An Introduction to Cold and Ultracold Chemistry

Jesús Pérez Ríos

An Introduction to Cold and Ultracold Chemistry

Atoms, Molecules, Ions and Rydbergs

 Springer

Jesús Pérez Ríos (iD)
Department of Molecular Physics
Fritz Haber Institute of the Max Planck Society
Berlin, Germany

ISBN 978-3-030-55938-0   ISBN 978-3-030-55936-6   (eBook)
https://doi.org/10.1007/978-3-030-55936-6

This Springer imprint is published by the registered company Springer Nature Switzerland AG
The registered company address is: Gewerbestrasse 11, 6330 Cham, Switzerland

To my nephew Dimas

Preface

This book is devoted to master's and graduate students interested in the most intimate nature of chemical reactions at very low collision energies where the inherent quantum mechanical nature of chemical processes emerges and becomes tangible. My approach to the subject in this book revolves around presenting novel and interesting theoretical methods for studying cold and ultracold chemistry reactions, going beyond the conventional techniques in the field. Moreover, I try to show the reader the broadness of the field through the study of Rydberg atoms, ions, and physics beyond the standard model with ultracold atoms and molecules.

In general, cold and ultracold chemistry are assumed to be disciplines focused on the control of chemical reactions involving atoms and molecules. However, in this book, it is shown that there is a lot more to cold and ultracold chemistry! For instance, we will see that ion–neutral collision leads to a new realm of chemical reactions that require the development of a chemical intuition different than that required for the neutral–neutral collision. In the same vein, the study of Rydberg–neutral collisions leads to a new set of chemical reactions that have acute effects on the different applications of Rydberg atoms for the development of quantum technologies. Moreover, throughout this book, the reader will see how cold and ultracold molecules are useful systems for understanding the Standard Model of particle physics and even the exploration of physics beyond the Standard Model. Finally, I want to make a statement: no matter what you are working on, if it involves atoms and molecules, chemistry will always be there. Therefore, better we understand the chemistry to design better machines and better quantum simulators.

This book is not just the result of my work, there are contributions from many people in different ways. First, I would like to thank all those who contributed or contribute to the field of cold and ultracold chemistry. Next, I would like to thank Prof. Sherwin Love for his support during the development of this book and for his valuable comments on Chap. 6, and Prof. Dr. Gerard Meijer for his understanding and support. Also, I would like to thank Dr. Christian Schewe for his help in some parts of this book, Dr. Rene Gerritsma for reading Chap. 9, Dr. David Cassidy for reading and commenting on Chap. 12, Dr. Marjan Mirahmadi for reading Chap. 10, Dr. Stefan Truppe for reading Chap. 3, Dr. Matthew T. Eiles for reading Chaps. 7 and 8, and Simon Hofsäss for reading Chaps. 1, 2, 3, and 4 of the book. Last but not least, I would like to express my gratitude to my family for their unconditional support, especially to my wife Anne for her constant help and motivation, to my sisters for the support, to my dad for being my number one fan, to my nephew Dimas, because without noticing, he makes everything funnier and more beautiful, and my mother María Teresa for her insistent attitude and for always being there for me.

Berlin, Germany   Jesús Pérez Ríos

Contents

List of Figures

List of Tables

The Realm of Cold and Ultracold

1

1.1 Why Ultracold Temperatures?

Chemistry, as we know and experience it every day, mostly occurs at room temperature ($T = 298$ K). At lower temperatures, the molecules in a gas have lower velocities, and hence, for the same density, collisions are less frequent. Then, one may think that as the temperature drops, the reaction rate will follow the same fate. This *classical* vision is partially true. However, one needs to keep in mind that as the temperature of a gas drops, quantum mechanics takes over, and phenomena insignificant at room temperature will dominate the physics and chemistry in this regime. One of these quantum phenomena is the onset of resonances [1–10], which are washed out at room temperature. Other interesting quantum phenomena are the threshold behaviors of elastic and inelastic collisions, also known as the Wigner threshold laws [11]. These predict that the elastic cross section tends toward a constant value as the temperature approaches zero, whereas the inelastic cross section increases with $1/\sqrt{T}$, where T is the temperature of the gas, as we will explain in this book.

Controlling chemical reactions is one of the main goals of modern chemistry. Indeed, many different techniques have been developed and proposed to that end [12, 13]. Maybe the most successful one is coherent quantum control [14, 15], which relies on using the coherence of external laser fields to control the fate of a reaction. However, at room temperature, frequent random collisions among the molecules induce decoherence, thus decreasing its efficiency. In the ultracold regime, the interaction of atoms and molecules with external fields is of the same order of magnitude as the collision energy, which can be exploited to establish precise control over the different degrees of freedom of a molecule [1, 7, 9, 16]. Thus, ultracold temperatures lead to a new realm for control and manipulation of reactions. This degree of control of internal degrees of freedom can be used to develop quantum simulators to study complex many-body problems [17, 18], to design new quantum information protocols [19, 20] or to develop high-precision

© Springer Nature Switzerland AG 2020
J. Pérez Ríos, *An Introduction to Cold and Ultracold Chemistry*,
https://doi.org/10.1007/978-3-030-55936-6_1

spectroscopy techniques to study time variation of fundamental constants [21–23] and physics beyond the standard model [24].

The wave-particle duality in quantum mechanics establishes that every particle also behaves as a wave with a wavelength given by $\Lambda = \hbar \sqrt{\frac{2\pi}{mk_B T}}$ – the de Broglie wavelength – where k_B is the Boltzmann constant, m the mass and $\hbar$ the reduced Planck constant. At room temperature, $\Lambda \sim 10^{-11}$ m that is ten times smaller than the typical interatomic distance of a molecule $\sim 1\text{Å} = 10^{-10}$ m. So, to observe the inherent wave nature of molecules at our energy scale, we would need to explore distances smaller than the size of a molecule. That is the reason why we do not see direct quantum effects in our daily life. However, as the temperature drops, the de Broglie wavelength increases and it may become of the same order of magnitude as the interparticle distance, and under appropriate circumstances leads to Bose–Einstein condensation (BEC), which is otherwise unobservable. The observation of BEC [25–27] revolutionized chemical physics, leading to the new paradigm of ultracold physics.

Chemistry, the topic of this book, is relevant at a vast range of temperatures. The set of rules that govern chemistry is not always the same, and they change as the temperature varies. To give a flavor of it, a sketch of the relevant chemical and physical processes characteristic of a given temperature range is presented in Fig. 1.1. The figure shows the suitable parameter space for chemistry, focusing on temperature, and the de Broglie wavelength. For temperatures below the ionization energy of the hydrogen atom (13.6 eV), electrons can be exchanged between

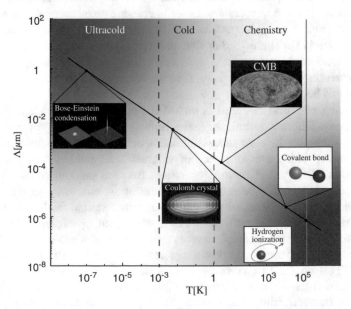

Fig. 1.1 De Broglie wavelength Λ (in μm) as a function of the temperature T (in K) covering the parameter space for chemistry. CMB stands for cosmic microwave background

different atoms, leading to the formation of the chemical bond, and hence leading to the beginning of chemistry. At lower temperatures, the cosmic microwave background (CMB) appears at $\sim$3 K, the lowest temperature in the universe.[1] The CMB consists of microwave photons filling the whole universe, a remnant of the de-coupling era, the time when the universe became transparent and photons could freely stream throughout the universe. Temperatures below 1 K are said to be in the cold regime. Different phenomena occur within this regime, but here, we emphasize the formation of Coulomb crystals owing to its relevance to different chapters of this book. A Coulomb crystal is a crystal-like structure that emerges as a consequence of the compensation between the repulsive Coulomb interaction (charge-charge interaction, since the ions have the same charge) and kinetic energy of the ions in the presence of a trap holding the ions in a given spatial region. This phenomenon is only possible to observe if there is a friction force acting on the ions, which is usually a laser inducing the cooling of the ions. Finally, below the mK range, the ultracold regime emerges. In particular, at $T \sim 100$ nK, BEC can be reached. A BEC is a state of matter in which a quantum state shows a macroscopic occupation, i.e., most of the particles of the system are in the same quantum state, leading to the manifestation of quantum effects at macroscopic scales [28]. It was predicted by Bose and Einstein [29, 30] in the 1920s, but it took 70 years for it to be observed in ultracold atoms [25–27].

1.2   Cold and Ultracold Chemistry

1.2.1   The Standard Approach

In general, the concept of cold chemistry refers to any chemical reaction happening at temperatures between 1 and 10^{-3} K, whereas the term ultracold chemistry is used for any chemical reaction occurring at temperatures $\lesssim 10^{-3}$ K [1–10]. These definitions correspond to a convention; however, they also contain a physical meaning, as we explain below.

1.2.2   A More Physical Approach

In chemical physics, the relevant physical systems are ensembles of atoms and molecules. The relevant energy scale is given by the interaction energy between colliding partners. For the sake of clarity, let us assume atoms in this section. Neutral atoms interact through the well-known van der Waals interaction given by

$$V_{\text{vdW}}(r) = -\frac{C_6}{r^6}, \tag{1.1}$$

---

[1]In this statement we do not consider the temperature reached in laboratories on earth.

where r stands for the interatomic distance and C_6 is the van der Waals coefficient [31]. From a classical perspective, a collision occurs when the interaction energy is similar to the collision energy, E_k, and hence a deviation from the uniform rectilinear motion predicted by Newton's first law of classical mechanics is observed [32, 33]. Its quantum mechanical counterpart is given by

$$E_k = V_{vdW} \rightarrow \frac{\hbar^2}{2\mu r^2} = \frac{C_6}{r^6}, \tag{1.2}$$

then solving for r and including by convention a factor $1/2$ yields

$$r_{vdW} = \frac{1}{2}\left(\frac{2\mu C_6}{\hbar^2}\right)^{1/4}. \tag{1.3}$$

r_{vdW} is called the van der Waals length [34]. The van der Waals length defines the characteristic length scale of neutral–neutral interactions, which if substituted back into the quantum kinetic energy leads to

$$E_{vdW} = \frac{\hbar^2}{2\mu r_{vdW}^2} = \left(\frac{2\hbar^6}{\mu^3 C_6}\right)^{1/2}, \tag{1.4}$$

known as van der Waals energy. For $Ek < E_{vdW}$, the scattering is dominated by the long-range tail of the underlying potential, which leads to the realm of ultracold physics. Another way of looking at it is that for collision energies $\lesssim E_{vdW}$ only a handful of partial waves contribute to the scattering; therefore, the scattering observables may show resonance effects and other pure quantum mechanical effects. It turns out that E_{vdW} for many atom–atom systems is of the order of 1 mK, as shown in Table 1.1. This is the reason why ultracold processes are generally those below 1 mK.

The concepts of van der Waals energy and length may be generalized to any long-range interaction as $\frac{C_n}{r^n}$, with $n \geq 2$. In particular, one finds

$$r_n = \left(\frac{2\mu C_n}{\hbar^2}\right)^{1/(n-2)},$$

$$E_n = \left(\frac{\hbar^2}{2\mu C_n^{2/n}}\right)^{n/(n-2)},$$

$$\tag{1.5}$$

where the factor $1/2$ on the length scale is omitted and the pertinent factor for the energy scale too. These equations establish that the scales associated with the ultracold realm strongly depend on the nature of the system at hand. Therefore, the rule $E \sim 1$ mK is not general.

Table 1.1 Length (in a_0) and energy scale (in mK) associated with the ultracold limit in different systems. The C_6 values are taken from [35], whereas the C_4 coefficients, given by half of the atom's polarizability, are taken from Refs. [36–38]

System	μ (a.m.u.)	C_6 ($\times 10^3$ a.u.)	C_4 (a.u.)	R (a_0)	E (mK)
Li-Li	3.47	1.39		32.4	47.55
Na-Na	11.49	1.47		44.3	7.67
K-K	19.98	3.81		64.5	2.08
Rb-Rb	42.73	4.43		81.02	0.62
Cs-Cs	66.45	6.33		98.9	0.27
Li-Li$^+$	3.47		82.0	1.02×10^3	0.05
Na-Na$^+$	11.49		81.3	1.85×10^3	4.40×10^{-3}
K-K$^+$	19.98		145.0	3.25×10^3	8.20×10^{-4}
Rb-Rb$^+$	42.73		160.0	4.99×10^3	1.63×10^{-4}
Cs-Cs$^+$	66.45		199.76	6.95×10^3	5.38×10^{-5}

In this book, we will cover ion–atom systems, or more generally, charged–neutral interactions. In that scenario, the long-range interaction reads as $-C_4/r^4$ [31], where the C_4 long-range coefficient only depends on the polarizability of the neutral system. Therefore, charged–neutral interactions are more attractive than neutral–neutral ones, as one would expect. The results for the characteristic length and energy scale for ultracold physics for ion–atom systems are shown in Table 1.1. In this table, we notice that the threshold energy characterizing the ultracold regime for charged–neutral systems is rather different from the one in neutral–neutral systems. In particular, the more attractive the long-range interaction is, the lower the energy threshold, and the larger the length scale becomes.

In the present book, we emphasize that the ultracold energy threshold depends on the system at hand. In particular, we will see that what is ultracold for neutral–neutral interaction is actually rather *hot*[2] for charged–neutral interactions.

1.3   Ultracold Physics and Fundamental Physics

In the present book, the reader finds a chapter dedicated to a field that is gaining more interest: the search for new physics[3] (NP) with ultracold atoms and molecules. This new field relies on using the properties of ultracold atoms and molecules to perform high precision spectroscopy to constrain new physics models [24,39,40] or to measure time-dependent variation of the fundamental constants of nature [21,22].

As presented in Sect. 1.1, at ultracold temperatures, the internal and external degrees of freedom of molecules are readily controllable. On top of that, in this

---

[2]Thermal is more precise, although we use hot here to emphasize its difference concerning the neutral–neutral case.

[3]Physics beyond the standard model.

temperature regime, the Doppler and collisional broadening are negligible. Thus, it is a unique playground for high-precision spectroscopy. Measuring high-resolution spectra and comparing these with standard model predictions will help to elucidate the possibility of new physics. In that way, atoms and molecules in the ultracold regime are a complementary tool for accelerator-based experiments to expand the parameter space for the quest of physics beyond the standard model. A chapter of this book is dedicated to this exciting field.

References

1. Krems RV (2008) Cold controlled chemistry. Phys Chem Chem Phys 10:4079. https://doi.org/10.1039/B802322K
2. Hutson JM, Soldán P (2007) Molecular collisions in ultracold atomic gases. Int Rev Phys Chem 26(1):1
3. Weck PF, Balakrishnan N (2006) Importance of long-range interactions in chemical reactions at cold and ultracold temperatures. Int Rev Phys Chem 25(3):283. https://doi.org/10.1080/01442350600791894
4. Quéméner G, Julienne PS (2012) Ultracold molecules under control! Chem Rev 112:4949
5. Carr LD, DeMille D, Krems RV Ye J (2009) Cold and ultracold molecules: science, technology and applications. New J Phys 11:055049
6. Dulieu O, Gabbanini C (2009) The formation and interactions of cold and ultracold molecules: new challenges for interdisciplinary physics. Rep Progress Phys 72(8):086401
7. Dulieu O, Osterwalder A (eds) (2018) Cold chemistry. Theoretical and computational chemistry series. The Royal Society of Chemistry. https://doi.org/10.1039/9781782626800
8. Weiner J (2003) Cold and ultracold collisions in quantum microscopic and mesoscopic systems. Cambridge University Press, Cambridge
9. Krems RV (2018) Molecules in electromagnetic fields. Wiley, New York
10. Krems RV, Stwalley WC, Friedrich B (eds) (2009) Cold molecules: theory, experiment, applications. CRC Press, Boca Raton
11. Wigner EP (1948) Phys Rev 73:1002
12. Prokhorenko VI, Nagy AN, Waschuk SA, Brown LS, Birge RR Miller RJD (2006) Coherent control of retinal isomerization in bacteriorhodopsin. Science 313:1257
13. Levin L, Skomorowski W, Rybak L, Kosloff R, Koch CP, Amitay Z (2015) Coherent control of bond making. Phys Rev Lett 114:233003
14. Shapiro M, Brumer P (2012) Quantum control of molecular processes. Wiley, New York
15. Kosloff R, Rice SA, Gaspar P, Tersegni S, Tannor DJ (1989) Wavepacket dancing: achieving chemical selectivity by shaping light pulses. Chem Phys 139:201
16. Marinescu M, You L (1998) Controlling atom-atom interaction at ultralow temperatures by dc electric fields. Phys Rev Lett 81:4596. https://doi.org/10.1103/PhysRevLett.81.4596
17. Bloch I, Dalibard J, Zwerger W (2008) Many-body physics with ultracold gases. Rev Mod Phys 80:885. https://doi.org/10.1103/RevModPhys.80.885
18. Aidelsburger M, Atala M, Lohse M, Barreiro JT, Paredes B, Bloch I (2013) Realization of the hofstadter hamiltonian with ultracold atoms in optical lattices. Phys Rev Lett 111:185301. https://doi.org/10.1103/PhysRevLett.111.185301
19. DeMille D (2002) Quantum computation with trapped polar molecules. Phys Rev Lett 88:067901. https://doi.org/10.1103/PhysRevLett.88.067901
20. Saffman M, Walker TG, Molmer K (2010) Quantum information with Rydberg atoms. Rev Mod Phys 82:2313
21. Chin C, Flambaum VV, Kozlov MG (2009) Ultracold molecules: new probes on the variation of fundamental constants. New J Phys 11(5):055048

22. Schiller S, Korobov V (2005) Tests of time independence of the electron and nuclear masses with ultracold molecules. Phys Rev A 71:032505. https://doi.org/10.1103/PhysRevA.71.032505
23. DeMille D, Sainis S, Sage J, Bergeman T, Kotochigova S, Tiesinga E (2008) Enhanced sensitivity to variation of m_e/m_p in molecular spectra. Phys Rev Lett 100:043202. https://link.aps.org/doi/10.1103/PhysRevLett.100.043202
24. Safronova MS, Budker D, DeMille D, Kimball DFJ, Derevianko A, Clark CW (2018) Search for new physics with atoms and molecules. Rev Mod Phys 90:025008. https://doi.org/10.1103/RevModPhys.90.025008
25. Davis KB, Mewes MO, Andrews MR, van Druten NJ, Durfee DS, Kurn DM, Ketterle W (1995) Bose-einstein condensation in a gas of sodium atoms. Phys Rev Lett 75:3969. https://doi.org/10.1103/PhysRevLett.75.3969
26. Bradley CC, Sackett CA, Tollett JJ, Hulet RG (1995) Evidence of bose-einstein condensation in an atomic gas with attractive interactions. Phys Rev Lett 75:1687. https://doi.org/10.1103/PhysRevLett.75.1687
27. Anderson MH, Ensher JR, Matthews MR, Wieman CE, Cornell EA (1995) Observation of bose-einstein condensation in a dilute atomic vapor. Science 269:198
28. Pethick CJ, Smith H (2002) Bose-Einstein condensation in dilute gases. Cambridge Unviersity Press, Cambridge
29. Bose (1924) Plancks gesetz und lichtquantenhypothese. Zeitschrift für Physik 26(1):178. https://doi.org/10.1007/BF01327326
30. Einstein A (2006) Quantentheorie des einatomigen idealen Gases. Wiley, pp 237–244
31. Stone A (2013) The theory of intermolecular forces, 2nd edn. Oxford University Press, UK
32. Levine RD, Bernstein RB (1987) Molecular reaction dynamics and chemical reactivity. Oxford University Press, New York
33. Landau LD, Lifshitz EM (1976) Mechanics. Elsevier Butterworth-Heinemann, Burlington
34. Jones KM, Tiesinga E, Lett PD, Julienne PS (2006) Ultracold photoassociation spectroscopy: long-range molecules and atomic scattering. Rev Mod Phys 78:483. https://doi.org/10.1103/RevModPhys.78.483
35. Marinescu M (1998) Dispersion coefficients for alkali-metal dimers. https://www.phys.uconn.edu/~rcote/Dispersion/AlkaliMetal.html
36. Molof R, Schwartz H, Miller T, Bederson B (1974) Measurements of electric dipole polarizabilities of the alkali-metal atoms and the metastable noble-gas atoms. Phys Rev A 10:1131. https://doi.org/10.1103/PhysRevA.10.1131
37. Ekstrom C, Schmiedmayer J, Chapman M, Hammond T, Pritchard D (1995) Measurement of the electric polarizability of sodium with an atom interferometer. Phys Rev A 51(5):3883
38. Gregoire MD, Hromada I, Holmgren WF, Trubko R, Cronin AD (2015) Measurements of the ground-state polarizabilities of Cs, Rb, and K using atom interferometry. Phys Rev A 92:052513. https://doi.org/10.1103/PhysRevA.92.052513
39. Ubachs W, Koelemeij J, Eikema K, Salumbides E (2016) Physics beyond the standard model from hydrogen spectroscopy. J Mol Spectrosc 320:1
40. Salumbides EJ, Schellekens AN, Gato-Rivera B, Ubachs W (2015) Constraints on extra dimension from precision molecular spectroscopy. New J Phys 17:033015

Quantum Scattering Theory

2

This chapter revolves around the theoretical minimum scattering theory to understand the main physical and chemical processes at cold and ultracold temperatures. The concept of scattering is introduced based on experimental grounds, followed by the classical and quantum definitions of the cross section. In this way, we hope to emphasize the main difference between the quantum and classical realms. Within the quantum theory of scattering, we pay special attention to the concept of scattering length and Wigner threshold laws owing to its importance in ultracold chemistry. In addition, we introduce some quantum mechanical effects on scattering observables such as Fano–Feshbach resonances and the glory effect, which is a very well-known phenomenon in chemical physics.

2.1 Collisions

The study of the subjacent interatomic interactions in molecular systems is one of the main driving forces of modern chemical physics. Information about how atoms and molecules interact is encoded in scattering, thermodynamics, and spectroscopy observables. Among these observables, the scattering ones allow an energy-dependent map of the underlying molecular interaction by studying collisions, characterized, generally, by the cross section. Therefore, it is fundamental to understand what a collision is and how to characterize it, which is the topic of this section.

2.1.1 When Does a Collision Occur?

Newton's first law establishes that in the absence of forces, a particle experiences a uniform rectilinear motion. However, if a particle is interacting with other particles

© Springer Nature Switzerland AG 2020
J. Pérez Ríos, *An Introduction to Cold and Ultracold Chemistry*,
https://doi.org/10.1007/978-3-030-55936-6_2

Fig. 2.1 Illustration of the Beer–Lambert law. A molecular beam of molecules type A (in orange) passes through a chamber containing molecules of type B (in blue), and as a consequence the molecular beam gets attenuated through collisions between molecules A and B

or feels any interaction, its trajectory will deviate from a straight line, as Newton's second law dictates. In that case, we can say that the particles undergo a collision.

2.1.2 Experimental Approach: The Mean Free Path

Let us assume a collimated molecular beam of molecules A propagating in the z direction with flux I_0 (number of molecules per unit time and surface). This beam hits a chamber filled with molecules of type B and constant density of n_B, as shown in Fig. 2.1. The walls of the chamber are transparent to molecules of type A but opaque to molecules B. As the molecules A go through the chamber, they experience collisions deflecting out its trajectories and, ultimately, an attenuation of the flux of the molecular beam. The average distance that an A molecule travels before a collision with a B molecule is the mean free path and is labeled as λ. Therefore, the probability of an A molecule colliding with a B molecule in a short distance dz is dz/λ, and, as a consequence, the flux will change in the same interval as

$$\frac{I(z+dz) - I(z)}{I(z)} = -\frac{dz}{\lambda}. \tag{2.1}$$

Solving this equation and keeping in mind that $I(0) = I_0$ one finds the so-called Beer–Lambert law

$$I(z) = I_0 e^{-z/\lambda}. \tag{2.2}$$

The collision probability depends on the number of scatterers. Therefore, λ will depend on n_B, and λ has units of length it should be given by [1]

$$\lambda = \frac{1}{n_B \sigma}, \tag{2.3}$$

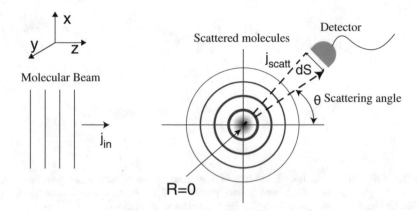

Fig. 2.2 Quantum mechanical interpretation of the experiment to measure the differential cross section. A molecular beam with flux $\mathbf{j}_{in}$ scatters off of a scattering center placed at $R = 0$. The flux of scattered particles at a given scattering angle θ is given by $\mathbf{j}_{scatt}$. The detector covers a given surface, which is small compared with the whole spherical shell and it is labeled dS. The expressions for the incoming and scattered flux are given by Eq. 2.6

where σ is the so-called cross section and has units of area. Thus, the concept of cross section emerges naturally from the mean free path, and it represents the effective collision area.

2.1.3 Quantum Mechanical Cross Section

Let us come back to the previous experiment, but in this case, we place the detector at a given angle θ with respect to the z axis. In this case, we will detect the number of molecules per unit of time that have been deflected toward the direction where the detector is located, as shown in Fig. 2.2. The ratio between the number of scattered molecules per unit of time, $\mathbf{j}_{scatt}$, within a surface element dS and the incident flux $\mathbf{j}_{in}$ is called the differential cross section and is given by

$$\frac{d\sigma}{d\Omega} = \frac{\mathbf{j}_{scatt} \cdot d\mathbf{S}}{|\mathbf{j}_{in}|}. \tag{2.4}$$

The differential cross section has units of length squared, and it reveals most of the information regarding the underlying interparticle interaction.

From a quantum mechanical perspective, a molecular beam is described as a plane wave propagating in the positive direction of the z. Here, $k = \sqrt{2\mu E/\hbar}$, which is known as the wave vector, where μ is the reduced mass of the colliding pair, E is the kinetic energy of the molecules, and $\hbar$ is the reduced Planck constant. The molecules deflected by the scattering center, far away from it, behave as a divergent spherical wave[1] with an amplitude depending on the scattering angle [4]. Therefore, the wave function as a function of the scattering coordinate R reads as

$$\Psi(R) \approx e^{ikz} + \frac{e^{ikR}}{R} f(\theta). \tag{2.5}$$

$f(\theta)$ is the amplitude of scattering, it has units of length and it is related to the probability of scattering a particle in a given angle, θ. Employing Eq. (2.5), the flux of the incoming particles and the scattered flux are given by

$$j_{\text{scatt}} = \frac{\hbar k}{\mu} \frac{|f(\theta)|^2}{R^2}, \tag{2.6}$$

$$j_{\text{in}} = \frac{\hbar k}{\mu},$$

respectively. Plugging Eqs. (2.6) into Eq. 2.4 we find

$$\frac{d\sigma}{d\Omega} = |f(\theta)|^2, \tag{2.7}$$

which is the quantum mechanical expression for the differential cross section, and performing the integration in Eq. (2.7) over the solid angle element, $d\Omega = 2\pi \sin\theta d\theta$, yields the scattering cross section as

$$\sigma = 2\pi \int |f(\theta)|^2 \sin(\theta) d\theta. \tag{2.8}$$

2.1.4 Classical Cross Section

Classically, the cross section is defined as the area of the plane perpendicular to the initial momentum of the incoming particle containing the scattering center [1, 5]. The cross section is determined through the impact parameter and energy. The impact parameter, b, is the component of the vector position perpendicular to the velocity of the incoming particle in the absence of interactions (this is the classical equivalent of asymptotic conditions in quantum mechanical theory). Equivalently, b is the component of the initial vector position lying in the perpendicular plane to the motion of the incoming particle, as shown in Fig. 2.3. In the same figure, it is noticed that for every impact parameter, there is a unique scattering angle θ, which implies that the area described by the impact parameter maps, $2\pi b db$, into the area described by the scattering angle $2\pi \sin(\theta) d\theta$. Therefore, the classical cross section reads as

---

[1]This is Green's function for the two-body Schrödinger equation in spherical coordinates [2, 3].

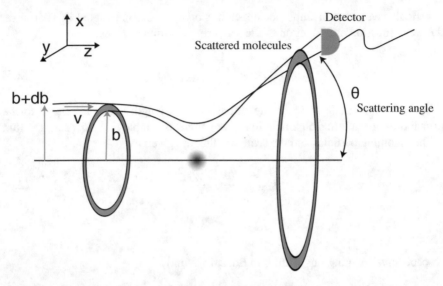

Fig. 2.3 In classical mechanics, each trajectory with a given impact parameter b leads to a given scattering angle θ. The area sustained by the blue circular shell in terms of the impact parameter is the same as the orange one for the scattering angle

$$\sigma = 2\pi \int P(b)b\,db, \tag{2.9}$$

where $P(b)$ is the opacity function and determines the probability that a given trajectory leads to the process under consideration. In this book, we introduce different opacity functions for different processes.

2.1.4.1 Hard-Sphere Collisions
Let us assume that two particles interact through the following potential,

$$V(R) = \begin{cases} \infty & R < r \\ 0 & R > r \end{cases} \tag{2.10}$$

which emulates the collision between two solid particles or radius $r/2$, as shown in Fig. 2.4. In the center of mass (CM) frame of reference, only in trajectories whose impact parameter is below the size of the potential will a collision occur; otherwise, the incoming particle will not feel any interaction. In this scenario, the opacity function has the following form

$$P(b) = \begin{cases} 1 & b < r \\ 0 & b > r. \end{cases} \tag{2.11}$$

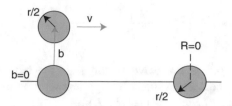

Fig. 2.4 Two particles of radius $r/2$ collide with a hard-sphere potential, i.e., only if the particles touch each other physically will a collision happen. This is like the collision between billiard balls

The opacity function is independent of the collision energy; thus, the probability of collision is energy independent. Then, plugging the opacity function into Eq. 2.9 we obtain

$$\sigma = 2\pi \int_0^r P(b)b\,db = 2\pi \int_0^r b\,db = \pi r^2. \tag{2.12}$$

This cross section is also known as the geometric cross section and often appears in this book. In particular, the geometric cross section gives an upper bound for the elastic scattering between two particles after the estimation of the range of the potential (Fig. 2.4).

2.2   Elastic Cross Section

2.2.1   Definition

Elastic collisions are those in which the internal states of the colliding partners do not change, in other words, the collision energy cannot be exchanged with internal degrees of freedom of the colliding partners. Elastic collisions are responsible for thermalization in gases; hence, they play an essential role in transport theory [6]. Let us assume two structureless particles colliding via a central potential $V(R)$, i.e., only elastic collisions occur. In that scenario, it is convenient to express the wave function in the CM coordinate system, and expand it in the complete basis of Legendre polynomials $P_l(\cos\theta)$ as, also known as partial wave decomposition,

$$\Psi(\mathbf{R}) = \sum_{l=0}^{\infty} (2l + 1) P_l(\cos\theta)\psi_l(R). \tag{2.13}$$

Here, R is the scattering coordinate, and θ is the azimuthal angle (introduced in Sect. 2.1.3). Plugging this expression into the Schrödinger equation in spherical coordinates leads to

$$\frac{1}{R}\frac{d^2}{dR^2}(R\psi_l) + \left(k^2 - \frac{l(l+1)}{R^2} - \frac{2\mu V(R)}{\hbar^2}\right)\psi_l = 0, \tag{2.14}$$

where $k = \sqrt{2\mu E/\hbar^2}$ is the wave vector, μ is the reduced mass, and E is the collision energy, i.e., the available energy that may be exchanged during the collision. The solution of Eq. (2.14) is [2, 4, 7, 8]

$$\psi_l(R) = A_l(k)j_l(kR) + B_l(k)n_l(kR), \tag{2.15}$$

where $j_l(.)$ and $n_l(.)$ are the spherical Bessel functions of the first and second kind respectively [9]. Since we are dealing with a scattering problem, we are interested in the asymptotic behavior of the wave function, and this is given by [9]

$$\psi_l(R)(R \to \infty) = A_l(k)\frac{\sin(kR - l\pi/2)}{kR} - B_l(k)\frac{\cos(kR - l\pi/2)}{kR}$$

$$= \tilde{A}_l(k)\frac{\sin(kR - l\pi/2 + \delta_l)}{kR}, \tag{2.16}$$

where

$$\tilde{A}_l(k) = \sqrt{A_l(k)^2 + B_l(k)^2} \tag{2.17}$$

and

$$\tan\delta_l(k) = -\frac{B_l(k)}{A_l(k)}. \tag{2.18}$$

Plugging Eq. (2.16) into Eq. (2.13) leads to the final form of the asymptotic scattering wave function as

$$\Psi(\mathbf{R}) \approx \sum_{l=0}^{\infty}(2l+1)A_l P_l(\cos\theta)\frac{\sin(kR - l\pi/2 + \delta_l)}{kR}, \tag{2.19}$$

which can be represented as an incoming wave plus an outgoing wave contribution as

$$\Psi(\mathbf{R}) \approx \sum_{l=0}^{\infty}(2l+1)A_l P_l(\cos\theta)\frac{\iota}{2kR}\left(-e^{(\iota kR - \iota l\pi/2 + \iota\delta_l)} + e^{-(\iota kR - \iota l\pi/2 + \iota\delta_l)}\right). \tag{2.20}$$

Equivalently, the scattering wave function of a central potential, given in Eq. (2.5), can be represented in terms of the Legendre polynomial basis set as [4]

$$\Psi(\mathbf{R}) \approx \sum_{l=0}^{\infty}(2l+1)P_l(\cos\theta)\left[\imath^l\frac{l}{2kR}\left(e^{-(\imath kR - \imath l\pi/2)} - e^{(\imath kR - \imath l\pi/2)}\right) + \frac{f_l}{R}e^{\imath kR}\right].$$

(2.21)

Here, we have used the decomposition of the scattering amplitude in partial waves as

$$f(\theta) = \sum_{l=0}^{\infty}(2l+1)P_l(\cos\theta)f_l,$$

(2.22)

and the plane wave expansion

$$e^{\imath kz} \approx \sum_{l=0}^{\infty}\imath^l(2l+1)P_l(\cos\theta)\frac{l}{2kR}\left(e^{-(\imath kR - \imath l\pi/2)} - e^{(\imath kR - \imath l\pi/2)}\right).$$

(2.23)

Equation (2.21) can be rewritten in terms of the incoming and outgoing waves as

$$\Psi(\mathbf{R}) \approx \sum_{l=0}^{\infty}\frac{\imath(2l+1)}{2kR}P_l(\cos\theta)\left((-1)^l e^{-\imath kR} - e^{\imath kR}(2\imath k f_l + 1)\right).$$

(2.24)

Comparing the terms proportional to $e^{-\imath kR}$ in Eqs. (2.20) and (2.24) leads to

$$A_l = \imath^l e^{\imath\delta_l}.$$

(2.25)

And comparing the terms regarding the outgoing wave, $e^{\imath kR}$, in Eqs. (2.20) and (2.24) and by means of Eq. (2.25) one gets

$$f_l = \frac{e^{2\imath\delta_l} - 1}{2\imath k},$$

(2.26)

where $S_l = e^{2\imath\delta_l}$ is the scattering matrix S [4, 8]. Equation (2.26) can be expressed in terms of the scattering matrix as

$$f_l = \frac{S_l - 1}{2\imath k},$$

(2.27)

Finally, keeping in mind the partial wave expansion of the scattering amplitude in Eq. (2.21) and the definition of cross section one gets

$$\sigma = \sum_{l=0}^{\infty}4\pi(2l+1)|f_l|^2,$$

(2.28)

which is better described in terms of the S-matrix through Eq. (2.27) as

$$\sigma = \frac{\pi}{k^2} \sum_{l=0}^{\infty} (2l + 1)|1 - S_l|^2. \tag{2.29}$$

This is the general expression for the quantal elastic cross section. Also, Eq. (2.29) can be recast in terms of the phase shift as

$$\sigma = \frac{4\pi}{k^2} \sum_{l} (2l + 1) \sin^2(\delta_l), \tag{2.30}$$

which is the well-known expression for the quantal elastic cross section for the single channel problem.

2.2.2 Wigner Threshold Law for the Elastic Cross Section

The threshold laws of Wigner determine the universal behavior of the elastic and inelastic cross sections as the collision energy goes to zero. These threshold laws are general and appear independently of the nature of the interaction between the colliding partners. At very low collision energies, $k \to 0$, it is convenient to study the solution of the two-body scattering problem at three different regimes:

- **Region I:** $R \leq R^*$. R^* stands for the range of the potential $V(R)$. In this region, the collision energy is negligible in comparison with the interaction potential and the two-body Schrödinger equation reads as

$$\frac{1}{R} \frac{d^2}{dR^2} (R\psi_l) - \left(\frac{l(l + 1)}{R^2} + \frac{2\mu V(R)}{\hbar^2} \right) \psi_l = 0. \tag{2.31}$$

  This region extends up to $100a_0$ in the case presented in Fig. 2.5.
- **Region II:** $R^* \gg R \ll 1/k$. This region covers distances between the range of the potential and the de Broglie wavelength, $1/k$. In this region, the collision energy and interaction potential are negligible and the Schrödinger equation reads as

$$\frac{1}{R} \frac{d^2}{dR^2} (R\psi_l) - \frac{l(l + 1)}{R^2} \psi_l = 0, \tag{2.32}$$

  and its solution is

$$\psi_l(R) = c_1 r^l + \frac{c_2}{r^{l+1}}. \tag{2.33}$$

  This region lies between 100 and 1,000 a_0 in Fig. 2.5, and in this region, the wave function shows linear behavior with respect to the scattering coordinate R.
- **Region III:** $kR \sim 1$ or asymptotic region. In this region we recover Eq. (2.14) and its solution can be expressed as (through Eqs. (2.17) and (2.18))

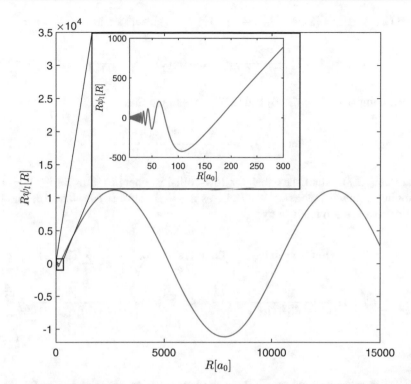

Fig. 2.5 Wave function for $E = 500\,\text{nK}$. The wave function shows the expected oscillatory behavior for large distances. The inset shows a zoom=in of the region $R \leq 300$ where the expected highly oscillatory behavior of the wave function transforms into a linear trend with R for $R \gtrsim 100 a_0$

$$\psi_l(R)(R \to \infty) = A_l(k)\,(j_l(kr) - \tan\,(\delta_l(k))n_l(kr))\,. \tag{2.34}$$

This region shows the characteristic oscillations associated with the asymptotic behavior of the wave function, as shown in Fig. 2.5.

At the interplay between regions II and III, the wave function follows Eq. (2.34) with $kr \ll 1$. In that case, taking into account that for small arguments the spherical Bessel functions behave as [10]

$$j_l(kr) \simeq \frac{(kr)^l}{(2l+1)!!},$$

$$n_l(kr) \simeq -\frac{(2l-1)!!}{(kr)^{l+1}},$$

$$\tag{2.35}$$

where $(2x + 1)!! = (2x + 1)(2x - 1)...$ Eq. (2.34) is written as

$$\psi_l(R) = A_l(k)\left(\frac{k^l}{(2l+1)!!}r^l + \tan{(\delta_l(k))}\frac{(2l-1)!!}{k^{l+1}r^{l+1}}\right). \tag{2.36}$$

Finally, comparing Eqs. (2.36) and (2.33), it is found that

$$\tan{(\delta_l(k))} = \frac{k^{2l+1}}{(2l+1)!!(2l-1)!!}\frac{c_2}{c_1}. \tag{2.37}$$

Equation (2.37) establishes that for low-collision energies $\lim_{k\to 0}\delta_l(k) \to 0$, following a power law as k^{2l+1}. Thus, as $k \to 0$, $\sin\delta_l(k) \approx \tan\delta_l(k)$ and the elastic cross section is given by

$$\lim_{k\to 0}\sigma = \lim_{k\to 0}\frac{4\pi}{k^2}\sum_l(2l+1)\sin^2(\delta_l(k)) \approx \frac{4\pi}{k^2}\sum_l(2l+1)\tan^2(\delta_l)$$

$$= \frac{4\pi}{k^2}\sum_l(2l+1)\frac{k^{4l+2}}{[(2l-1)!!]^2[(2l+1)!!]^2}\frac{c_2^2}{c_1^2} = 4\pi\sum_l(2l+1)\frac{k^{4l}}{[(2l-1)!!]^2[(2l+1)!!]^2}\frac{c_2^2}{c_1^2}. \tag{2.38}$$

The elastic cross section goes to zero for low-collision energies except for the s-wave scattering $(l = 0)^2$ that tends to a constant value as

$$\lim_{k\to 0}\sigma = 4\pi\frac{c_2^2}{c_1^2}. \tag{2.39}$$

This is the Wigner threshold law for the elastic cross section and establishes that for low-collision energies, only the s-wave scattering contributes to the cross section [11]. Besides, the cross section is constant at low energies, and its value is an intrinsic property of the underlying two-body potential, as explained in the next section.

2.2.3   Collision Between Two Identical Particles

From a quantum mechanical perspective, the electrons of an atom are equivalent; in other words, they have no intrinsic property to be distinguished, and there is no unique way to label each of these electrons. However, we also know that energy cannot depend on the labeling of the particles of a system. Then, the exchange of particles will not affect energy. In other words, the exchange operator commutes

---

[2]This is only true for bosons. For fermions the lowest partial wave is $l = 1$.

with the Hamiltonian of the system. However, the scattering properties will depend on the behavior of the wave function with respect to the exchange operator.

The wave function of a system can be either symmetric or antisymmetric with respect to the exchange of two particles; the first are bosons and the second fermions. In particular, for the two-body problem, we find

$$P_{12}\psi(\mathbf{r}_1, \mathbf{r}_2) = \psi(\mathbf{r}_2, \mathbf{r}_1) = \pm\psi(\mathbf{r}_1, \mathbf{r}_2), \tag{2.40}$$

where $\pm$ is for bosons/fermions.

The scattering of two identical particles in the CM frame of reference is sketched in Fig. 2.6 and given by the following asymptotic wave function

$$\Psi(R) = e^{\imath kz} \pm e^{-\imath kz} + \frac{e^{\imath kR}}{R}\left(f(\theta) \pm f(\pi - \theta)\right). \tag{2.41}$$

Here, we notice two plane waves along the direction of the z axis, but with opposite orientations, which correspond to the motion of the particles in the CM frame. Since both particles can be scattered, we must take that into account, which is the reason behind the two scattering amplitudes for θ and $\pi - \theta$ in Eq. (2.41). Then, taking into account the definition of differential cross section in Eqs. (2.6) and Eq. (2.41), it is possible to get the elastic cross section as

Fig. 2.6 The role of the quantal indistinguishability on collisions. Two different processes (in the CM frame) associated with the same collision experiment are shown. A detector will record as a scattering event the processes shown in (**a**) and (**b**) without distinction at the same scattering angle θ

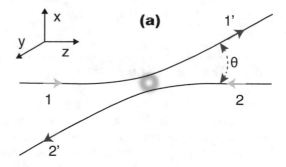

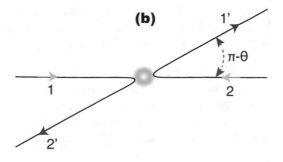

$$\sigma = \pi \int |f(\theta) \pm f(\pi - \theta)|^2 \sin(\theta) d\theta, \tag{2.42}$$

where one needs to account for the proper incoming flux of particles (twice the one in the case of distinguishable particles). Using Eq. (2.22) and taking into account that $P_l(-x) = (-1)^l P_l(x)$ yields

$$\sigma = \pi \int | \sum_l (2l+1) P_l(\cos\theta) f_l \left(1 \pm (-1)^l\right) |^2 \sin(\theta) d\theta$$

$$= 2\pi \sum_l (2l+1) \left(1 \pm (-1)^l\right)^2 |f_l|^2. \tag{2.43}$$

Finally, using Eq. (2.26) we find the ultimate expression for the elastic quantum mechanical scattering of two identical particles as

$$\sigma = \frac{8\pi}{k^2} \sum_{l=0}^{l=\text{even}} (2l+1) \sin^2(\delta_l) \tag{2.44}$$

for bosons, and for fermions it reads as

$$\sigma = \frac{8\pi}{k^2} \sum_{l=1}^{l=\text{odd}} (2l+1) \sin^2(\delta_l). \tag{2.45}$$

2.3   The Scattering Length

In this section, we introduce one of the most critical concepts in ultracold physics: the scattering length. This magnitude is an inherent property of any two-body interaction potential, and it reveals essential information for scattering observables. In most textbooks, the scattering length is presented as a result of a particular interaction potential, e.g., the square well, and then generalized, or it is introduced as a definition. For this reason, we prefer to present a version of the original derivation by E. Fermi in its seminal paper about the anomalous shift of the Rydberg lines in the presence of a dense background of inert gases [12]. Indeed, the concept of scattering length was born in the field of Rydberg physics, which is covered in this book.

In 1934, Amaldi and Segrè observed that the absorption lines for Rydberg states in alkali atoms showed a shift proportional to the density of the background gas [13]. At the density that Emaldi and Segrè performed their experiment, around 10^4 atoms of the background gas, also called perturbers, were within the orbit of the Rydberg electron, as sketched in Fig. 2.7. In this scenario, Fermi suggested that the electron

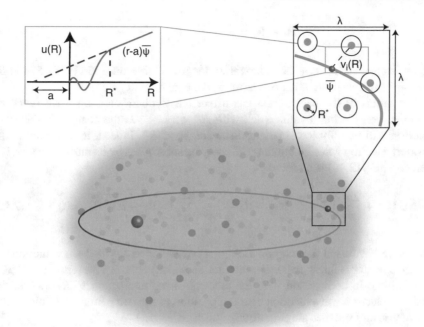

Fig. 2.7 A Rydberg atom in a high dense background gas. Many atoms of the background (orange balls) gas lie within the Rydberg orbit interacting with the Rydberg electron. The upper right panel shows a volume of a cube with side λ, the de Broglie wavelength of the electron. In this region, the Rydberg electron wave function is characterized by $\bar{\psi}$, except when it interacts with the atoms of the background gas via the potential $V_i(R)$ with size R^*. In the upper left panel it is shown how the electron wave function looks within a sphere of radius R^* around each of the atoms of the background gas

colliding with the background atoms would be the primary source of the shift of the Rydberg energy levels.

In the model of Fermi, the Rydberg electron is perturbed by the presence of the background atoms, and as a consequence the Rydberg electron wave function satisfies the following Schrödinger equation

$$-\Big[\frac{\hbar^2}{2m_e}\Delta + U(\mathbf{R}) + \sum_i V_i(\mathbf{R}_i - \mathbf{R})\Big]\psi(\mathbf{R}) = E\psi(\mathbf{R}), \qquad (2.46)$$

where

$$U(\mathbf{R}) = -\frac{e^2}{4\pi\epsilon_0}\frac{1}{R}, \qquad (2.47)$$

stands for the electron-ionic core Coulomb interaction, ϵ_0 for the vacuum permittivity, m_e is the electron mass, and e is the electron charge. The interaction between the atoms of the background and the electron is given by

$$V_i(\mathbf{R}_i - \mathbf{R}) = -\frac{1}{(4\pi\epsilon_0)^2}\frac{\alpha e^2}{2|\mathbf{R}_i - \mathbf{R}|}, \tag{2.48}$$

which is characterized by the polarization length R^* introduced in Eq. (1.5) and $\mathbf{R}_i$ represents the position of the i-th atom of the background gas.

The atoms of the background gas affect the Rydberg electron wave function; however, considering a volume L^3, with $R^* \ll L \ll \lambda$, the change on the wave function will be only local, i.e., at distances $R < R^*$. Indeed, in this volume the Rydberg electron wave function, $\psi(\mathbf{R})$, may be substituted by its mean value, $\bar{\psi}(\mathbf{R})$; therefore, Eq. (2.46) is recast as

$$\Delta\bar{\psi}(\mathbf{R}) + \frac{2m_e}{\hbar^2}[E - U(\mathbf{R})]\bar{\psi}(\mathbf{R}) - \frac{2m_e}{\hbar^2}\overline{\sum_i V_i(\mathbf{R}_i - \mathbf{R})\psi(\mathbf{R})} = 0, \tag{2.49}$$

where we have used the relation $\Delta\bar{\psi} = \overline{\Delta\psi}$, which is fulfilled since the kinetic energy operator is linear. Inside the interaction region, $R < R^*$, the term $[E - U(\mathbf{R})]$ is negligible in comparison with the electron–atom interaction $V(R)$; hence, the Rydberg electron wave function near the position of the i-th perturber is spherically symmetric and satisfies the equation

$$\Delta\psi(R) - \frac{2m_e}{\hbar^2}V(R)\psi(R) = 0, \tag{2.50}$$

which is further simplified by assuming, $\psi(R) = u(R)/R$, yielding

$$u''(R) = \frac{2m_e}{\hbar^2}V(R)u(R), \tag{2.51}$$

being $u''(R) \equiv d^2u(R)/dR^2$. For large distances, $V(R) \to 0$; therefore, $u(R)$ must be a linear function of R. In particular, we know that for $R \to R^*$, $u(R)/R = \bar{\psi}$ and hence the solution of Eq. (2.51) for $R > R^*$ reads as

$$u(R) = (R - a)\bar{\psi}, \tag{2.52}$$

where a is the scattering length for the electron–perturber interaction. To understand in more detail the meaning of scattering length it is worth comparing Eq. (2.52) with Eq. (2.33) assuming $l = 0$, which results in $c_1 = 1$ and $c_2 = -a$. Plugging these results into Eq. (2.37) a relation between phase shift and scattering length is obtained as

$$a = -\lim_{k \to 0}\frac{\tan\delta_0(k)}{k}. \tag{2.53}$$

Next, employing the definition of the elastic cross section in Eq. (2.30) and taking into account that for $k \to 0$, $\delta_0 \to 0$ and hence $\tan\delta_0 \approx \sin\delta_0$, one finds that for low-collision energies the elastic cross section is constant and is given by

$$\sigma = 4\pi a^2. \tag{2.54}$$

However, for identical particles it would be

$$\sigma = 8\pi a^2. \tag{2.55}$$

The threshold behavior for the elastic cross section was first observed in nuclear physics. In particular, Fermi and Marshall observed that neutron-nuclei scattering at low energies is isotropic [14]. That is, the differential cross section is constant independently of the scattering angle and hence the cross section.

Let us go back to the energy shift of the Rydberg electron in the presence of perturbing atoms. Assuming a single perturber atom per unit of volume and taking into account Eqs. (2.51) and (2.52), the effect of the perturber atom on the Rydberg electron is given by

$$-\frac{2m_e}{\hbar^2}\overline{V_i(\mathbf{R}_i - \mathbf{R})\psi(\mathbf{R})} = -\frac{2m_e}{\hbar^2}\int V(R)\psi(R)4\pi R^2 dR =$$

$$-4\pi \int u''(R)R\,dR = -4\pi[u'(R)R - u(R)]_0^R = 4\pi a\bar{\psi}. \tag{2.56}$$

Therefore, for a density ρ of perturbers, Eq. (2.46) will read as

$$-\frac{\hbar^2}{2m_e}\Delta\bar{\psi}(\mathbf{R}) - \left[E - U(\mathbf{R})\right]\bar{\psi}(\mathbf{R}) + \frac{2\pi\rho\hbar^2}{m_e}\bar{\psi}(\mathbf{R}) = 0, \tag{2.57}$$

explaining the shift proportional to the density that Amaldi and Segrè observed.

2.4   Inelastic Cross Section

In inelastic collisions the internal states of one or of both colliding partners change within a scattering event. Each of the different internal states of the colliding partners are labeled as fragments, whereas each possible combination between fragments and partial waves[3] is called a channel. In other words, in the asymptotic region, the fragments stand for the internal levels of the colliding partners, since there is no interaction. However, as the two particles approach each other, the fragments split into different channels associated with different partial waves, as shown in Fig. 2.8. The fragments are related to good quantum numbers that describe the internal state of the partners and the scattering observables.

---

[3]A partial wave is each of the centrifugal quantum numbers, l, that contribute to the scattering process at a given collision energy.

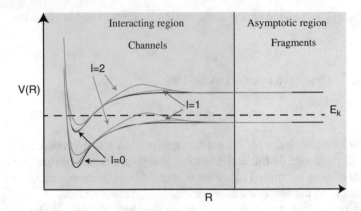

Fig. 2.8 Inelastic collision. The fragments are represented by the red lines and define the asymptotic energy of the internal states of the system under consideration. In the region of interaction, each fragment correlates with channels associated with each value of l, or different partial waves. Therefore, every channel feels a different interaction energy. E_k stands for the collision energy. In this case there are three open channels and three closed ones

Let us assume a collision between a particle with internal states labeled by i and a particle without structure. In that case, Eq. (2.29) can be generalized to include the dependence on the internal states leading to the state-to-state cross section defined as [15]

$$\sigma_{ij}(k) = \frac{\pi}{k_i^2 g_i} \sum_J \sum_l \sum_{l'} (2J+1)|\delta_{i,j}\delta_{l,l'} - S_{il;jl'}^J|^2, \tag{2.58}$$

where g_i indicates a degeneracy factor associated with the nature of the internal states of the particle, and k_i is the wave vector for the fragment i. J is the total angular momentum of the system that replaces l in Eq. (2.29), whereas l and l' are the labels for the centrifugal quantum number associated with the entrance and outgoing channels respectively. The state-to-state cross section is crucial to understanding inelastic and reactive processes.

From the state-to-state cross section, the elastic cross section is given by

$$\sigma_{ii}(k) = \frac{\pi}{k_i^2 g_i} \sum_J \sum_l (2J+1)|1 - S_{il;il}^J|^2, \tag{2.59}$$

whereas the inelastic cross section is defined as

$$\sigma_i^r(k) = \frac{\pi}{k_i^2 g_i} \sum_{j \neq i} \sigma_{ij}(k) = \frac{\pi}{k_i^2 g_i} \sum_{j \neq i} \sum_J \sum_l \sum_{l'} (2J+1)|S_{il;jl'}^J|^2. \tag{2.60}$$

But, taking into account the unitarity of the S-matrix (see Sect. 2.4.1, Eq. (2.60)) yields

$$\sigma_i^r(k) = \frac{\pi}{k_i^2 g_i} \sum_J \sum_l (2J+1) \left(1 - |S_{il;il}^J|^2\right); \tag{2.61}$$

therefore, the total cross section for the fragment i reads as [4]

$$\sigma_i^T(k) = \sigma_{ii}(k) + \sigma_i^r(k) = \frac{2\pi}{k_i^2 g_i} \sum_J \sum_l (2J+1) \left(1 - \Re(S_{il;il}^J)\right), \tag{2.62}$$

where $\Re()$ stands for the real part of the element within parenthesis. Equation (2.62) is a remarkable result: the total cross section only depends on the diagonal elements of the S-matrix [4]. In other words, the total cross section is dominated by the isotropic component of the two-body potential.

2.4.1 Unitarity of the S-Matrix

We have seen in Sect. 2.2.1 that the scattering wave function is always a combination of a convergent wave toward the scattering center (the term proportional to e^{-ikR} in Eq. (2.24)) and a divergent wave as the outgoing wave (the term proportional to e^{ikR} in Eq. (2.24)). The number of particles has to be conserved; therefore, the flux associated with the divergent and convergent waves must be the same. In a single-channel problem, the S-matrix is a complex scalar with unit magnitude. However, in the presence of different channels, the S-matrix becomes a matrix, and the conservation of particles implies that it has to be a unitary matrix, i.e.,

$$S^\dagger S = \mathcal{I}, \tag{2.63}$$

where $\mathcal{I}$ is the identity matrix, or

$$\sum_j |S_{ij}|^2 = 1. \tag{2.64}$$

The unitarity of the S-matrix is also related to the time-reversal symmetry of the scattering problem [16].

2.4.2 Wigner Threshold Law for Inelastic Scattering

We have seen in Sect. 2.4 that the presence of inelastic channels affects the properties of the elastic cross section owing to the unitarity of the S-matrix. In particular, the scattering length will become an imaginary magnitude, i.e.,

$$a = \alpha - \iota\beta. \tag{2.65}$$

Therefore, the Wigner threshold law derived in Sect. 2.2.2 will read as

$$\sigma = 4\pi(\alpha^2 + \beta^2), \tag{2.66}$$

and the s-wave phase shift as

$$\delta_0(k) = -k(\alpha - \iota\beta), \tag{2.67}$$

which comes from Eq. (2.53).

We have already defined the total inelastic cross section in terms of the S-matrix through Eq. (2.61). However, for the sake of clarity, in this discussion, we neglect the degeneracy factor g_i and assume that the collision occurs in the s-wave regime ($l = 0$). Therefore, the total inelastic cross section reads as

$$\sigma^r(k) = \frac{\pi}{k_i}\left(1 - |S_{ii}^0|^2\right). \tag{2.68}$$

The elastic S-matrix has the same functional form as in Sect. 2.2.1; thus,

$$S_{ii}^l = e^{2\iota\delta_l^i(k)}, \tag{2.69}$$

but, in the case of inelastic channels the low-energy phase shift for the fragment i reads as $\delta_0^i(k) = k(-\alpha_i + \iota\beta_i)$, which is a generalization of Eq. (2.67). Therefore, the zero-energy limit of the total inelastic cross section is given by [17]

$$\lim_{k \to 0} \sigma^r(k) \approx \frac{4\pi\beta}{k}, \tag{2.70}$$

which is the so-called Wigner threshold law for inelastic collisions [11]. This threshold law predicts that at ultracold temperatures, temperatures below the van der Waals energy ~ 1 mk (see Chap. 1), the inelastic processes increase as $1/k$ or $1/\sqrt{E_k}$, which is very different from the constant behavior predicted for the elastic cross section.

So far, we have obtained the Wigner threshold laws for ultracold temperatures. However, we do not know when these threshold laws start to describe the elastic and inelastic cross sections adequately. In Chap. 1 we have introduced, in Sect. 1.2.2 the concept of van der Waals energy as the energy threshold for the ultracold regime. Indeed, this energy also defines the beginning of the region where the Wigner threshold may be applicable.

2.5   Scattering Resonances

The presence of resonances in scattering observables is one of the most prominent features of the intrinsic quantal nature of atomic and subatomic systems. Resonances, in general, emerge as a consequence of interference effects or couplings between discrete and continuum states. At ultracold temperatures, the scattering observables can be tuned through external fields that control the strength of the interaction between discrete and continuous states [18]. In this section, we briefly expose the main types of resonances relevant for cold and ultracold chemistry.

2.5.1   Shape Resonances or Orbiting

Shape resonances (in physics) or orbiting (in chemistry) are single-channel resonances that occur when the scattering state is resonant with a quasi-bound state of the same channel. The centrifugal barrier in the collision modifies the potential interaction through the creation of a potential barrier (see Fig. 2.9). When this barrier is sufficiently high, the effective potential (potential + centrifugal barrier) may show bound states with positive energy E_{nl}. However, these states will tunnel out the barrier, as shown in Fig. 2.9, and decay in a given time, Γ_{nl}. Therefore, they are quasi-bound states. The presence of quasi-bound states will modify the scattering phase shift taking the Breit–Wigner form

$$\delta_l(k) = \delta_l^0(k) + \arctan\left(\frac{\Gamma_{nl}}{2(E_{nl} - \frac{\hbar^2 k^2}{2\mu})}\right), \tag{2.71}$$

where $\delta_l^0(k)$ varies smoothly with the collision energy $E_k = \frac{\hbar^2 k^2}{2\mu}$ and hence the elastic cross section yields

$$\sigma(k) = \frac{4\pi}{k^2} \sum_l (2l + 1)$$

$$\left[\sin^2(\delta_l^0(k)) + \frac{\Gamma_{nl}^2 \cos(2\delta_l^0(k)) + 2\Gamma_{nl}(\frac{\hbar^2 k^2}{2\mu} - E_{nl})\sin(2\delta_l^0(k))}{4(\frac{\hbar^2 k^2}{2\mu} - E_{nl})^2 + \Gamma_{nl}^2}\right]. \tag{2.72}$$

Equation (2.72) shows that the shape of the cross section strongly depends on the value of $\delta_l^0(k)$, which is an inherent property of the two-body potential.

Fig. 2.9 Orbiting or shape resonance. A scattering state with collision energy E_k is resonant with a quasi-bound state of energy E_{nl} and lifetime Γ_{nl}. The potential $V_l(R) = V(R) + \frac{\hbar^2 l(l+1)}{2\mu R^2}$, i.e., is the result of the two-body interaction potential plus the centrifugal barrier for quantum angular momentum l

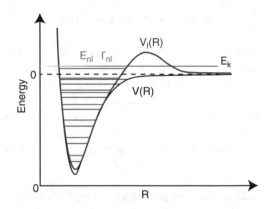

2.5.2 Fano–Feshbach Resonances

A Fano–Feshbach resonance, in brief, occurs when a scattering state is close in energy to a bound state of a closed channel. Therefore, it is a pure quantum mechanical phenomenon without classical analog. The first evidence for this resonant behavior was found and studied in the context of nuclear physics by E. Feshbach [19, 20]. However, U. Fano developed a more general description of the nature of these novel resonances [21]. For that reason, we believe that it is more accurate to use the term Fano–Feshbach resonance rather than Feshbach resonance.

Since the first observation of Fano–Feshbach resonances in an ultracold atomic system [22], interest in them has increased exponentially. The presence of Fano–Feshbach resonances allows atomic and molecular interactions to be controlled and tailored that have applications in the study of the stability of BECs and the generation of solutions in ultracold systems [23–29], as well as the formation of ultracold molecules [30–32].

When a scattering state is close to a bound state of a closed channel, the interaction between these two states dramatically changes the scattering state if the collision energy is low enough [33]. Let us assume a two-channel scattering problem such as the one shown in Fig. 2.10. The scattering occurs at ultracold temperatures, and one of the channels is open, and another is closed, labeled as $V_{bg}(r)$ and $V_c(r)$ respectively. These channels are coupled to some external field, either magnetic or electric. The channels show a different response to variations on the external field. In this case, varying the strength of the external field, one may control the energy difference between the scattering state and the bound state of the closed channel (see Fig. 2.10). For concreteness, in the case of atoms and molecules in an external magnetic field the scattering length is given by [34]

$$a(B) = a_{bg}\left(1 - \frac{\Delta}{B - B_0}\right). \tag{2.73}$$

Fig. 2.10 Two-level
Fano–Feshbach resonance
model. A Fano–Feshbach
resonance appears when a
scattering state (continuum)
couples to a bound state, E_c,
of a closed channel $V_c(R)$
(discrete). An external field
can be employed to shift the
relative energy between the
channels, which changes the
strength of the coupling
between the discrete and the
scattering state. Therefore,
the scattering length may be
modified by external fields

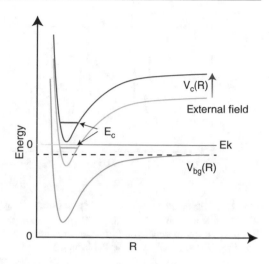

where a_{bg} is the scattering length in the absence of resonances, B_0 is the resonance
position (the magnetic field that satisfies that $E_c = E_k$), and Δ is the width of the
resonance.

Equation (2.73) shows how the scattering length changes as a function of
the external field strength in the proximity of a Fano–Feshbach resonance. As
the external field keeps further changing, eventually, another bound state of a
closed channel may be resonant with the collision energy, thus appearing as a
new resonance. Therefore, through the scan of an external field, one may gain
information about the bound states[4] of the system at hand. As an example, in
Fig. 2.11 we show the elastic cross section for $^{17}O_2(^3\Sigma_g^-)$–$^{17}O_2$ $(^3\Sigma_g^-)$ collisions
in the lowest energy state, for the range of magnetic fields considered, and at $1\,\mu K$
of collision energy. As a result, it is observed that the elastic cross section changes
in more than two orders of magnitude through a Fano–Feshbach resonance, which
shows the capabilities of these resonances to tailor O_2–O_2 interactions. Also, the
shape of the cross section is readily described by Eq. (2.73).

The density of Fano–Feshbach resonances per unit of the magnetic field (in G)
depends on the atoms and molecules of the system, collision energy, and entrance
channel. For instance, for ultracold alkali atoms, the density of Fano–Feshbach
resonances is $\sim 10^{-3}\,G^{-1}$. For molecules, the density of Feshbach resonances
is $\sim 10^{-2}\,G^{-1}$ for O_2–O_2 collisions [35, 36], which is a little bit higher than
alkali–alkali collision due to the more anisotropic molecule–molecule interaction.
However, for atoms with sizeable magnetic dipole moments like erbium (Er) or
dysprosium (Dy), the density is $\sim 10\,G^{-1}$ [37, 38]. This striking difference in the
density of resonances is due to the sizeable atom–atom anisotropy interaction and

---

[4]We should talk about quasi-bound states, since they are above the threshold of the entrance and
channel, and hence they are not truly bound states.

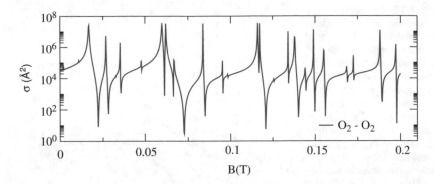

Fig. 2.11 Elastic cross section (in Å^2) for O_2–O_2 collisions as a function of the magnetic field in T. The state is $|NM_NM_S\rangle = |00-1\rangle$ (this is an approximate labeling) and the collision energy is $1\,\mu\text{K}$. For more detailed information look at Ref. [35]. (Figure adapted with permission from Ref. [35] Copyright (2020) (AIP Publishing))

dipole–dipole interaction that Er and Dy show. Surprisingly enough, the distribution of resonances in these systems is universal, and it is predicted by random matrix theory [39], leading to the field of quantum chaos [40–42]. The same is applicable to ultracold dipolar gases, i.e., a gas showing dipole–dipole interaction (either magnetic or electric) at ultracold temperatures [37, 38, 43].

2.5.3 The Glory Effect

In general, for collisions, some trajectories with small impact parameters experience the attractive and repulsive part of the inter-particle potential, leading to a small scattering angle. These trajectories may interfere with those with large impact parameters that do not feel the potential. This interference effect is known as the glory effect, which leads to oscillations in the cross section known as glory undulations [44, 45]. Therefore, the glory effect does not have a classical analog.

Let us assume a two-body collision at a given collision energy E_k in a central potential $V(R)$. Therefore, the local kinetic energy is given by $E_k - V(R)$. However, for particles showing a significant impact parameter, the local kinetic energy is better described by the collision energy. In this scenario, taking into account that $E_k = \hbar^2 k^2/(2\mu)$, it is possible to introduce the refraction index as[5]

$$n(R,k) = \sqrt{\frac{k^2}{k^2 - V(R)}}. \tag{2.74}$$

---

[5]The refraction index in optics is analogous to the role of potential on classical mechanics [46,47].

Fig. 2.12 An optical analog
to the glory effect. A
trajectory with an impact
parameter b_g scatters with a
small scattering angle owing
to the attractive nature of the
underlying potential
interaction. This trajectory
will interfere with trajectories
with a large impact parameter.
The potential interaction is
characterized by the well
depth of the interaction, ϵ,
and its position, R_m, as well
as the classical turning point
at zero energy, σ^*

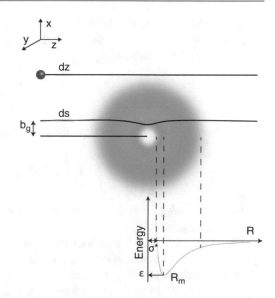

In this context, the refraction index is the ratio between the velocity of the colliding
particles without any interaction and its local velocity. From the refraction index it is
possible to establish the change on the path length between trajectories with a large
impact parameter and those with a small impact parameter (feeling the potential)
as [46–48]

$$\Delta(k) = k \int_{-\infty}^{\infty} dz - k \int_{-\infty}^{\infty} n(R, k) ds, \tag{2.75}$$

where λ is the wavelength associated with the vector wave $k = 2\pi/\lambda$, and dz
indicates that the trajectories are moving in the z direction, as shown in Fig. 2.12. In
the second term of Eq. (2.75), the integral is over the arc length ds, since we need to
keep in mind that the trajectory is deflected owing to the presence of the interaction
potential. When the collision energy is larger than the depth of the potential ϵ,
Eq. (2.75) reads as

$$\Delta(k) \approx -\frac{\mu}{\hbar^2 k} \int_{-\infty}^{\infty} V(R) dz. \tag{2.76}$$

Keeping in mind that $\Delta(k) = 2\delta_g(k)$ [49] and $R^2 = z^2 + b^2$ we find

$$\delta_g(k) \approx -\frac{\mu}{\hbar^2 k} \int_b^{\infty} \frac{V(R) R dR}{\sqrt{R^2 - b^2}}, \tag{2.77}$$

which is equivalent to the Massey–Mohr expression [50]. Finally, assuming that v is the velocity of the trajectory, and R_m the position of the potential minimum, Eq. (2.77) can be expressed as

$$\delta_g(k) = -\frac{\epsilon R_m}{\hbar v} F, \tag{2.78}$$

where F is a factor that depends on the area of the potential, which is defined as

$$\int_{\sigma*}^{\infty} V^*(R^*) dR^*, \tag{2.79}$$

where $\sigma^* = \sigma/R_m$, σ is the classical turning point at zero kinetic energy, $V(\sigma) = 0$, $R^* = R/R_m$, and $V^* = V/\epsilon$.

The cross section shows an oscillatory behavior as a consequence of the phase $\delta_g(k)$, which are known as glory undulations. In particular, the anti-nodes of the glory undulations will appear every time there is a constructive interference in Eq. (2.78), i.e., $\delta_g(k) = 2m\pi$, with $m = 0, 1, 2, \ldots$. Therefore, at higher collision energies, the maxima of the glory undulations will be closer than at low collision energies. In the same vein, for deeper potentials, the maxima of the undulations will be closer than for shallow potentials, and the same physics hold for the value of R_m.

Since glory undulations are a consequence of the interference between trajectories, its amplitude will be modulated by the product of the amplitude of the interfering trajectories. Following this argument, it can be shown that the amplitude of the glory undulations is given by [49]

$$A_g \propto \sqrt{\frac{-2\pi}{k\theta'_g}}, \tag{2.80}$$

where $\theta'_g = (d\theta/db)_{b_g}$. From Eq. (2.80), it is clearly seen that for higher collision energies, the smaller the amplitude of the undulations. Another interesting aspect of the glory undulations is that the number of glory undulations is equal to the number of bound states sustained by the isotropic term of the interaction potential, as Bernstein showed based on Levinson's theorem [51].

References

1. Levine RD, Bernstein RB (1987) Molecular reaction dynamics and chemical reactivity. Oxford University Press, New York
2. Schwabl F (2007) Quantum mechanics. Springer, Berlin
3. Taylor JR (2000) Scattering theory. Dover, New York
4. Landau LD, Lifshitz EM (1958) Quantum mechanics. Butterworth-Heinemann. Oxford, UK
5. Levine RD (2005) Molecular reaction dynamics. Cambridge University Press, Cambridge
6. Zhdanov VM (2002) Transport processes in multicomponent plasma. Taylor and Francis, London

7. Schiff LI (2010) Quantum mechanics. Mc Graw-Hill, New York
8. Galindo A, Pascual P (1990) Quantum mechanics, vols. I and II. Springer, Berlin
9. Abramowitz M, Stegun IA (1972) Handbook of mathematical functions. Dover, New York
10. Whittaker ET, Watson GN (1927) A course of modern analysis. Cambridge University Press, New York
11. Wigner EP (1948) Phys Rev 73:1002
12. Fermi E (1934) Sopra lo spostamento per pressione delle righe elevate delle serie spettrali. Nouvo Cimento 11:157
13. Amaldi E, Segrè E (1934) Effetto della pressione sui termini elevati degli alcalini. Il Nuovo Cimento (1924–1942) 11(3):145. https://doi.org/10.1007/BF02959828
14. Fermi E, Marshall L (1947) On the interaction between neutrons and electrons. Phys Rev 72:1139. https://doi.org/10.1103/PhysRev.72.1139
15. Arthurs AM, Dalgarno A, Bates DR (1960) The theory of scattering by a rigid rotator. Proc R Soc Lond Ser A Math Phys Sci 256(1287):540. https://doi.org/10.1098/rspa.1960.0125
16. Mott NF, Massey H (1965) The theory of atomic collisions. Oxford University Press, New York
17. Balakrishnan N, Kharchenko V, Forley RC, Dalgarno A (1997) Complex scattering lengths in multi-channel atom-molecule collisions. Chem Phys Lett 280:5
18. Krems RV (2018) Molecules in electromagnetic fields. Wiley, New York
19. Feshbach H (1958) Unified theory of nuclear reactions. Ann Phys 5(4):357
20. Feshbach H (1962) A unified theory of nuclear reactions. II. Ann Phys 19(2):287
21. Fano U (1961) Effects of configuration interaction on intensities and phase shifts. Phys Rev 124:1866. https://doi.org/10.1103/PhysRev.124.1866
22. Inouye S, Andrews MR, Stenger J, Miesner HJ, Stamper-Kurn MD, Ketterle W (1998) Nature 392:151
23. Donley EA, Clausen NR, Cornish SL, Roberts JL, Cornell EA, Wieman CE (2001) Nature 412:295
24. Roberts JL, Clausen NR, Burke JP Jr, Greene CH, Cornell EA, Wieman CE (2001) Phys Rev Lett 86:010403
25. Cornish SL, Thompsom ST, Wieman CE (2006) Phys Rev Lett 96:1740401
26. Khaykovich L, Schreck F, Ferrari G, Bourdel T, Cubizolles J, Carr LD, Castin Y, Salomon C (2002) Science 296:1290
27. Strecker KE, Partdrige GB, Truscott AG, Hullet RG (2002) Nature 417:150
28. Weber T, Herbig J, Nägerl HC, Grimm R (2003) Phys Rev Lett 91:123201
29. Rychtarik D, Engeser B, Nägerl HC, Grimm R (2004) Phys Rev Lett 92:173003
30. Ni KK, Ospelkaus S, de Miranda MHG, Pe'er A, Neyenhuis B, Zirbel JJ, Kotochigova S, Julienne PS, Jin DS, Ye J (2008) A high phase-space-density gas of polar molecules. Science 322(5899):231. http://science.sciencemag.org/content/322/5899/231.abstract
31. Danzl JG, Haller E, Gustavsson M, Mark MJ, Hart R, Bouloufa N, Dulieu O, Ritsch H, Nägerl H-C (2008) Quantum gas of deeply bound ground state molecules. Science 321(5892):1062. https://science.sciencemag.org/content/321/5892/1062.full.pdf, https://science.sciencemag.org/content/321/5892/1062
32. Danzl JG, Mark MJ, Haller E, Gustavsson M, Hart R, Adegunde J, Hutson JM, Nägerl HC (2010) An ultracold high-density sample of rovibronic ground-state molecules in an optical lattice. Nat Phys 6:265
33. Chin C, Grimm R, Julienne PS, Tiesinga E (2010) Feshbach resonances in ultracold gases. Rev Mod Phys 82:1225
34. Moerdijk AJ, Verhaar BJ, Axelsson A (1995) Resonances in ultracold collisions of ^{6}Li, ^{7}Li, and ^{23}Na. Phys Rev A 51:4852
35. Pérez-Ríos J, Campos-Martínez J, Hernández MI (2011) Ultracold o2 + o2 collisions in a magnetic field: on the role of the potential energy surface. J Chem Phys 134(12):124310. https://doi.org/10.1063/1.3573968
36. Tscherbul TV, Suleimanov YV, Aquilanti V, Krems RV (2009) Magnetic field modification of ultracold molecule–molecule collisions. New J Phys 11(5):055021

37. Maier T, Kadau H, Schmitt M, Wenzel M, Ferrier-Barbut I, Pfau T, Frisch A, Baier S, Akiawa K, Chomaz L, Mark MJ, Ferlaino F, Makrides C, Tiesinga E, Petrov A, Kotochigova S (2015) Emergence of chaotic scattering in ultracold Er and Dy. Phys Rev X 5:041029
38. Frish A, Mark M, Aikawa K, Ferlaino F, Bohn JL, Makrides C, Petrov A Kotochigova S (2014) Quantum chaos in ultracold collisions of gas-phase erbium atoms. Nature (London) 507:475
39. Metha ML (1967) Random matrices and the statistical theory of energy levels. Academic Press, New York
40. Gutzwiller MC (1990) Chaos in classical and quantum mechanics. Springer, New York
41. Izrailev FM (1990) Simple models of quantum chaos: spectrum and eigenfunctions. Phys Pep 196:299
42. Bohigas O, Giannoni MJ, Schmidt C (1984) Characterization of chaotic quantu spectra and universality of level fluctuation laws. Phys Rev Lett 52:1
43. Yang BC, Pérez-Ríos J, Robicheaux F (2017) Classical fractals and quantum chaos in ultracold dipolar collisions. Phys Rev Lett 118:154101. https://doi.org/10.1103/PhysRevLett.118.154101
44. Child MS (1974) Molecular collision theory. Academic Press, New York
45. Ford KW, Wheeler JA (1959) Semiclassical description of scattering. Ann Phys 7(3):259
46. Guenther RD (1990) Modern optics. Wiley, New York
47. Born M, Wolf E (1999) Priciples of optics. Cambridge University Press, New York
48. Audouard P, Duplàa E, Vigué V (1995) Europhys Lett 32:397
49. Kong P, Mason EA, Munn RJ (1969) Am J Phys 38:294
50. Massey HSW, Mohr CBO (1934) Proc R Soc Lond A 144:188
51. Bernstein RB (1962) J Chem Phys 37:1880

Ultracold Gases

3

3.1 Bose–Einstein Condensation

The observation of Bose–Einstein condensation (BEC) in atomic gases [1–3] has
changed the paradigm of chemical physics (or atomic, molecular, and optical
physics, in brief, AMO) with the birth of a new field: ultracold physics (see Fig. 3.1).
Therefore, we believe that we should provide a basic, but necessary, introduction to
the physics of ultracold gases to pave the way for ultracold chemistry.

There is a vast body of literature on the topic of BEC owing to its importance
in condensed matter physics, and hence most of the work in the field treats BEC
from a condensed matter physicist's point of view. However, in this chapter we will
follow an atomic physics approach, which is closely related to chemical physics and
physical chemistry.

3.1.1 The Ideal Bose Gas

The ideal Bose gas is best described within the grand canonical ensemble. The grand
canonical potential characterizes the grand canonical ensemble, $\Omega = E - TS - \mu N$, where the temperature T and chemical potential μ are the natural variables.
S is the entropy, E is the internal energy of the gas, and N is the total number
of particles. The chemical potential appears as a consequence that the number of
particles remains constant in the grand canonical ensemble. The grand canonical
potential reads as [4,5]

$$\Omega = -k_B T \ln Z, \tag{3.1}$$

where k_B is the Boltzmann constant, and Z is the partition function defined as [4,5]

© Springer Nature Switzerland AG 2020
J. Pérez Ríos, *An Introduction to Cold and Ultracold Chemistry*,
https://doi.org/10.1007/978-3-030-55936-6_3

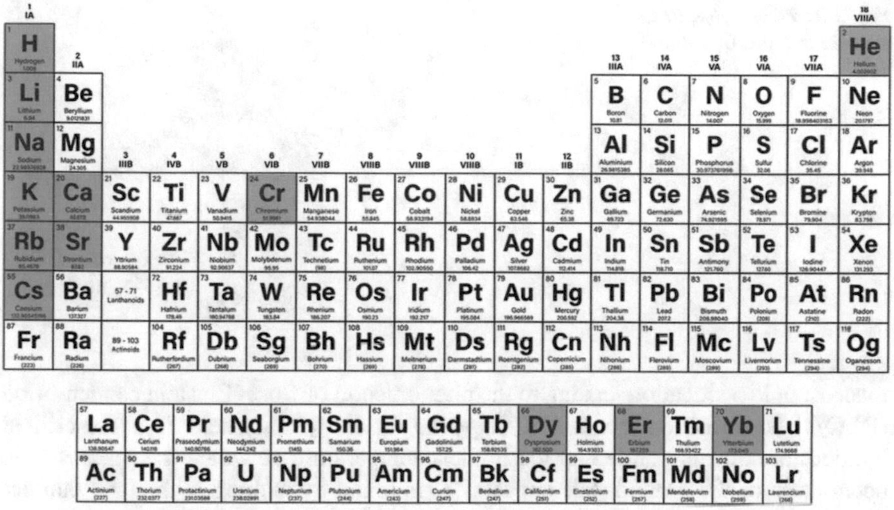

1 IA												13 IIIA	14 IVA	15 VA	16 VIA	17 VIIA	18 VIIIA
H	2 IIA																He
Li	Be											B	C	N	O	F	Ne
Na	Mg	3 IIIB	4 IVB	5 VB	6 VIB	7 VIIB	8 VIIIB	9 VIIIB	10 VIIIB	11 IB	12 IIB	Al	Si	P	S	Cl	Ar
K	Ca	Sc	Ti	V	Cr	Mn	Fe	Co	Ni	Cu	Zn	Ga	Ge	As	Se	Br	Kr
Rb	Sr	Y	Zr	Nb	Mo	Tc	Ru	Rh	Pd	Ag	Cd	In	Sn	Sb	Te	I	Xe
Cs	Ba	57-71 Lanthanoids	Hf	Ta	W	Re	Os	Ir	Pt	Au	Hg	Tl	Pb	Bi	Po	At	Rn
Fr	Ra	89-103 Actinoids	Rf	Db	Sg	Bh	Hs	Mt	Ds	Rg	Cn	Nh	Fl	Mc	Lv	Ts	Og

La	Ce	Pr	Nd	Pm	Sm	Eu	Gd	Tb	Dy	Ho	Er	Tm	Yb	Lu
Ac	Th	Pa	U	Np	Pu	Am	Cm	Bk	Cf	Es	Fm	Md	No	Lr

Fig. 3.1 Atomic species that have been cooled down until reaching the BEC. Helium has been only achieved in its metastable state, since the ground state is closed shell. Dy, Er, and Cr are highlighted because of their magnetic dipole moment. Indeed, the BEC of these atoms show magnetic dipole interactions and are called dipolar BECs

$$Z = \sum_i \frac{1}{e^{\beta(\epsilon_i - \mu)} - 1}. \tag{3.2}$$

Here, ϵ_i stands for the i-th energy state of the system and $\beta = (k_B T)^{-1}$. Finally, taking into account that the total number of particles is the conjugate variable of μ

$$N = -\frac{\partial \Omega}{\partial \mu} = \sum_i n_i, \tag{3.3}$$

with $i = 0, 1, \ldots, \mathcal{N}$, where $\mathcal{N}$ is the number of states, and by means of Eqs. (3.1)–(3.3), we find

$$n_i = \frac{1}{e^{\beta(\epsilon_i - \mu)} - 1}. \tag{3.4}$$

The chemical potential can be defined as a function of N and T by taking into account that the total number of particles must be equal to the sum of the occupancies of the individual levels given by Eq. (3.4). Moreover, $\mu \leq \epsilon_0$, otherwise the occupation of the ground state will be negative.

For bosons, there is no constraint regarding the occupation number of a given state ϵ_i. For a large number of bosons N the zero point energy of the system may be neglected, and hence $\epsilon_0 = 0$, which is the minimum energy of a boson in a box. In fact, under appropriate conditions, the ground state of N bosons shows a

Fig. 3.2 Phase space of a particle in a box of volume L^3

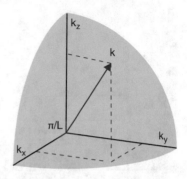

macroscopic occupation leading to the phenomenon of Bose–Einstein condensation (BEC). The transition into a BEC happens for $T \leq T_c$, where T_c is the critical temperature and is defined as the highest temperature at which the macroscopic occupation of the ground state appears. T_c is found by making N the total number of particles in the excited states when $\mu = \epsilon_0$, which in the present case reads as

$$N = N_T(\mu = \epsilon_0, T_c) = \int_0^\infty \frac{g(E)dE}{e^{\beta E} - 1}, \tag{3.5}$$

where the summation has been substituted by an integral in the energy and the density of states $g(E)$ needs to be included to account for the number of states properly. $g(E)$ is a system-dependent magnitude, and hence the BEC will depend on the system at hand. Here, for the sake of simplicity, we are studying the BEC of N bosons in a box, i.e, $\epsilon_0 = 0$.

Let us assume N non-interactive bosons of mass m in a box with side L and volume $V = L^3$. In this scenario the single-particle Hamiltonian is

$$H = -\frac{\hbar^2}{2m}\nabla^2, \tag{3.6}$$

and its eigenvalues are

$$E = \frac{\hbar^2\pi^2}{2mL^2}\left(n_x^2 + n_y^2 + n_z^2\right) \equiv \frac{\hbar^2}{2m}k^2. \tag{3.7}$$

The number of states with energy equal to or less than E will be equivalent to the number of states in k-space with magnitude less than or equal to $k = \sqrt{2mE}/\hbar$, which is given by the octave of the sphere of radius k in k-space, shown in Fig. 3.2, as

$$\mathcal{N}(k) = \frac{1}{8}\frac{4\pi}{3}k^3\left(\frac{L}{\pi}\right)^3. \tag{3.8}$$

Here, we have kept in mind that the unit of volume in k-space is $(\pi/L)^3$. Thus, the density of states for the ideal Bose gas in a box reads as

$$g(E) = \frac{d\mathcal{N}(k)}{dk}\frac{dk}{dE} = \frac{Vm^{3/2}}{2^{1/2}\pi^2}\sqrt{E}. \tag{3.9}$$

Now, we are ready to calculate the critical temperature for BEC following Eq. (3.5) that through Eq. (3.9) reads as

$$N = \int_0^\infty \frac{g(E)dE}{e^{\beta E}-1} = \frac{Vm^{3/2}}{2^{1/2}\pi^2}\int_0^\infty \frac{\sqrt{E}dE}{e^{\beta E}-1} = \frac{V}{2^{1/2}\pi^2}\left(\frac{m}{\beta}\right)^{3/2}\zeta(3/2)\Gamma(3/2), \tag{3.10}$$

where $\zeta(x)$ is the Riemann zeta function evaluated at x and $\Gamma(x)$ is the Euler gamma function of argument x. Solving Eq. (3.10) for T and taking into account $\Gamma(3/2) = \pi/\sqrt{2}$, it is found [6, 7]

$$k_B T_c = \frac{2\pi}{m}\hbar^2\left(\frac{\rho}{\zeta(3/2)}\right)^{2/3}, \tag{3.11}$$

where ρ is the density of the Bose gas. From Eq. (3.11) it is noticed that the critical temperature depends on the density of the gas. In particular, the lower the density, the lower the temperature needs to be to reach the BEC region. Introducing the thermal de Broglie wavelength as $\lambda_T = \sqrt{\frac{2\pi\hbar^2}{mk_B T}}$ it is possible to describe the BEC condition expressed by Eq. (3.11) as

$$\lambda_T^3\rho = \zeta(3/2). \tag{3.12}$$

This equation shows the relationship between the inter-particle distance and their thermal wavelength to reach the BEC. Indeed, one notices that only when $\lambda_T \sim \rho^{-3}$ will the BEC occur since $\zeta(3/2) = \mathcal{O}(1)$. In other words, only when the interparticle distance is comparable with the intrinsic wavelength of the quantum system will BEC be possible.

The range of densities and temperatures for a BEC depends on the nature of the system at hand. However, it is good to have an idea about the temperature and density required to reach a BEC. In Fig. 3.3 the density (panel (a)) and critical temperature (panel (b)) for different atomic masses are shown. In panel (a), a temperature of 500 nK is assumed, whereas, in panel (b), a density of 10^{13} cm^3 is taken. In Fig. 3.3, it is noticed that both temperature and density have a strong dependence on the mass of the bosons, as Eq. (3.11) shows. In the same vein, the density shows a slighter steeper dependence on the mass of the atoms, according to Eq. (3.12).

The last part of this section is devoted to the study of the temperature dependence of the condensed fraction. The number of bosons in the ground state is $N_0 = N - N_T$, and in virtue of Eqs. (3.10) and (3.11), one gets

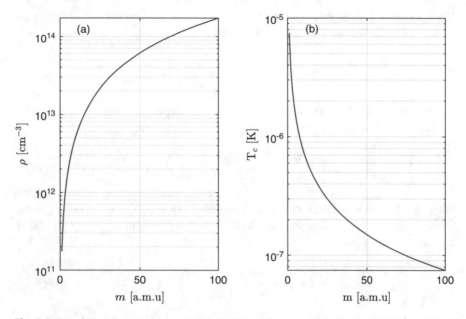

Fig. 3.3 Density and temperature needed for Bose–Einstein condensation. Panel (**a**): critical density in cm^{-3} as a function of the mass of the atoms in a.m.u. assuming a temperature of the gas of 500 nK. Panel (**a**): critical temperature (in K) as a function of the mass of the atoms (in a.m.u.) assuming a gas density of 10^{13} cm^3

$$N_0(T) = N\left[1 - \left(\frac{T}{T_c}\right)^{3/2}\right]. \tag{3.13}$$

This equation shows that at $T = 0$, all the bosons will be in the ground state. However, as the temperature increases, the ground state is depleted, and the excited states are populated until reaching the critical temperature, in which all the bosons are occupying excited states.

3.2 Interacting Bose Gas

Here, we study how the interparticle interactions affect the structure of a Bose–Einstein condensate. It is known that particles do interact, and the assumption of non-interaction particles is an idealization that may be accurate under specific circumstances. In particular, we introduce the Gross–Pitaevskii equation, which governs the dynamics of interacting bosons at $T = 0$, and the Thomas–Fermi approximation, which follows the same logic as the well-known Thomas–Fermi approximation in quantum chemistry.

3.2.1   Gross–Pitaevskii Equation

In a BEC, at $T = 0$, all the bosons occupy the same quantum state. This scenario, following a Hartree formalism, may be described as a symmetrized product of single-particle wave functions (or orbitals, in electronic structure theory) $\phi(\mathbf{r})$ as

$$\psi(\mathbf{r}_1, \ldots, \mathbf{r}_N) = \prod_{i=1}^{N} \phi(\mathbf{r}_i), \qquad (3.14)$$

where N is the number of bosons and $\phi(\mathbf{r}_i)$ is a unit normalized eigenstate of the single-particle Hamiltonian,

$$H_i^{(1)} = \left[\frac{\mathbf{p}_i^2}{2m} + V(\mathbf{r}_i) \right] \phi(\mathbf{r}_i) = E^{(1)} \phi(\mathbf{r}_i), \qquad (3.15)$$

where $V(\mathbf{r}_i)$ is the external potential, thus

$$\int |\phi(\mathbf{r}_i)|^2 d^3 \mathbf{r}_i = 1. \qquad (3.16)$$

The BEC wave function given by Eq. (3.14) only contains single-particle physics and ignores any possible correlation that may emerge as a conscquence of two-body collisions. In Sect. 2.3, we have seen that at ultracold temperatures, the elastic cross-section leads to a hard-sphere collision of radius a, the scattering length. The two-body interaction is represented through the effective potential $U_0\delta(\mathbf{r}_i - \mathbf{r}_j)$, where $U_0 = 4\pi\hbar^2 a/m$.[1] Thus, it is possible to build the effective BEC Hamiltonian

$$H = \sum_{i=1}^{N} H_i^{(1)} + \sum_{i>j} U_0\delta(\mathbf{r}_i - \mathbf{r}_j), \qquad (3.17)$$

with expectation value

$$E = \langle \psi|H|\psi \rangle = \int \prod_{i'} \phi^*(\mathbf{r}_{i'}) \sum_{i=1}^{N} H_i^{(1)} \prod_{j'} \phi(\mathbf{r}_{j'}) d^3\mathbf{r}_1 \ldots d^3\mathbf{r}_N + \quad (3.18)$$

---

[1]This is the same effective potential that Fermi derived for a Rydberg electron colliding with a neutral atom (but taking into account that for the BEC problem the electron mass has to be substituted by the reduced mass $m/2$, as we presented in Sect. 2.3, and it is explicitly explained in [8].

$$\int \prod_{i'} \phi^*(\mathbf{r}_{i'}) \sum_{i>j} U_0 \delta(\mathbf{r}_i - \mathbf{r}_j) \prod_{j'} \phi(\mathbf{r}_{j'}) d^3\mathbf{r}_1 \dots d^3\mathbf{r}_N, \quad (3.19)$$

is the energy of N bosons in a BEC, including two-body interactions characterized by the scattering length a. The energy can be finally expressed as

$$E = N \left[\int \left(-\frac{\hbar^2}{2m} \nabla^2 \phi(\mathbf{r}) + V(\mathbf{r}) \right) d^3\mathbf{r} + \binom{N}{2} \int U_0 |\phi(\mathbf{r})|^4 d^3\mathbf{r} \right], \quad (3.20)$$

where $\binom{N}{2}$ stands for the number of different ways in which a couple of bosons can be selected out of N bosons.

The inter-particle interaction may push some of the bosons to excited states, leading to the so-called *depletion* of the condensate. This phenomenon limits our description of the energy of the BEC by Eq. (3.20). However, luckily enough, the depletion of the condensate is $(\rho a^3)^{1/3}$[6, 7], where ρ is the density of particles, is $\lesssim 1\%$ in most of the experimental conditions. Therefore, unless a is of the same order of magnitude as the inter-particle distance, Eq. (3.20) should be a realistic description of the condensate state. At this point, it is useful to introduce the wave function of the condensate as

$$\Psi(\mathbf{r}) = \sqrt{N}\phi(\mathbf{r}). \quad (3.21)$$

The number of particles through Eq. (3.16) will be $N[\Psi(\mathbf{r})] = \int |\Psi(\mathbf{r})|^2 d^3\mathbf{r}$; therefore,

$$\rho(\mathbf{r}) = |\Psi(\mathbf{r})|^2, \quad (3.22)$$

and the energy becomes a function of the wave function,

$$E[\Psi(\mathbf{r})] = \int \left[-\frac{\hbar^2}{2m} \Psi^*(\mathbf{r}) \nabla^2 \Psi(\mathbf{r}) + V(\mathbf{r})|\Psi(\mathbf{r})|^2 \right] d^3\mathbf{r} + \binom{N}{2} \int U_0 |\Psi(\mathbf{r})|^4 d^3\mathbf{r}, \quad (3.23)$$

where terms $\mathcal{O}(1/N)$ have been neglected. The ground-state wave function is calculated by minimizing the energy, as in the variational principle. However, it is also necessary to impose the conservation of the number of particles N. This optimization problem with a constraint is solved through the method of Lagrange multipliers by minimizing the variation

$$\delta E[\Psi(\mathbf{r})] - \mu \delta N[\Psi(\mathbf{r})] = 0. \quad (3.24)$$

In particular, its variation with respect to $\Psi^*(\mathbf{r})$ read as

$$\frac{\delta E[\Psi(\mathbf{r})]}{\delta \Psi^*(\mathbf{r})} - \mu \frac{\delta N[\Psi(\mathbf{r})]}{\delta \Psi^*(\mathbf{r})} = 0, \tag{3.25}$$

leads to

$$-\frac{\hbar^2}{2m}\nabla^2 \Psi(\mathbf{r}) + V(r)\Psi(\mathbf{r}) + U_0|\Psi(\mathbf{r})|^2\Psi(\mathbf{r}) = \mu\Psi(\mathbf{r}). \tag{3.26}$$

This is the Gross–Pitaevskii equation and describes the dynamics of the BEC in the presence of two-body interactions. Equation (3.26) resembles the equation governing the dynamics of a Rydberg electron in a high dense media, as shown Sect. 2.3. The Gross–Pitaevskii equation establishes the role of inter-particle interaction in BEC, and hence it is an essential equation regarding the dynamics of an interacting BEC.

3.2.2   Thomas–Fermi Approximation

For a BEC with a large number of atoms, N, and repulsive interactions (positive scattering length), it can be shown that the kinetic energy term is smaller than the rest of energies in the Gross–Pitaevskii equation [6,7]; thus, Eq. (3.26) is approximated as

$$\left(V(r)\Psi(\mathbf{r}) + U_0|\Psi(\mathbf{r})|^2\right)\Psi(\mathbf{r}) = \mu\Psi(\mathbf{r}), \tag{3.27}$$

or

$$\rho(\mathbf{r}) = |\Psi(\mathbf{r})|^2 = \frac{\mu - V(r)}{U_0}. \tag{3.28}$$

Within this approximation, the energy per particle is the same in every point of the BEC. The energy per particle is the result of the external potential plus a mean-field interaction $\propto U_0\rho(\mathbf{r})$, which is the chemical potential of an interacting ultracold uniform gas with a local density equal to $\rho(\mathbf{r})$.

In the 1920s, Thomas and Fermi proposed a model for electronic structure calculations in which the electrons of an atom behave as a uniform electron gas, i.e., $\rho(r) \propto k_F^3(r)$ [9,10], where $k_F(r)$ is the local Fermi momentum [4,5,11]. The local Fermi momentum depends on the local energy due to the nuclei and electron–electron repulsion interaction. Within these assumptions it is possible to establish a direct relationship between local density and interaction energy, as in Eq. (3.28); that is the reason why Eq. (3.27) is known as the Thomas–Fermi approximation for a BEC.

The Thomas–Fermi approximation states that the density profile of a BEC is the inverse of the trapping potential, as it is noticed from Eq. (3.28). This is better illustrated through an example. Let us assume an ultracold gas of N bosons at $T =$

0, with inter-particle scattering length a immersed in a spherical harmonic trap, i.e.,

$$V(r) = \frac{1}{2}m\omega^2 r^2. \tag{3.29}$$

Therefore, through Eq. (3.28) one finds

$$\rho(r) = \frac{\left(\mu - \frac{1}{2}m\omega^2 r^2\right)}{U_0}. \tag{3.30}$$

And integrating Eq. (3.30) over the volume of the trap yields

$$N = \int_0^{\sqrt{\frac{2\mu}{m\omega^2}}} \left(\frac{\mu - \frac{1}{2}m\omega^2 r^2}{U_0}\right) 4\pi r^2 dr = \frac{2}{15}\frac{\pi 2^{5/2}\mu^{5/2}}{U_0 m^{3/2}\omega^3}. \tag{3.31}$$

It is essential to notice that the upper limit of integration goes up to the radius at which the external potential is equal to the chemical potential, i.e., $\mu = V(r)$. From Eq. (3.31) we find that the chemical potential for a spherical harmonic trap is

$$\mu = \left(\frac{15U_0 m^{3/2}\omega^3}{\pi 2^{9/2}}\right)^{2/5}. \tag{3.32}$$

This equation shows the dependency of chemical potential on the scattering length of the atomic species as well as on the properties of the trapping potential. In particular, for systems showing a larger scattering length, the chemical potential will be larger since the atomic interaction will raise the energy per particle in the BEC. Therefore, tight traps induce larger chemical potential than shallow traps. Figure 3.4, shows the results for $N = 10^4$ ^{7}Li atoms in a spherical harmonic trap with $\omega = 2\pi \times 1\,\text{kHz}$. This figure shows that within the Thomas–Fermi approximation, the trapping potential tailors the density of the BEC, as stated by Eq. (3.28).

3.2.3   Healing Length

The healing length, ξ, of a BEC is the length scale in which the BEC wave function tends to its bulk value in the presence of a localized perturbation. ξ is usually called the coherence length of the BEC. The healing length is defined as the distance for which the kinetic energy of the BEC is equal to its interaction energy [6, 7]. Thus, following the Gross–Pitaevskii equation, Eq. (3.26), we find

$$\frac{\hbar^2}{2m\xi^2} = \rho U_0, \tag{3.33}$$

Fig. 3.4 Thomas–Fermi approximation for an ultracold gas in a spherical harmonic trap. Top panel: density (in cm^{-3}) as a function of the distance (in nm) to the center of the BEC along one of the trap axes. Bottom panel: potential (in kHz) as a function of the distance (in μm) to the center of the BEC along one of the trap axes. This calculation is for 10^4 ^{7}Li atoms and $\omega = 2\pi \times 1$ kHz

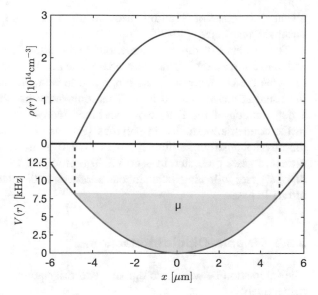

or

$$\xi = \frac{\hbar}{\sqrt{\rho m U_0}}. \tag{3.34}$$

We have seen in Sect. 2.1 that a collision occurs when the kinetic energy is similar to the interaction energy, and surprisingly enough, this is the same condition as stated in Eq. (3.33). Therefore, the healing length defines the distance in which the particles within a BEC experience a classical collision.

Under typical experimental conditions, the distance between atoms is much larger than the scattering length, and therefore the healing length is larger than the interatomic separation.

3.3   Ultracold Dipolar Gases

The possibility of realizing ultracold gases showing dipole–dipole interactions is revolutionizing the field of ultracold physics owing to their unique properties and the new possibilities in relation to those. Namely, long-range anisotropic interactions and highly tunable scattering observables, that may show novel quantum phases and an intriguing dynamics [12–14]. Dipolar interactions can be achieved through highly magnetic atoms [15–17], polar molecules [18–20], and Rydberg atoms [21, 22], and even with ultra-long-range Rydberg molecules [23]. Among these candidates, the most successful to date is the use of highly magnetic atoms thanks to the pioneering achievements for Cr [15, 24], and recently for Er and Dy. In particular, with Er and Dy, it has been possible to reveal unique characteristics of quantum ferrofluids and the emergence of supersolid phases [25–29]. Moreover, these ultracold gases revealed the onset of quantum chaos owing to its large number of magnetic Fano–Feshbach resonances [30–32].

The presence of dipolar interactions in an ultracold gas modifies the dynamics and behavior of the condensate phase drastically. The inter-particle interaction includes the characteristic long-range and anisotropic interaction term apart from the usual scattering length term. This long-range interaction leads, ultimately, to a deformation of the BEC cloud and the emergence of instabilities [26], absent in the condensates studied in previous sections of this chapter. However, most of the physics regarding a dipolar gas at $T = 0$ is similar to the one for the regular ultracold gases presented in Sect. 3.2. Therefore, we will use this scenario to study dipolar gases following an analogous approach to the one presented in Sect. 3.2 for ultracold gases.

3.3.1 Dipole–Dipole Interactions

The interaction between two dipoles with orientations $\mathbf{e}$ and $\mathbf{e}'$ placed at $\mathbf{r}$ and $\mathbf{r}'$ respectively is

$$U_{dd}(\mathbf{r}) = \frac{C_{dd}}{4\pi} \left[\frac{\mathbf{e} \cdot \mathbf{e}'}{|\mathbf{r} - \mathbf{r}'|^3} - 3 \frac{(\mathbf{e} \cdot (\mathbf{r} - \mathbf{r}'))(\mathbf{e}' \cdot (\mathbf{r} - \mathbf{r}'))}{|\mathbf{r} - \mathbf{r}'|^5} \right], \tag{3.35}$$

where C_{dd} is the dipole–dipole coupling constant, which reads as $C_{dd} = \mu_0 \mu_M^2$ for magnetic dipoles, where μ_0 is the vacuum permeability and μ_M is the magnetic dipole of the particle. For electric dipoles $C_{dd} = d^2/\epsilon_0$ where d is the electric dipole moment of the particle and ϵ_0 is the permittivity of vacuum. Equation (3.35) can be further simplified for the case of polarized dipoles, i.e., the dipoles are oriented in the same direction, leading to

$$U_{dd}(\mathbf{r}) = \frac{C_{dd}}{4\pi} \left[\frac{1}{|\mathbf{r} - \mathbf{r}'|^3} - 3 \frac{(z - z')^2}{|\mathbf{r} - \mathbf{r}'|^5} \right]. \tag{3.36}$$

Equation (3.36) shows the anisotropy of the dipolar interaction since it depends on the angle of the dipoles with respect to the vector linking the two dipoles in space. Indeed, noticing in Eq. (3.36) $(z - z')^2 \equiv \cos^2 \theta |\mathbf{r} - \mathbf{r}'|^2$, it is possible to find an angle θ_m for which $U_{dd}(\mathbf{r})$ vanishes, and it is $\theta_m = \arccos(1/\sqrt{3})$. θ_m is called the magic angle as in the nuclear magnetic resonance (NMR) community [33, 34]. However, in the case of atomic and molecular systems the magic angle is not the same as in NMR owing to the role of long-range coefficients, as has been shown for ultra-long Rydberg molecules [23].

The dipole–dipole coupling constant depends drastically on the magnetic or electric nature of the interacting dipoles or on the atomic/molecular nature of the system at hand. Magnetic moments are of the order of the Bohr magneton μ_B, which in atomic units is $\mu_B = \hbar e/2m_e$, where e is the electric charge of the electron and

Table 3.1 Dipole moments for different molecular and atomic species. The electric dipoles are in units of Debyes (D); $1D = 0.39$ ea$_0$. The magnetic moments are in units of Bohr magneton μ_B. The value for Rb$_2^*$ stands for a ultra-long-range Rydberg molecular state of Rb, in particular, a butterfly state with $n = 25$ [35]

Species	Dipole	$r_{dd}(a_0)$
^{87}Rb	$1.0\,\mu_B$	0.7
^{52}Cr	$6.0\,\mu_B$	16
^{168}Er	$7.0\,\mu_B$	69.9
^{164}Dy	$10.0\,\mu_B$	132.3
^{40}K^{87}Rb	0.57D	4.0×10^3
^{23}Na^{40}K	2.72D	4.3×10^4
Rb$_2^*$	525D	7.6×10^8

m_e is the electron mass. On the other hand, electric dipoles are of the order of ea_0, where a_0 is the Bohr radius. Therefore, the ratio of the coupling constants C_{dd} for the electric and magnetic dipoles goes as $\frac{(ea_0)^2}{\epsilon_0\mu_0(\mu_B)^2} = 4/\alpha^2 \approx 7.5 \times 10^4$, showing that the electric dipoles interact more strongly than magnetic ones. Therefore, to observe more extreme dipolar effects, it is preferable to have polar molecules rather than magnetic atoms, as shown in Table 3.1.

Dipolar interactions can be characterized by the effective range $r_{dd} = \frac{C_{dd}m}{12\pi\hbar^2}$ [12–14]. The effective range for different dipole–dipole interactions are shown in Table 3.1, where it is noted that polar molecules show in general an effective range $\sim 10^4$ larger than magnetic atoms, as expected based on the ratio for C_{dd} between electric dipoles and magnetic ones.

3.3.2 Theory of the Dipolar Bose–Einstein Condensation

The wave function of a dipolar BEC at $T = 0$ fulfils the Gross–Pitaevskii equation, which reads as

$$-\frac{\hbar^2}{2m}\nabla^2\Psi(\mathbf{r}) + V(r)\Psi(\mathbf{r}) + U_0|\Psi(\mathbf{r})|^2\Psi(\mathbf{r}) + \left[\int U_{dd}(\mathbf{r} - \mathbf{r}')|\Psi(\mathbf{r}')|^2 d^3\mathbf{r}'\right]\Psi(\mathbf{r})$$

$$= \mu\Psi(\mathbf{r}), \tag{3.37}$$

which is very similar to Eq. (3.26) but includes the dipolar interaction term. The interaction term will induce some anisotropies and instabilities on the BEC wave function, which will depend on the relative strength of the dipolar interaction with respect to the scattering length as

$$\epsilon_{dd} = \frac{r_{dd}}{a} = \frac{C_{dd}m}{12\pi\hbar^2 a}. \tag{3.38}$$

To understand the effects of dipolar interactions in an ultracold gas is preferable to work within the Thomas–Fermi approximation. Then, the density distribution $\rho(\mathbf{r})$ satisfies the following equation

$$\mu = V(\mathbf{r}) + U_0 \rho(\mathbf{r}) + \left[\int U_{dd}(\mathbf{r} - \mathbf{r}')\rho(\mathbf{r}')d^3\mathbf{r}' \right]. \tag{3.39}$$

Let us assume a dipolar gas in a harmonic trap and treat the dipolar interactions as a perturbation. Then, the unperturbed density is given by Eq. (3.30) as $\rho^0(\mathbf{r}) = \frac{\left(\mu - \frac{1}{2}m\omega^2 r^2\right)}{U_0}$, but the dipolar BEC density reads as

$$\rho(\mathbf{r}) = \frac{\mu - V(\mathbf{r}) - V_{dd}^0(\mathbf{r})}{U_0}, \tag{3.40}$$

with

$$V_{dd}^0(\mathbf{r}) = \int U_{dd}(\mathbf{r} - \mathbf{r}')\rho^0(\mathbf{r}')d^3\mathbf{r}'. \tag{3.41}$$

The integral in Eq. (3.41) is simplified by noticing that

$$\frac{1}{|\mathbf{r} - \mathbf{r}'|^3} - \frac{(z - z')^2}{|\mathbf{r} - \mathbf{r}'|^5} = -\frac{4\pi}{3}\delta(\mathbf{r} - \mathbf{r}') - \frac{\partial^2}{\partial z^2}\left(\frac{1}{|\mathbf{r} - \mathbf{r}'|}\right), \tag{3.42}$$

and hence

$$V_{dd}^0(\mathbf{r}) = \frac{C_{dd}}{4\pi}\left[-\frac{4\pi\rho^0(\mathbf{r})}{3} - \frac{\partial^2}{\partial z^2}\int \frac{\rho^0(\mathbf{r}')}{|\mathbf{r} - \mathbf{r}'|}d^3\mathbf{r}' \right]. \tag{3.43}$$

The integral in Eq. (3.43) is determined by dividing the integration region into two: a first region for $r' < r$ and a second for $r' > r$. In the first region $|\mathbf{r} - \mathbf{r}'| = r$, whereas in the second region $|\mathbf{r} - \mathbf{r}'| = r'$, leading to

$$\int \frac{\rho^0(\mathbf{r}')}{|\mathbf{r} - \mathbf{r}'|}d\mathbf{r}' = 4\pi \int_0^r \frac{\rho^0(\mathbf{r}')}{r}r'^2dr' + 4\pi \int_r^{\sqrt{\frac{2\mu}{m\omega^2}}} \frac{\rho^0(\mathbf{r}')}{r'}r'^2dr'$$

$$= \frac{4\pi}{U_0}\left[\frac{1}{r}\int_0^r \left(\mu - \omega^2 r'^2\right)r'^2dr' + \int_r^{\sqrt{\frac{2\mu}{m\omega^2}}} \left(\mu - m\omega^2 r'^2\right)r'dr' \right]$$

$$= \frac{4\pi}{U_0}\left(-\frac{\mu r^2}{6} + \frac{\mu^2}{2m\omega^2} - \frac{m\omega^2 r^4}{40} \right). \tag{3.44}$$

Next, plugging the solution of Eq. (3.44) into Eq. (3.43) yields

$$V_{dd}^0(\mathbf{r}) = \frac{C_{dd}m\omega^2}{15U_0}(r^2 - 3z^2) \equiv \frac{\epsilon_{dd}m\omega^2}{5}(r^2 - 3z^2), \qquad (3.45)$$

where we have taken into account the ratio of the dipolar force to the scattering length given by Eq. (3.38). Therefore, the density of the dipolar gas, following Eq. (3.40), is given by

$$\rho(\mathbf{r}) = \frac{1}{U_0}\left[\mu - \frac{1}{2}m\omega^2(1 - 2\epsilon_{dd}/5)(x^2 + y^2) - \frac{1}{2}m\omega^2(1 + 4\epsilon_{dd}/5)z^2\right]. \qquad (3.46)$$

Dipolar interactions modify the shape of the BEC cloud. In particular, as observed in Eq. (3.46), the presence of dipolar interaction causes an elongation of the BEC cloud toward the direction in which the dipoles are aligned. This effect occurs because the dipolar interaction is attractive when the vector joining two dipoles lies along the z axis. In contrast, it is repulsive in a perpendicular plane to it. In chemical physics language, the dipolar interaction is attractive for collision in the L geometry, whereas it is repulsive for the H geometry.

3.4   Spinor Bose–Einstein Condensates

A spinor Bose–Einstein condensate is a BEC in which the atoms are in different spin states; in other words, the total spin becomes an internal degree of freedom of the condensate. Below the critical temperature, a spinor BEC may show different ground states associated with different phases [36–42], which are controlled via an external magnetic field. In particular, the spin ordered phases may be ferromagnetic, cyclic or nematic [36,41,43]. The atoms in a spinor BEC show interesting collisional properties that can be exploited to study state-selective chemical reactions, as shown recently [44].

In a spinor BEC, all the atoms share the same spatial wave function but have different magnetic substates, m_F, which are called the components of the spinor BEC. The atoms in a spinor BEC are in a given spin state $|F_i m_{F_i}\rangle$ (in this case the total angular momentum of the atom corresponds to $F_i = S_i + I_i + l_i$). Therefore, the total spin for a two-body process is $\hat{\mathbf{F}} = \hat{\mathbf{F}}_1 + \hat{\mathbf{F}}_2$. The interactions are rotationally invariant, and hence F is the proper quantum number to describe the interaction. Therefore, the contact interaction of the Gross–Pitaevskii equation for a spinor BEC reads as [42]

$$U(\mathbf{r}_1 - \mathbf{r}_2) = \delta(\mathbf{r}_1 - \mathbf{r}_2)\sum_F U^{(F)}\mathcal{P}_F, \qquad (3.47)$$

with

$$\mathcal{P}_F = \sum_{M_F}|F_1 F_2 F M_F\rangle\langle F M_F F_1 F_2|. \qquad (3.48)$$

Here, M_F is the projection of the total spin on the quantization axis and

$$U^{(F)} = \frac{4\pi\hbar^2 a^{(F)}}{m}, \tag{3.49}$$

is the contact interaction for a given F state, where $a^{(F)}$ is the appropriate scattering length.

Let us assume the s-wave scattering of two atoms with $F_i = 1$; thus, the total spin would be $F = 0, 1, 2$. The two-body scattering wave function is an eigenfunction of the exchange operator with eigenvalue $(-1)^{F+l}$, where l stands for the centrifugal quantum number. Keeping in mind that we are working with bosons and that $l = 0$ for s-wave scattering we find that $F = 0, 2$. Therefore, $U(\mathbf{r}_1 - \mathbf{r}_2) = \delta(\mathbf{r}_1 - \mathbf{r}_2)(U^{(0)}\mathcal{P}_0 + U^{(2)}\mathcal{P}_2)$. However, it is convenient to express the interaction energy in terms of the spin of the atoms $\hat{\mathbf{F}}_1, \hat{\mathbf{F}}_2$ by means of the following relations:

$$\mathcal{P}_0 = \frac{1 - \hat{\mathbf{F}}_1 \cdot \hat{\mathbf{F}}_2}{3}$$

$$\mathcal{P}_2 = \frac{2 + \hat{\mathbf{F}}_1 \cdot \hat{\mathbf{F}}_2}{3}. \tag{3.50}$$

And finally obtaining

$$U(\mathbf{r}_1 - \mathbf{r}_2) = \delta(\mathbf{r}_1 - \mathbf{r}_2)\left[\frac{U^{(0)} + 2U^{(2)}}{3} + \frac{U^{(2)} - U^{(0)}}{3}\hat{\mathbf{F}}_1 \cdot \hat{\mathbf{F}}_2\right], \tag{3.51}$$

showing that spin–spin interactions affect the contact energy term on the wave function of the BEC, and hence they will affect the shape of the BEC. Moreover, the presence of the spin–spin interaction allows spin–exchange transitions. Spin–exchange interaction induces the change of m_F state of the atoms before and after the collision takes place. Therefore, in a spinor BEC, the number of atoms in a given component may change owing to the cited spin interactions.

References

1. Davis KB, Mewes MO, Andrews MR, van Druten NJ, Durfee DS, Kurn DM, Ketterle W (1995) Bose-einstein condensation in a gas of sodium atoms. Phys Rev Lett 75:3969. https://doi.org/10.1103/PhysRevLett.75.3969
2. Anderson MH, Ensher JR, Matthews MR, Wieman CE, Cornell EA (1995) Observation of bose-einstein condensation in a dilute atomic vapor. Science 269:198
3. Bradley CC, Sackett CA, Tollett JJ, Hulet RG (1995) Evidence of bose-einstein condensation in an atomic gas with attractive interactions. Phys Rev Lett 75:1687. https://doi.org/10.1103/PhysRevLett.75.1687
4. Pathria RK (1996) Statistical mechanics. Butterworht Heinemann, Oxford
5. Feynman R (1998) Statistical mechanics: a set of lectures. Advanced books classics. Avalon Publishing. https://books.google.de/books?id=pJ4_BAAAQBAJ

6. Pethick CJ, Smith H (2002) Bose-Einstein condensation in dilute gases. Cambridge Unviersity Press, Cambridge
7. Pitaevskii L, Stringari S (2018)   Bose-Einstein condensation and superfluidity. Oxford University Press, New York
8. Liebisch TC, Schlagmüller M, Engel F, Nguyen H, Balewski J, Lochead G, Böttcher F, Westphal KM, Kleinbach KS, Schmid T, Gaj A, Löw R, Hofferberth S, Pfau T, Pérez-Ríos J, Greene CH (2016) Controlling rydberg atom excitations in dense background gases. J Phys B Atomic Mol Opt Phys 49(18):182001. https://doi.org/10.1088/0953-4075/49/18/182001
9. Thomas LH (1927) The calculation of atomic fields. Math Proc Camb Philos Soc 23(5):542. https://doi.org/10.1017/S0305004100011683
10. Fermi E (1927) Un metodo statistico per la determinazione di alcune propietà dell'atomo. Rend Accad Naz Lincei 6:602
11. Ashcroft N, Mermin N (2011) Solid state physics. Cengage Learning https://books.google.de/books?id=x_s_YAAACAAJ
12. Baranov M, Dobrek L, Góral K, Góral K, Santos L, Lewenstein M (2002) Ultracold dipolar gases? A challenge for experiments and theory. Phys Scr T102(1):74
13. Lahaye T, Menotti C, Santos L, Lewenstein M, Pfau T (2009) The physics of dipolar bosonic quantum gases. Rep Prog Phys 72:126401
14. Baranov MA, Dalmonte M, Pupillo G, Zoller P Condensed matter theory of dipolar quantum gases. Chem Rev 112:5012 (2012)
15. Griesmaier A, Werner J, Hensler S, Stuhler J, Pfau T (2005) Bose-Einstein condensation of chromium. Phys Rev Lett 94:160401. https://doi.org/10.1103/PhysRevLett.94.160401
16. Aikawa K, Frisch A, Mark M, Baier S, Rietzler A, Grimm R, Ferlaino F (2012) Bose-Einstein condensation of erbium. Phys Rev Lett 108:210401
17. Lu M, Burdick NQ, Youn SH, Lev BL (2011) Strongly dipolar Bose-Einstein condensate of dysprosium. Phys Rev Lett 107:190401
18. Ni K-K, Ospelkaus S, Nesbitt DJ, Ye J Jin DS (2009) A dipolar gas of ultracold molecules. Phys Chem Chem Phys 11:9626
19. Deiglmayr J, Repp M, Grochola A, Dulieu O, Wester R, Weidemüller M (2011) Dipolar effects and collisions in an ultracold gas of LiCs molecules. J Phys Conf Ser 264:012014
20. Park JW, Will SA, Zwierlein MW (2015)   Ultracold dipolar gas of fermionic $^{23}Na^{40}K$ molecules in their absolute ground state. Phys Rev Lett 114:205302
21. Löw R, Weimer H, Nipper J, Balewski JB, Butscher B, Büchler HP, Pfau T (2012)   An experimental and theoretical guide to strongly interacting rydberg gases. J Phys B Atomic Mol Opt Phys 45(11):113001
22. Balewski JB, Krupp AT, Gaj A, Hofferberth S, Löw R, Pfau T (2014)   Rydberg dressing: understanding of collective many-body effects and implications for experiments. New J Phys 16(6):063012
23. Eiles MT, Lee H, Pérez-Ríos J, Greene CH (2017)   Anisotropic blockade using pendular long-range rydberg molecules. Phys Rev A 95:052708. https://doi.org/10.1103/PhysRevA.95.052708
24. Lahaye T, Koch T, Fröhlich B, Fattori M, Metz J, Griesmaier A, Giovnazzi S, Pfau T (2007) Strong dipolar effects in a quantum ferrofluid. Nature (London) 448:672
25. Ferrier-Barbut I, Kadau H, Schmitt M, Wenzel M, Pfau T (2016)   Observation of quantum droplets in a strongly dipolar bose gas. Phys Rev Lett 116:215301. https://doi.org/10.1103/PhysRevLett.116.215301
26. Kadau H, Schmitt M, Wenzel M, Wink C, Maier T, Ferrier-Barbut I, Pfau T (2016) Observing the rosensweig instability of a quantum ferrofluid. Nature 530(7589):194. https://doi.org/10.1038/nature16485
27. Böttcher F, Schmidt J-N, Wenzel M, Hertkorn J, Guo M, Langen T, Pfau T (2019) Transient supersolid properties in an array of dipolar quantum droplets. Phys Rev X 9:011051. https://doi.org/10.1103/PhysRevX.9.011051
28. Chomaz L, Petter D, Ilzhöfer P, Natale G, Trautmann A, Politi C, Durastante G, van Bijnen RMW, Patscheider A, Sohmen M, Mark MJ, Ferlaino F (2019).   Long-lived and transient

supersolid behaviors in dipolar quantum gases. Phys Rev X 9:021012. https://doi.org/10.1103/PhysRevX.9.021012

29. Natale G, van Bijnen RMW, Patscheider A, Petter D, Mark MJ, Chomaz L, Ferlaino F (2019) Excitation spectrum of a trapped dipolar supersolid and its experimental evidence. Phys Rev Lett 123:050402. https://doi.org/10.1103/PhysRevLett.123.050402

30. Frish A, Mark M, Aikawa K, Ferlaino F, Bohn JL, Makrides C, Petrov A, Kotochigova S (2014) Quantum chaos in ultracold collisions of gas-phase erbium atoms. Nature (London) 507:475

31. Maier T, Kadau H, Schmitt M, Wenzel M, Ferrier-Barbut I, Pfau T, Frisch A, Baier S, Akiawa K, Chomaz L, Mark MJ, Ferlaino F, Makrides C, Tiesinga E, Petrov A, Kotochigova S (2015) Emergence of chaotic scattering in ultracold Er and Dy. Phys Rev X 5:041029

32. Yang BC, Pérez-Ríos J, Robicheaux F (2017) Classical fractals and quantum chaos in ultracold dipolar collisions. Phys Rev Lett 118:154101. https://doi.org/10.1103/PhysRevLett.118.154101

33. Mehring M (1983) Principles of high resolution NMR in solids. Springer, Berlin

34. Sichter CP (1996) Principles of magnetic resonances. Springer, Berlin

35. Niederprüm T, Thomas O, Eichert T, Pérez-Ríos J, Greene CH, Ott H (2016) Observation of pendular butterfly Rydberg molecules. Nat Commun 7:12820

36. Demler E, Zhou F (2002) Spinor bosonic atoms in optical lattices: symmetry breaking and fractionalization. Phys Rev Lett 88:163001. https://doi.org/10.1103/PhysRevLett.88.163001

37. Ho T-L (1998) Spinor bose condensates in optical traps. Phys Rev Lett 81:742. https://doi.org/10.1103/PhysRevLett.81.742

38. Song JL, Semenoff GW, Zhou F (2007) Uniaxial and biaxial spin nematic phases induced by quantum fluctuations. Phys Rev Lett 98:160408. https://doi.org/10.1103/PhysRevLett.98.160408

39. Ohmi T, Machida K (1998) Bose-einstein condensation with internal degrees of freedom in alkali atom gases. J Phys Soc Jpn 67(6):1822. https://doi.org/10.1143/JPSJ.67.1822

40. Koashi M, Ueda M (2000) Exact eigenstates and magnetic response of spin-1 and spin-2 Bose-Einstein condensates. Phys Rev Lett 84:1066. https://doi.org/10.1103/PhysRevLett.84.1066

41. Ciobanu CV, Yip S-K, Ho T-L (2000) Phase diagrams of $f = 2$ spinor bose-einstein condensates. Phys Rev A 61:033607. https://doi.org/10.1103/PhysRevA.61.033607

42. Kawaguchi Y, Ueda M (2012) Spinor Bose–Einstein condensates. Phys Rep 520(5):253. https://doi.org/10.1016/j.physrep.2012.07.005. http://www.sciencedirect.com/science/article/pii/S0370157312002098

43. Zhou F, Semenoff GW (2006) Quantum insulating states of $f = 2$ cold atoms in optical lattices. Phys Rev Lett 97:180411. https://doi.org/10.1103/PhysRevLett.97.180411

44. Blasing DB, Pérez-Ríos J, Yan Y, Dutta S, Li C-H, Zhou Q, Chen YP (2018) Observation of quantum interference and coherent control in a photochemical reaction. Phys Rev Lett 121:073202. https://doi.org/10.1103/PhysRevLett.121.073202

Cooling and Trapping of Molecules

4

The advancement in ultracold physics is attached to the development of cooling, manipulation, and trapping techniques for atoms and molecules. Therefore, it is necessary to understand some of the techniques that make ultracold physics. Efficient cooling and trapping techniques for atoms have been available since the mid-1980s and there is an extensive range of literature on that topic; it has even been beautifully introduced in several books [1, 2]. However, cooling and trapping techniques for molecules are generally discussed in specialized journals, and certainly, they are not discussed in books about ultracold gases. For this reason, in this chapter, we introduce most of the cooling and trapping techniques for molecules, since molecules are the workhorse of chemistry and the purpose of this book. Moreover, throughout this chapter, the reader may gain some insights into the complications that molecules offer with respect to atoms owing to the presence of internal degrees of freedom, that ultimately can be overcome.

Ultracold molecules are achieved through two different strategies:

- Direct methods: use of external fields to control the translational degrees of freedom of molecules to cool them down directly from a source or beam, e.g., Stark decelerator, Zeeman decelerator, or buffer gas cooling, among others.
- Indirect methods: ultracold molecules are assembled from ultracold atoms by manipulation of the atom–atom interaction via a laser field (photoassociation) or a magnetic field (magnetoassociation)

In this chapter we present the main direct and indirect cooling methods, covering the basic physical aspects of each of them. But, before diving into the respective methods we would like to reflect on the physics behind cooling techniques for molecules.

Supersonic jets or supersonic expansions are within the most well-known techniques for creating molecular beams with a well-defined velocity and internal state population of a molecule, and hence one of the most important techniques

© Springer Nature Switzerland AG 2020
J. Pérez Ríos, *An Introduction to Cold and Ultracold Chemistry*,
https://doi.org/10.1007/978-3-030-55936-6_4

for studying gas-phase reactions in chemical physics. In a supersonic expansion the initial internal energy of the molecular gas is transformed into kinetic energy of the molecules as a consequence of the conservation of enthalpy and the isentropic nature of the expansion (see Appendix B). As a consequence, the energy stored in the internal degrees of freedom of the molecule is reduced and molecules at the end of the expansion are very fast but *internally cold*. In particular, the molecules are in the ground electronic and vibrational state and only a few rotational states are populated. This phenomenon is a consequence of the rupture of equilibrium among different degrees of freedom of the molecules (electronic, vibrational, rotational, and translational) owing to the different relaxation mechanisms associated with those degrees of freedom [3, 4].

After a supersonic expansion the molecules show a population of a few internal states but moving extremely fast, leading to a large mean kinetic energy. Therefore, to reach ultracold temperatures something else needs to be done to remove the extra kinetic energy of the molecules, which is the role of the cooling methods in this chapter. In particular, cooling is achieved through the application of dissipative forces based on external fields or through thermalization of the system at hand when it is brought in contact with a colder one.

4.1   Stark Deceleration

Stark deceleration, like many other techniques in this chapter, was inspired by the groundbreaking experiment of Otto Stern and Walther Gerlach in 1922, known as the Stern–Gerlach experiment [5]. The Stern–Gerlach experiment showed that it is possible to deflect a beam of silver atoms by applying an inhomogeneous magnetic field, owing to the interaction between the magnetic field and the magnetic moment of the atoms. In other words, by external fields it is possible to control and manipulate molecules, which is the core idea behind the Stark decelerator. In particular, Stark deceleration exploits the force exerted by an inhomogeneous electric field on polar molecules to manipulate their translational degrees of freedom. This technique was experimentally demonstrated for the first time in 1999 [6], and since then has been implemented in many laboratories around the world [7–13]; it has even been extended to Rydberg atoms and Rydberg molecules [9, 10, 14].

The quantum states of polar molecules in an external electric field **E**, owing to their permanent dipole moment, experience an energy shift according to the Hamiltonian

$$H_{\text{Stark}} = -\mathbf{d} \cdot \mathbf{E}. \tag{4.1}$$

Here, **d** stands for the dipole moment of the molecule. The shift in energy as a consequence of the Hamiltonian given by Eq. (4.6) is known as the Stark effect. In Fig. 4.1 the Stark effect on aluminum monofluoride (AlF) is shown, where it is noted that the quantum states of the molecule mainly respond in two different ways to the external electric field:

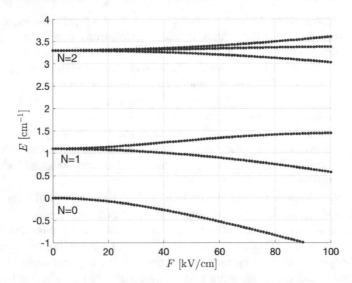

Fig. 4.1 Stark effect in AlF. Energy levels of AlF ($X^1\Sigma^+$) as a function of the magnitude of the external electric field F (in kV/cm) N denotes the rotational quantum number. The dipole moment, rotational constant, and centrifugal distortion term are taken from Truppe et al. [15]

- Low-field seeker (lfs) states: increase their energy with the strength of the electric field.
- High-field seeker (hfs) states: lower their energy with the strength of the electric field.

A lfs states prefer regions of small electric field magnitude to minimize its energy. In contrast, a hfs states reduce its energy in regions where the magnitude of the electric field is large. In regions of high electric field, a lfs states experiences a sizeable Stark energy shift (see Fig. 4.1), thus lowering its kinetic energy. In other words, lfs states have to climb the potential hill owing to the Stark effect in regions of high electric field. Therefore, by making a molecule in a lfs to climb the hill several times, via proper timing for switching on and off different electric field configurations, its kinetic energy will be lessened accordingly. This is the main idea behind the concept of Stark deceleration. In fact, if the electric field configuration is kept constant in time the molecule does not lose kinetic energy, since every time that the molecule climbs the potential hill it climbs it down.

In Fig. 4.2 the electric field configuration for Stark deceleration is shown, and let us visualize a bunch of molecules traveling through the electrodes. At point A, the vertical electrodes (blue electrodes) are switched on. Molecules in a lfs state will gain potential energy as they move toward point B, which translates into a reduction of its kinetic energy and deceleration. If at B, the horizontal configuration (red electrodes) is switched on, the molecules in lfs will lose kinetic energy again. In panel (a) of Fig. 4.3 the electric field between two pairs of electrodes, which is

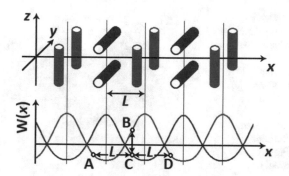

Fig. 4.2 Electric field configuration for Stark deceleration. The blue electrodes are the vertical ones whereas the red electrodes represent the horizontal ones, the potential of the electrodes is the same but opposite in sign (within the same color), and the distance between pairs of electrodes is L. $W(x) \equiv H_{\text{Stark}}(x)$ is the potential that a molecule experiences as a function of its position in x axis. The points A, B and C denote different stages during the Stark decelerator and they are explained in the text. (Figure courtesy of Dr. Christian Schewe)

generally called a stage of the decelerator, is shown. Reproducing this combination of voltages several times, it is possible to decelerate a molecular beam effectively.

The time sequence for switching on and off the different electrodes is chosen according to a synchronous molecule traveling with a given velocity. That is, a molecule in a given lfs state and particular velocity. The phase angle, ϕ, specifies the position of the synchronous molecule when the configuration of the electrodes changes. Molecules coming from a molecular beam show a velocity distribution. Thus, the molecules are placed at different positions, x_{ns}, when the configuration of electrodes changes, which is better characterized by the phase angle relative to a stage $\phi = \frac{x_{ns}\pi}{L}$ [16]. $\phi = 0$ is when the molecule is at the same distance between two consecutive pairs of electrodes, whereas $\phi = \pm\pi/2$ stands for a molecule at the grounded and charged electrode pair respectively. Deceleration occurs when $\phi < 0$ and the optimal situation is when $\phi = -\pi/2$. The phase space is defined from the velocity of the molecules and ϕ, and it is shown in panel (b) of Fig. 4.3. In particular, the stable region of the phase space for a Stark decelerator is shown. This region defines the range of velocities and phase angles of the molecules that will be effectively decelerated, it is called *acceptance region*, and it can be shown that it depends on ϕ [17, 18].

4.2   Optical Stark Deceleration

The optical Stark deceleration uses the optical dipole force to trap and decelerate molecules during a given pulse duration. The main idea is to make use of the AC Stark shift on molecules, induced by a high-intensity pulsed laser field, to reduce the kinetic energy of the molecules. The technique follows the same logic as the Stark decelerator, although in the present case, the deceleration of the molecules is due

Fig. 4.3 (a) Normalized electric field configuration in the $x - y$ plane, which corresponds to the configuration where the horizontal electrodes are charged and the vertical ones are grounded. (b) "Acceptance" are for a Stark decelerator for different phase-angles ϕ. (Figure courtesy of Dr. Christian Schewe)

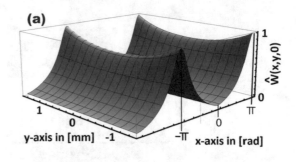

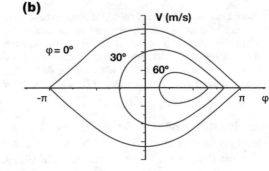

Fig. 4.4 Scheme of the experimental setup for optical Stark deceleration. The molecular beam is locally trapped owing to the intense far-off resonant laser field interacting with the molecules

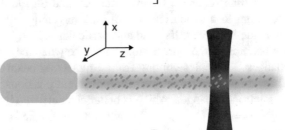

to light–matter interaction. The optical Stark deceleration technique is extensible to any molecule [19–23].

The potential energy of a molecule in a far detuned laser field is [24]

$$U(\mathbf{r}, t) = -\frac{\alpha}{2\epsilon_0 c} I(\mathbf{r}, t), \qquad (4.2)$$

where $I(\mathbf{r}, t)$ is the time and spatial dependent intensity of the laser, α, is the effective polarizability of the molecule, ϵ_0, is the electric constant, and c is the speed of light in a vacuum. A laser field far detuned to the red for any electronic transition guarantees that all the states of the molecule behave as hfs states [25]. Therefore, the molecules try to move toward regions with a higher intensity. Assuming that the molecules are in a collimated beam moving along the z axis, the laser is applied in a direction perpendicular to the beam, as shown in Fig. 4.4. The far-off resonant laser light induces a force, and hence the motion of the molecules is described as

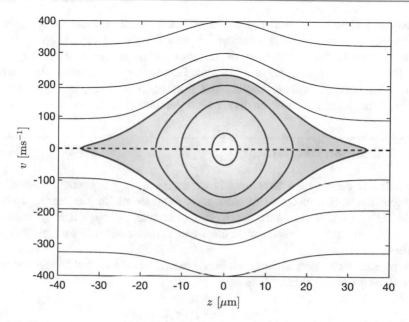

Fig. 4.5 Phase space portrait for molecules under the action of a far-off resonant light field. The velocity of benzene molecules (in m/s) as a function of their position with respect the center of the laser beam (in μm). The stable region is shown as the shaded blue region, and it is a consequence of the far-off laser–matter interaction. The *separatrix* is depicted by the solid blue line

$$\frac{dz}{dt} = v \tag{4.3}$$

$$\frac{dv}{dt} = -\frac{2\alpha I(z)z}{m\epsilon_0 c w_0^2}. \tag{4.4}$$

These equations are derived applying second Newton's law and taking into account Eq. (4.2). In Eq. (4.3), it has been assumed that the intensity profile of the laser is given by

$$I(z, y, t) = I_0 e^{-[2(z^2+y^2)/w_0^2]}, \tag{4.5}$$

where I_0 is the peak intensity of the laser beam and w_0 is its radius.

In Fig. 4.5 the phase space associated with Eq. (4.3) for benzene is shown. The results for this figure have been obtained by taking $\alpha = 1.16 \times 10^{-39}$ cm^{2}V^{-1}, $I_0 = 1.6 \times 10^{16}$ W/m^2, and employing different initial conditions. The phase portrait in Fig. 4.5 shows a stability region highlighted in blue, which ends at the *separatrix*, i.e., the curve that divides the phase space into stable and unstable orbits. As a result, only molecules in a certain region and with a particular velocity may be trapped and decelerated. However, despite this complication, it was shown that the net effect is a reduction in 25 m/s in a single state[1] for a pulse duration of 15 ns [19].

In conclusion, the optical Stark deceleration technique seems to be an *efficient* tool for decelerating all sorts of molecules. However, it only handles the external degrees of freedom without cooling the internal ones. Thus, one may have a molecule that is cold (translationally) but hot (internally) [26].

4.3   Zeeman Decelerator

The Zeeman deceleration technique is based on using the Zeeman shift of the quantum states of a molecule or atom to manipulate its kinetic energy. Thus, it is similar in spirit to the Stark decelerator technique, although it is germane for molecules with unpaired electrons, in other words, molecules with a magnetic dipole moment [27–32].

A paramagnetic atom or molecule in an external magnetic field experiences an interaction given by the Hamiltonian

$$H_{\text{Zeeman}} = -\mu_B g_s \mathbf{B} \cdot \hat{\mathbf{S}}, \tag{4.6}$$

where $\mathbf{S}$ is the electronic spin, μ_B is the Bohr magneton, and g_s is the spin g-factor of the molecular state. An example of lfs and hfs states is depicted in Fig. 4.6, where the energy levels for the two lowest rotational states ($N = 1$, $N = 2$) of $^{16}O_2$ as a function of the magnetic field (in T) are shown. In this figure, it is noted, as we expected, that lfs states prefer regions of small magnetic field magnitude (blue lines) to minimize their energy. In contrast, the hfs states reduce their energy in regions where the magnitude of the magnetic field is significantly higher.

A Zeeman decelerator consists of a number of identical solenoids that are equidistantly placed, as shown in Fig. 4.7. As the molecules move toward the center of the solenoid, molecules in lfs states will experience a positive potential energy through Eq. (4.6), hence lowering its kinetic energy. As the molecules get closer to the center of the solenoid, the intensity flowing in the solenoid is switched off, and simultaneously the next solenoid is turned on with the same current as the previous solenoid. In that way, the molecules in lfs states will experience a reduction of their kinetic energy in each of the solenoids, which are commonly called Zeeman stages. Therefore, the Zeeman decelerator is similar to the Stark decelerator. However, the electrodes are substituted by solenoids and polar molecules by paramagnetic ones.

Although the core idea behind the Stark and Zeeman decelerators is the same, it is crucial to remark two main differences between these two direct cooling techniques [28].

---

[1] Based on the velocity of the molecules in the beam, this reduction would be equivalent to 8%.

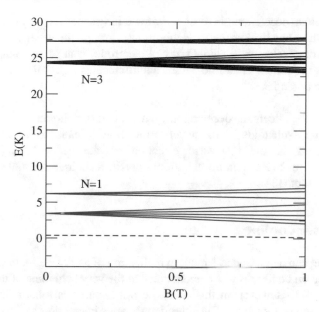

Fig. 4.6 Zeeman levels for $^{16}O_2$ as a function of the magnetic field (in T). The hfs states are shown in red, whereas the lfs states are presented in blue. The spin-rotation constant of $^{16}O_2$ is taken as $\gamma_{SR} = -0.0089 \, cm^{-1}$, the spin–spin coupling constant as $\lambda_{SS} = 1.985 \, cm^{-1}$ and the rotational constant as $B_e = 1.438 \, cm^{-1}$ [33]

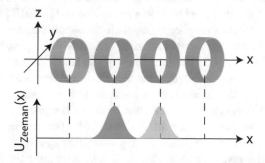

Fig. 4.7 Magnetic field configuration for a Zeeman decelerator. Identical solenoids are placed at the same distance from each other, as a consequence the atoms or molecules will feel the interaction $U_{Zeeman}(x)$ (dark blue-marine area). Next, this solenoid is switched off when the molecules are at a given x, losing kinetic energy. At the same time, the nearest solenoid is switched on (light blue-marine area) inducing a reduction of the kinetic energy of the molecule or atom, as in the previous solenoid. Thus, repeating this process in each of the solenoids, the molecules or atoms will show a prominent reduction of its kinetic energy

- The Zeeman decelerator has a cylindrical symmetry to the beam axis, which affects the trapping properties in the traverse axis. This symmetry is absent in the Stark decelerator.

- In the Zeeman decelerator, in most of the switching schemes, only one solenoid is active at the same time as the molecules move through the decelerator. However, in the Stark decelerator, a whole array of electrodes is at some voltage, whereas the rest is grounded. Therefore, atoms and molecules in a Stark decelerator feel a periodic potential.

Recently, the Zeeman decelerators have been ameliorated by alternating hexapoles and solenoids. This novel arrangement leads to a major degree of confinement in the transverse direction of the beam [34]. These new improvements make Zeeman deceleration a useful technique for molecular cross-beam experiments [34].

4.4   Laser Cooling

Laser cooling can be viewed as a modern realization of a very well-known fact: light exerts a force on bodies, as was demonstrated at the very beginning of the twentieth century [35–37]. However, in this case, the light source is a laser field, and the body is an atom or molecule. Thus, the light–matter interaction has to be analyzed from a quantum mechanical perspective [38–40]. Laser cooling is the most common technique for cooling down atoms owing to its straightforward implementation[2] and general application to any atom (although the complexity of the setup changes for each atom). Laser cooling is extensible to molecules [41], despite the presence of internal degrees of freedom, as DeMille's group showed in 2010 [42].

Let us assume a two-level molecule[3] with transition frequency ω_0 interacting with two identical counter-propagating laser beams with frequency ω, as shown in panels (a) and (b) of Fig. 4.8. The frequency of the lasers is tuned below resonance with a detuning $\Delta = \omega - \omega_0$.[4] If the molecule is moving with velocity $\mathbf{v}$ owing to the Doppler effect, the molecule sees photons with frequencies $\omega \pm \mathbf{k} \cdot \mathbf{v}$. Thus, the molecule in the ground state may absorb one of the photons, receiving a kick of momentum $\hbar k$ in the direction of the absorbed photon. Photon absorption is followed by the isotropic emission of a photon through spontaneous emission. After many absorptions, on average, the emission does not have any effect on the momentum of the molecule. Therefore, the molecule loses kinetic energy with the number of absorptions.

---

[2] As a theoretician; I never implemented this technique by myself. However, my experimentalist colleagues continuously implement this setup without too many technical issues. That is the reason why I used the word straightforward. Having said this, I do not mean that its implementation is trivial.

[3] A two-level molecule is analogous to a two-level atom, although the two states in this cases are represented by rovibrational states of the molecule.

[4] The laser frequency is red-detuned from the atomic transition.

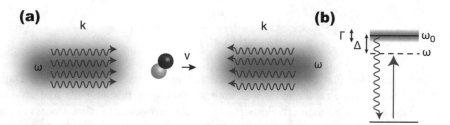

Fig. 4.8 Laser cooling of a molecule. Panel (**a**), a molecule with velocity v is exposed to two counter-propagated beams of equal frequency ω. The frequency of the lasers is detuned from the molecular transition ω_0 by Δ, as shown in panel (**b**). The molecule will absorb a photon in the direction of the laser fields, owing to the Doppler shift (see text for details), and it will de-excite through spontaneous emission in a random direction. As a net effect, the molecule reduces its kinetic energy in the direction of the lasers. This is a 1D cooling scheme. By applying three counter-propagating lasers, it is possible to cool down a molecule to temperature $\sim100\,\mu$K

The photon absorption probability from a laser, whose frequency is detuned by Δ with respect a given transition reads as [1, 43, 44]

$$\rho_{ee} = \frac{I}{2I_0}\frac{1}{\left(1 + \frac{I}{I_0} + \frac{4\Delta^2}{\Gamma^2}\right)}, \tag{4.7}$$

where I is the laser intensity, Γ is the spontaneous emission rate or natural linewidth of the transition, and I_0 is the saturation intensity. $I_0 = 2\pi^2\hbar c/(3\Gamma\lambda_0)$, where λ_0 stands for the transition wavelength.[5] The absorption of a photon leads to the following force

$$\mathbf{F} = \hbar\mathbf{k}\rho_{ee}. \tag{4.8}$$

Therefore, a molecule interacting with two counter-propagating beams experiences the total force

$$\mathbf{F}_\pm(\mathbf{v}) = \pm\hbar k\frac{I}{2I_0}\frac{1}{\left(1 + \frac{I}{I_0} + \frac{4(\Delta\mp\mathbf{k}\cdot\mathbf{v})^2}{\Gamma^2}\right)}, \tag{4.9}$$

where $\pm$ stands for photon absorption from the parallel/anti-parallel beam; therefore, the net force reads as $\mathbf{F}_T = \mathbf{F}_+(\mathbf{v}) + \mathbf{F}_-(\mathbf{v})$. Next, for low molecular speeds, the force can be written as [41, 45, 46]

---

[5]It is worth noting that ρ_{ee} stands for the population of the excited state, which is the second diagonal term of the density matrix [43, 44]. This is the typical notation for the optical Bloch equations.

$$F_T = -\frac{8\hbar k^2 \Delta I}{I_0 \Gamma \left(1 + \frac{I}{I_0} + 4\frac{\Delta^2}{\Gamma^2}\right)^2} v, \tag{4.10}$$

where it is noted that the force is proportional and opposite to the velocity. Thus, the net force acts as a damping mechanism, ultimately bringing the molecule to rest. However, owing to the randomness on the spontaneous emission process of the molecule, it experiences a random motion that limits the velocity reachable by laser cooling as

$$v_D = \sqrt{\hbar \Gamma / (2m)}. \tag{4.11}$$

And hence, the lowest temperature by means of laser cooling is given by

$$k_B T_D = \hbar \Gamma / 2, \tag{4.12}$$

known as the Doppler limit of the temperature, which is $\sim$100–10 μK for most of the atoms and molecules.

The previous model, assuming a two-level molecule interacting with a radiation field, is readily extensible to multilevel atoms and molecules, with the complication that a few extra lasers, called repump lasers, are needed to close the transition. Surprisingly enough, and despite that a priori one may think that internal states prevent laser cooling, after the impressive and pioneering work in the field of laser cooling of molecules [41,42,47–54], it is possible to identify molecules that can be laser cooled similar to an atom, such as CaF [52–54] or AlF [15].

In Fig. 4.9 is shown the laser cooling scheme for AlF. This molecule shows a perfect overlap between the ground vibrational states in the $A^1\Pi$ and $X^1\Sigma$ states, which translates into an excellent Franck–Condon factor. In other words, the laser cooling of AlF can be treated as the laser cooling of an atom. However, a repump laser for $A^1\Pi(v' = 0) \leftarrow X^1\Sigma(v = 1)$ is needed to close the transition and ensure the efficiency of the laser cooling.

Figure 4.10 displays the force experienced by an AlF molecule when exposed to two identical counter-propagating laser beams with a frequency detuned by $\Delta = -\Gamma$ with respect to the $A^1\Pi(v' = 0) \leftarrow X^\Sigma(v = 0)$ transition. For the $A^1\Pi$ state is known that $\Gamma = 2\pi \times 83$ MHz [15]. The force showed in Fig. 4.10 shows the linear behavior as a function of the velocity for low velocities, as expected from Eq. (4.10). It is also noticed that the most efficient cooling (or largest force) occurs for molecules with a velocity $\approx$20 ms^{-1}, which depends on Δ and the intensity of the laser beams.

Fig. 4.9 Potential energy curves of the three electronic states of AlF relevant for laser cooling. The inset shows the transition wavelength and calculated Franck–Condon factors for the X-A transition [15]. (Figure adapted with permission from Truppe et al. [15] Copyright (2020) (American Physical Society))

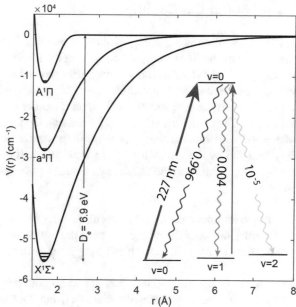

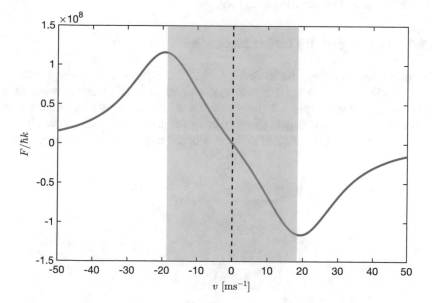

Fig. 4.10 Force exerted by two identical counter-propagating laser beams on AlF as a function of the velocity (in m/s). The force is given in units of $\hbar k$, which is the unit of momentum transfer for the laser beams. The laser beams detuned from the $A^1\Pi(v'=0) \leftarrow X^1\Sigma(v=0)$ transition by $\Delta = -\Gamma$, and the intensity is equal to the saturation intensity

4.5   Buffer Gas Cooling

Buffer gas cooling is a cooling method in which a buffer gas (a cold monoatomic gas, generally helium) cools down a hotter gas via thermalization, i.e., through elastic collisions between atoms of the buffer gas and the particles of the hot gas. The technique is general and applicable to any atom [55], molecule [54], and molecular ion [56], as long as the species survives on the order of 10 to 100 collisions with the buffer gas. The technique is realized by loading the target molecules into a chamber where the cold monoatomic gas is flowing. As the target molecules cool down, they will follow the monoatomic gas flow and will leave the chamber, as shown in Fig. 4.11. Therefore, the efficiency of buffer gas cooling relies on three different points:

- How to introduce the target molecules in the buffer gas.
- Thermalization of the molecules with the buffer gas.
- Diffusion of the cold molecules through the buffer gas.

The density of the monoatomic gas has to be high enough to thermalize the target molecules before they leave the chamber, but low enough to avoid three-body recombination and cluster formation [56].

4.5.1   Loading of the Target Molecules

The target molecules can be loaded into the buffer gas source through different methods [57], but the most frequently employed is laser ablation owing to its versatility, as we will explain. In laser ablation, a high-intensity laser beam hits the surface of the precursor target. The surface of the precursor is volatilized after the laser hit, thus ejecting atoms and molecules of the target species, although usually

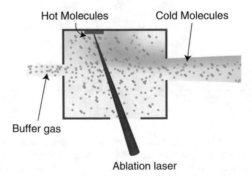

Fig. 4.11   A simplified version of a buffer gas cooling source. The buffer gas gets into the chamber from the orifice in the left of the chamber, and it flows toward the exit hole at the right of the chamber. The ablation laser liberates the target molecules that thermalize through elastic collisions with the buffer gas. As the target molecules thermalize they diffuse in the chamber until they reach the exit nozzle and leave the chamber at the same temperature (ideally) as the buffer gas

accompanied by other species (see Fig. 4.11). The efficiency of the ablation process depends on different factors such as the laser power and the angle of the ablation laser with respect to the solid precursor, among other physics variables [58, 59]. Indeed, the microscopic processes governing ablation are unknown. For instance, it is not understood the different chemical reactions and how the ions and electrons created during the ablation affect the total number of target molecules and its internal state.

Ablation is better understood by means of an example. In this case, we chose the creation of AlF. This molecule is stable, but it has to be created by a given chemical reaction. In this case, the solid precursor target is just an aluminum cylinder. However, into the chamber, apart from the buffer gas flowing, a gas is injected that triggers the reaction of interest. The gas of choice is SF_6, which reacts with aluminum atoms coming from the ablation, $Al + SF_6 \longrightarrow AlF + SF_5$, to give AlF [60], which is cooled down through elastic collisions until it thermalizes with the buffer gas. Ablation allows the production of all sorts of molecules utilizing different precursors and gases during ablation. Thus, laser ablation is the most versatile method of generating molecules within a buffer gas chamber.

4.5.2 Thermalization

Thermalization is a consequence of the second law of thermodynamics, from which the heat can only flow from a hot body into a cold one until the exchange of energy ceases, thus reaching equilibrium. In the case at hand, the heat flow is due to the exchange of energy between particles of the warm system (target molecules) and the cold one (buffer gas atoms) through collisions.

In buffer gas cooling, since one deals with gases, thermalization is better understood as the result of elastic collisions between the particles of the different gases. Let us suppose that particles with mass m_h of the hot gas at temperature T_h collide with a particle of mass m_b of the cold gas at temperature T_b. Assuming that the particles behave as hard spheres, imposing the conservation of energy and momentum during a collision, and keeping in mind that $1/2m\langle v^2 \rangle = 3/2k_B T$, the mean change in the temperature of the hot gas as a consequence of a single collision with the particles of the cold gas reads as [61–63]

$$\Delta T_h = -\frac{(T_h - T_b)}{\kappa}, \tag{4.13}$$

with

$$\kappa = \frac{(m_h + m_b)^2}{2m_h m_b}. \tag{4.14}$$

Therefore, after N collisions the temperature of the hot particles is

$$T_h(N) - T_h(N - 1) = -\frac{(T_h(N - 1) - T_b)}{\kappa}, \tag{4.15}$$

and if N is large and the change in the temperature per collision is small, Eq. (4.15) can be expressed as a first-order differential equation as follows:

$$\frac{dT_h(N)}{N} = -\frac{(T_h(N) - T_b)}{\kappa}, \tag{4.16}$$

whose solution is

$$T_h(N) = T_b + (T_h - T_b)e^{-N/\kappa}$$
$$\frac{T_h(N)}{T_b} \approx 1 + \frac{T_h}{T_b}e^{-N/\kappa}. \tag{4.17}$$

The second part of the equation is approximately valid as long as $T_h \gg T_b$, where T_h is the initial temperature of the particles of the hot gas. From Eq. (4.17) it is straightforward to estimate the number of collisions needed to thermalize a given target mass. For instance, assuming AlF at 1,000 K in a buffer gas of He at 4K, i.e., $m_h = 46$ a.m.u. and $m_b = 4$ a.m.u., an AlF molecule needs 29 collisions with He to reach $T_h(29) = 4.04$ K, which is 1% more of the temperature of the buffer gas. Indeed, most of the molecules in a buffer gas thermalize within a time scale ~ 1 ms.

The presence of the buffer gas thermalizes not only the translational degrees of freedom, as we have shown, but also the rotational degrees of freedom. In particular, the He molecule cross section for rotational relaxation is $\approx 10^{-15} - 10^{-16}\,\text{cm}^2$, which is one to two orders of magnitude smaller than the typical value for the He molecule elastic cross section. Therefore, rotational degrees of freedom may thermalize in the buffer gas chamber if it is long enough to allow sufficient inelastic rotational collisions before the molecules leave the buffer gas cell. For vibrational degrees of freedom, the story is quite different. It turns out that monoatomic gases do not efficiently relax the vibrational degrees of freedom. As a consequence, some of the molecules after leaving the buffer gas source may be in excited vibrational states [64–66].

4.5.3   Diffusion of the Cold Target Molecules

After the species of interest is thermalized, it has to diffuse until reaching the exit hole of the source. The diffusion of the target molecules in the buffer gas source is of capital interest to estimate how long it takes them to abandon the source, and how many of them collide and freeze with the walls of the source. The diffusion constant of a molecular gas, whose molecules have a mass m_{mol}, diffusing in a buffer gas, with atoms of mass m_b satisfying $m_b \ll m_{mol}$, is [63, 67]

$$D = \frac{3\pi \langle v_b \rangle}{32 \rho_b \sigma_{\text{b-mol}}}, \tag{4.18}$$

where it has been supposed that the density of the buffer gas is much higher than the density of the target molecule, i.e., $\rho_b \gg \rho_{\text{mol}}$. The diffusion process can be viewed as Brownian motion in 3D space, and in that case, the mean-squared displacement of the target molecules (from its initial position) within the buffer gas is

$$\langle r^2 \rangle (t) = 6Dt = \frac{9\pi \langle v_b \rangle t}{16 \rho_b \sigma_{\text{b-mol}}}. \tag{4.19}$$

Thus, it is possible to define the diffusion time as

$$\tau_{\text{diff}} = \frac{16 A_{\text{cell}} \rho_b \sigma_{\text{b-mol}}}{9\pi \langle v_b \rangle}, \tag{4.20}$$

where A_{cell} is the buffer gas cell cross section or diameter of the cell. In general, the diffusion time is of the order 1–10 ms for most of the buffer gas sources.

As a concluding remark it is worth mentioning that the buffer gas cooling method is available many molecular systems, which makes this technique one of the most convenient for starting the cooling pathway to the ultracold regime. The downside of buffer gas cooling is that the lowest temperature reachable depends on the temperature of the buffer gas. In general, using ^{4}He as a buffer gas it is possible to have the molecules at 4K and using the ^{3}He it is possible to reach a temperature of 200 mK for molecules. Therefore, with buffer gas cooling it is possible to bring any molecule to the cold regime.

4.6   Optical Trapping: Dipole Traps

Optical dipole traps use far-detuned laser light with respect to a molecular or atomic transition to enhance the induced dipole–electric field interaction, $U_{\text{dip}} \propto \mathbf{d} \cdot \mathbf{E}$, in comparison with the absorption of light. Optical dipole traps show a trap depth $\lesssim 1$ mK, depending on the intensity of the laser light and the from a given transition. Atoms and molecules within an optical dipole trap have lifetimes ~ 10 s, which is more than enough to study all sorts of internal dynamics, many-body processes, or chemical processes [24, 68].

We have seen in Sect. 4.2 that a molecule in a far detuned laser field will experience an interaction given by Eq. (4.2). In that equation, only the average polarizability was relevant for the optical Stark deceleration. However, for the optical dipole trap it is vital to account explicitly for the dependence of the polarizability with the laser frequency ω. Thus, the dynamic polarizability of a given molecular state, $\alpha_i(\omega)$, is needed to calculate the interaction energy of a molecule in a given state i in the presence of a laser field. Indeed, the interaction energy reads as

$$U_{\text{dip}}(\mathbf{r}) = -\frac{\mathcal{R}(\alpha_i(\omega))I(\mathbf{r})}{2\epsilon_0 c}, \tag{4.21}$$

where $\mathcal{R}(x)$ denotes the real part of x and

$$\alpha_i(\omega) = 2\sum_f \frac{\omega_{if} - \imath\gamma_f/2}{\left(\omega_{if} - \imath\gamma_f/2\right)^2 - \omega^2}|\langle f|d(R)\hat{\mathbf{R}} \cdot \hat{\epsilon}|i\rangle|^2. \tag{4.22}$$

In this equation, the summation is performed over all the accessible electronic and rovibrational states. $\omega_{if} = \omega_f - \omega_i$ stands for the excitation frequency from the initial state to the state $|f\rangle$, which has a scattering rate γ_f. $d(R)\hat{\mathbf{R}}$ is the transition dipole moment, R is the inter-atomic distance and $\hat{\mathbf{R}}$ is its orientation, and $\hat{\epsilon}$ is the polarization of the laser field.[6] A molecule in a laser field will absorb photons followed by the spontaneous emission process, as has been shown in Sect. 4.4. The scattering rate for spontaneous emission is

$$\Gamma_{\text{sr}}(r) = \frac{1}{\hbar\epsilon_0 c}\mathcal{I}(\alpha_i(\omega))I(\mathbf{r}). \tag{4.23}$$

Assuming a two-level model for the molecule and the Lorentz model for the dynamic polarizability [24], which is fairly accurate for the case of far-detuned transitions in atoms, the dipole potential and scattering rate read as

$$U_{\text{dip}}(\mathbf{r}) = -\frac{3\pi c^2}{2\omega_0^3}\left(\frac{\Gamma}{\omega - \omega_0} + \frac{\Gamma}{\omega + \omega_0}\right)I(\mathbf{r})$$

$$\Gamma_{\text{sr}}(r) = \frac{3\pi c^2}{2\hbar\omega_0^3}\left(\frac{\omega}{\omega_0}\right)^3\left(\frac{\Gamma}{\omega - \omega_0} + \frac{\Gamma}{\omega + \omega_0}\right)^2 I(\mathbf{r}), \tag{4.24}$$

where ω_0 is the transition frequency of the molecule, and Γ is the spontaneous decay rate from the excited state. Equations (4.24) emerge for two-level systems; in the case of molecules, the numerical computation of the dynamic polarizability is necessary [70,71]. Next, assuming that the detuning $\Delta = \omega_0 - \omega \gg \Gamma$ but $\Delta \ll \omega_0$, which is equivalent to $\omega/\omega_0 \approx 1$ Eqs. (4.24) are further simplified to

$$U_{\text{dip}}(\mathbf{r}) = -\frac{3\pi c^2}{2\omega_0^3}\frac{\Gamma}{\Delta}I(\mathbf{r})$$

---

[6]It is worth noting that the transition dipole moment is expressed with respect to the reference frame fixed to the molecule. In contrast, the electric field is defined in the lab frame. Therefore, a proper transformation from the molecular frame to the lab frame has to be accomplished for the proper evaluation of the matrix elements [69–71].

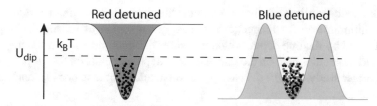

Fig. 4.12 Red and blue detuned optical dipole traps. The molecules are trapped in the maxima of the intensity of the red detuned trap, although in the minimum of the intensity for the blue detuned trap. A Gaussian beam is employed for the red detuned trap in which the molecules are trapped in the maximum of the intensity of the beam. For the blue detuned trap, the molecules are trapped in the minimum of the intensity of the laser field; hence, it needs to work with the Laguerre–Gaussian LG_{01} mode

$$\Gamma_{sr}(r) = \frac{3\pi c^2}{2\hbar \omega_0^3} \left(\frac{\Gamma}{\Delta} \right)^2 I(\mathbf{r}). \tag{4.25}$$

These are the equations that govern the possibility of trapping atoms and molecules for a given transition, intensity, and detuning. From Eq. (4.25) it is noticed that

$$\hbar \Gamma_{sr}(r) = \frac{\Gamma}{\Delta} U_{dip}(\mathbf{r}), \tag{4.26}$$

where it is clear that for larger detuning, fewer photons will be absorbed; and hence, a more stable dipole force will emerge.

In Eqs. (4.25), it is noticed that for red-detuned laser beams, the minimum of the potential will be reached in the regions of higher intensity, whereas for the case of blue-detuned light it is the opposite, as shown in Fig. 4.12. Therefore, to optically trap atoms and molecules is enough with a red-detuned Gaussian beam focused on a region of space. It is also possible to trap atoms and molecules with blue-detuned light. However, use of Laguerre–Gaussian modes is necessary, instead, as shown in Fig. 4.12.

4.7   Magnetic Trapping

In Sect. 4.3, it has been shown that paramagnetic atoms and molecules experience a potential due to the interaction between an external magnetic field and the electronic spin. As a result, some states will behave as lfs or hfs states (notation introduced in Sect. 4.1).

In principle, to trap atoms and molecules, one would need a maximum or minimum magnetic field. However, the theorem of Earnshaw establishes that it is only possible to generate magnetic field minima in the absence of currents [72]. Therefore, only lfs states are trappable states. The original formulation of Earnshaw dealt with a static electric field. In particular, he established that only with the presence of electrostatic forces is it not possible to find a stationary equilibrium [73].

A magnetic trap is a particular configuration of coils and conductors to design a local minimum on the magnetic field in a given region of space. The traps are characterized by the trap depth, which measures the maximum kinetic energy of the atoms and molecules held on the trap. The trap depth depends on the property of the system under study, and for the case of $^3\Sigma$ molecules, for instance O_2, reads as

$$U_{\text{trap}} = 2\mu_B|\vec{B}|, \tag{4.27}$$

where μ_B is the Bohr magneton. However, for alkali atoms the trap depth is given by [74]

$$U_{\text{trap}} = \mu_B|\vec{B}|. \tag{4.28}$$

There are magnetic traps based on different designs and configurations. One of the most common configurations is two coils in anti-Helmholtz configuration [75–77]. In this configuration, two coils experience the same current but in opposite directions, leading to a minimum of the magnetic field at the center of the segment drawn by the two centers of the coils. The minimum created by these coils is null; hence, the atoms and molecules may be lost from the trap at the minimum, which is known as Majorana losses. However, this issue can be solved by using a time-dependent magnetic field or by using four extra conductors leading to the Ioffe–Pritchard configuration. Discussion of the different configurations for magnetic traps is beyond the scope of this chapter. However, we encourage the reader to take a look at reference [77] for a comprehensive study of the different magnetic traps for ultracold physics.

4.8 Evaporative Cooling

Evaporative cooling is the most effective cooling method available for atoms. Hess proposed it in the context of dilute gases [78] in the late 1980s and it was realized by other scientists such as W. Ketterle [79]. The evaporative cooling method can be better understood by explaining how a cup of coffee cools down. In a cup of coffee, hotter molecules leave the cup as steam; thus, the remainder molecules are less energetic and, therefore, will be colder (thermalization). Evaporative cooling is very similar, but instead of having a cup, we have a trap (either magnetic or optical), and instead of coffee, we have atoms or molecules. Evaporative cooling is based on forcing atoms or molecules with higher kinetic energy to leave the trap by adiabatically lowering the trap depth, as shown in Fig. 4.13.

Losses in a trap, either optical or magnetic, occur for two reasons:

- The kinetic energy of a particle is greater than the depth of the trap.
- Inelastic collisions: collisions that induce a change in the internal quantum state of the colliding partners leading to a final state that it is not trappable.

Fig. 4.13 Evaporative
cooling. The trap depth is
adiabatically lowered to allow
the most energetic particles to
leave the trap. Thus, the
particles remaining will be
colder owing to
thermalization, with the less
energetic particles in the trap

Table 4.1 Evaporative
cooling efficiency as a
function of γ, i.e., the ratio
between the probability of
elastic collisions with respect
to inelastic ones

γ	Efficiency
10	Bad
100	Fine
1000	Good
10,000	Excellent

The efficiency of evaporative cooling depends on the relationship between the thermalization (elastic collisions) and losses (inelastic processes) through the following parameter

$$\gamma = \frac{\sigma_{el}}{\sigma_{ine}}. \tag{4.29}$$

The efficiency of evaporative cooling is summarized in Table 4.1, where it is noted that we need to have at least 100 elastic collisions per inelastic collision to ensure a decent evaporative cooling scheme for the molecule or atom at hand.[7]

The success of the evaporative cooling for atoms and molecules is contingent on controlling the kinetic energy of the particles before switching on the trap [80]. The maximum kinetic energy that a trappable particle can have is determined by the parameter,

$$\eta = \frac{U_{trap}}{k_B T}, \tag{4.30}$$

where U_{trap} is the trap depth. To ensure successful trapping, and hence a good chance of making evaporative cooling efficient is to guarantee that $\eta > 5$ [80].

---

[7]It also depends on the quantum state of the atomic or molecular system under consideration.

4.9 Assembly of Molecules at Ultracold Temperatures

It is possible to obtain ultracold molecules in a well-defined quantum state using so-called indirect methods. These methods rely on forming the molecules out of an ultracold gas of atoms. In particular, the molecular assembly is performed through an external laser (photoassociation) or magnetic field (magnetoassociation). Indirect methods have the benefit that the initial atoms are already in the ultracold regime; hence, no further cooling technique is necessary. However, elaborate, although efficient, preparation techniques for manipulation of the internal degrees of freedom such as the stimulated Raman adiabatic passage (STIRAP) [81] are required. The downside of indirect methods is that they are only applicable to a modest number of molecules, since the atomic species that are available at ultracold temperatures are somewhat restricted. Prevailing experiments show that indirect methods are the most efficient experimental approach to have an ensemble of ultracold molecules [82–87], although recently the application of direct techniques has been spreading around the ultracold physics community [41, 42, 50, 52, 88].

4.9.1 Photoassociation

Photoassociation (PA) is a quantum process in which two colliding atoms absorb a photon to form a molecule in an excited electronic state, that is,

$$A + B + \gamma \rightarrow AB^*, \tag{4.31}$$

as shown in Fig. 4.14. Then, the excited molecule will decay through spontaneous emission to one of the vibrational states of the ground electronic state of the molecule as

$$AB^* \rightarrow AB + \gamma, \tag{4.32}$$

or it can decay back to the two free atoms state as

$$AB^* \rightarrow A + B + \gamma. \tag{4.33}$$

The decay to the two free atoms leads to atoms with kinetic energy different than that they had in (4.31). The processes (4.32) and (4.33) are illustrated in Fig. 4.14, and both induce a loss of a pair of atoms.[8] Thus, monitoring the trap loss as a function of the frequency of the applied laser field is possible to extract the binding energy of the rovibrational states of the excited molecular state. Indeed,

---

[8]In the case of (4.33) it may be the case that the atoms have lower kinetic energy than before and hence they remain trapped. However, this possibility is unlikely [89].

Fig. 4.14 Photoassociation. Two colliding atoms at collision energy E_k may observe a photon resonant to a rovibrational state of the AB* electronic state of the molecule (1) that correlates with the A + B* atomic asymptote. Then, this molecular state may decay through (2) to a rovibrational state of the electronic ground state of AB, or through (3) to two-free atoms with the consequent kinetic energy release

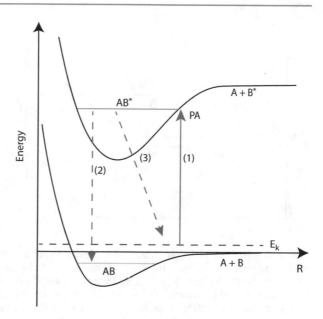

this method allows measuring absolute binding energies, as opposed to conventional spectroscopy techniques, where only the relative energy can be measured [89].

Photoassociation has been the preferred technique for achieving a high-density ensemble of ultracold homonuclear molecules [90–92], and it was the first technique to show that heteronuclear molecules can be efficiently formed in its absolute rovibrational ground state [93]. Moreover, PA has been applied in Bose–Einstein condensates, leading to coherent oscillations between a condensate of atoms and molecules [94], in spinor BECs [95, 96] and more recently in BECs with a synthetic spin–orbit interaction [97].

4.9.2 Magnetoassociation

In Sect. 2.5.2, it has been shown that Fano–Feshbach resonances emerge as a consequence of a coupling between the scattering state and a closed vibrational state, as shown in Fig. 4.15 and in more detail in Sect. 2.5.2. If there is a coupling between these states, it must be a way of transferring population from one state to the other, and that is the idea behind magnetoassociation. Indeed, a Fano–Feshbach resonance can be visualized as a coupled two-level system in the dressed state picture, where one of the states is a continuum instead of a bound one, as depicted in the inset of Fig. 4.15. Therefore, by sweeping the magnetic field adiabatically across resonance is possible to transform a pair of atoms into a bound molecule. The binding energy of the molecule would be [98]

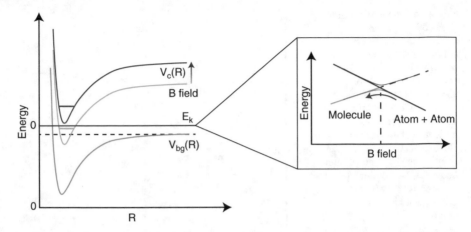

Fig. 4.15 Magnetoassociation. Schematic representation of a Fano–Feshbach resonance similar to the one presented in Fig. 2.10. The inset shows a different perspective of a Fano–Feshbach resonance, in which the initial scattering state is decoupled from the molecular state. However, as the magnetic field is reduced, the molecular state and scattering state strongly interact, and by means of an adiabatic ramp of the magnetic field it is possible to transfer a pair of atoms into a bound molecular state

$$E_B = \frac{\hbar^2}{2\mu a(B)^2}, \tag{4.34}$$

where $a(B)$ is the scattering length as a function of the magnetic field (see Eq. (2.73)). The molecules formed through magnetoassociation are commonly called Feshbach molecules; they are characterized as being weakly bound and show a large bond length[9] [99, 100].

Magnetoassociation has been successfully applied to homonuclear systems [91, 92, 101–105] as well as to heteronuclear systems [82–87, 106, 107]. Indeed, magnetoassociation is the world's leading technique in obtaining ultracold polar molecules. Nevertheless, the use of direct methods may bring more interesting molecules into play, improving our understanding of the ultracold realm.

References

1. Metcalf HJ, van der Straten P (1999) Laser cooling and trapping. Springer, Utrecht
2. Letokhov VS (2007) Laser control of atoms and molecules. Oxford University Press, New York
3. Hirschfelder JO, Curtiss CF, Bird RB (1954) Molecualr theory of transport in gases. Wiley, New York

---

[9]Here we use bond length in the physicists' sense, i.e., we are not referring to a covalent, ionic or van der Waals binding mechanism.

4. Montero S, Pérez-Ríos J (2014) Rotational relaxation in molecular hydrogen and deuterium: theory versus acoustic experiments. J Chem Phys 141:2014
5. Gerlach W, Stern O (1922) Der experimentelle nachweis der richtungsquantelung im magnetfeld. Zeitschrift für Physik 9(1):349
6. Bethlem HL, Berden G, Meijer G (1999) Decelerating neutral dipolar molecules. Phys Rev Lett 83(8):1558
7. van de Meerakker SYT, Bethlem HL, Meijer G (2008) Taming molecular beams. Nat Phys 4(8):595. https://doi.org/10.1038/nphys1031
8. van de Meerakker SYT, Bethlem HL, Vanhaecke N, Meijer G (2012) Manipulation and control of molecular beams. Chem Rev 112(9):4828. https://doi.org/10.1021/cr200349r
9. Hogan SD, Seiler C, Merkt F (2009) Rydberg-state-enabled deceleration and trapping of cold molecules. Phys Rev Lett 103(12):123001
10. Vliegen E, Hogan SD, Schmutz H, Merkt F (2007) Stark deceleration and trapping of hydrogen rydberg atoms. Phys Rev A 76:023405. https://doi.org/10.1103/PhysRevA.76.023405
11. Sawyer BC, Stuhl BK, Lev BL, Ye J, Hudson ER (2008) Mitigation of loss within a molecular stark decelerator. Eur Phys J D 48(2):197. https://doi.org/10.1140/epjd/e2008-00097-y
12. Hudson ER, Bochinski JR, Lewandowski HJ, Sawyer BC, Ye J (2004) Efficient stark deceleration of cold polar molecules. Eur Phys J D 31(2):351. https://doi.org/10.1140/epjd/e2004-00138-7
13. Wall TE, Tokunaga SK, Hinds EA, Tarbutt MR (2010) Nonadiabatic transitions in a stark decelerator. Phys Rev A 81:033414. https://doi.org/10.1103/PhysRevA.81.033414
14. Hogan SD (2016) Rydberg-stark deceleration of atoms and molecules. EPJ Tech Instrum 3(1):2. https://doi.org/10.1140/epjti/s40485-015-0028-4
15. Truppe S, Marx S, Kray S, Doppelbauer M, Hofsäss S, Schewe HC, Walter N, Pérez-Ríos J, Sartakov BG, Meijer G (2019) Spectroscopic characterization of aluminum monofluoride with relevance to laser cooling and trapping. 1908.11774
16. Bethlem HL, FMH Crompvoets, Jongma RT, SYT van de Meerakker SYT, Meijer G (2002) Deceleration and trapping of ammonia using time-varying electric fields. Phys Rev A 65:053416. https://doi.org/10.1103/PhysRevA.65.053416
17. van de Meerakker SYT, Vanhaecke N, Bethlem HL, Meijer G (2005) Higher-order resonances in a stark decelerator. Phys Rev A 71:053409. https://doi.org/10.1103/PhysRevA.71.053409
18. Gubbels K, Meijer G, Friedrich B (2006) Analytic wave model of stark deceleration dynamics. Phys Rev A 73:063406. https://doi.org/10.1103/PhysRevA.73.063406
19. Fulton R, Bishop AI, Barker PF (2004) Optical stark decelerator for molecules. Phys Rev Lett 63:243004
20. Fulton R, Bishop AI, Barker PF (2005) An optical decelerator for cold molecules. In: 2005 quantum electronics and laser science conference, vol. 1, pp. 490–492. https://doi.org/10.1109/QELS.2005.1548826
21. Fulton R, Bishop AI, Shneider MN, Barker PF (2006) Controlling the motion of cold molecules with deep periodic optical potentials. Nat Phys 2:465
22. Yang Y, Shunyong H, Deng L (2018) Optical stark deceleration of neutral molecules from supersonic expansion with a rotating laser beam. Chin Phys B 27:053701
23. Bishop AI, Wang L, Barker PF (2010) Creating cold stationary molecular gases by optical stark deceleration. New J Phys 12(7):073028. https://doi.org/10.1088/1367-2630/12/7/073028
24. Grimm R, Weidemüller M, Ovchinnikov YB, Bederson B, Walther H (2000) Adv Atom, Mol Opt Phys, vol. 42 Academic Press, pp. 95–170
25. Seideman T (1999) New means of spatially manipulating molecules with light. J Chem Phys 111(10):4397. https://doi.org/10.1063/1.479204
26. Friedrich B, Doyle JM (2009) Why are cold molecules so hot? ChemPhysChem 10(4):604
27. Dulitz K, Vanhaecke N, Softley TP (2015) Model for the overall phase-space acceptance in a zeeman decelerator. Phys Rev A 91:013409. https://doi.org/10.1103/PhysRevA.91.013409

28. Wiederkehr AW, Hogan SD, Merkt F (2010) Phase stability in a multistage zeeman decelerator. Phys Rev A 82:043428. https://doi.org/10.1103/PhysRevA.82.043428
29. Narevicius E, Parthey CG, Libson A, Narevicius J, Chavez I, Even U, Raizen MG (2007) An atomic coilgun: using pulsed magnetic fields to slow a supersonic beam. New J Phys 9(10):358. https://doi.org/10.1088/1367-2630/9/10/358
30. Hogan SD, Wiederkehr AW, Andrist M, Schmutz H, Merkt F (2008) Slow beams of atomic hydrogen by multistage zeeman deceleration. J Phys B Atomic Mol Opt Phys 41(8):081005
31. Hogan SD, Wiederkehr AW, Schmutz H, Merkt F (2008) Magnetic trapping of hydrogen after multistage zeeman deceleration. Phys Rev Lett 101:143001. https://doi.org/10.1103/PhysRevLett.101.143001
32. Hogan SD, Sprecher D, Andrist M, Vanhaecke N, Merkt F (2007) Zeeman deceleration of H and D. Phys Rev A 76:023412. https://doi.org/10.1103/PhysRevA.76.023412
33. Mizushima M (1975) The theory of rotating diatomic molecules. Wiley, New York
34. Cremers T, Janssen N, Sweers E van de Meerakker SYT (2019) Design and construction of a multistage zeeman decelerator for crossed molecular beams scattering experiments. Rev Sci Instrum 90(1):013104
35. Lebedev P (1901) Untersuchungen über die druckkräte des lichtes. Ann Phys (Leipzig) 6:433
36. Nichols EF, Hull GF (1901) A preliminary communication on the pressure of heat and light radiation. Phys Rev 13:307
37. Nichols EF, Hull GF (1903) The pressure due to radiation. Phys Rev 17:26
38. Phillips WD (1998) Nobel lecture: laser cooling and trapping of neutral atoms. Rev Mod Phys 70:721. https://doi.org/10.1103/RevModPhys.70.721
39. Chu S (1998) Nobel lecture: the manipulation of neutral particles. Rev Mod Phys 70:685. https://doi.org/10.1103/RevModPhys.70.685
40. Cohen-Tannoudji CN (1998) Nobel lecture: manipulating atoms with photons. Rev Mod Phys 70:707. https://doi.org/10.1103/RevModPhys.70.707
41. Tarbutt MR (2018) Laser cooling of molecules. Contemp Phys 59(4):356. https://doi.org/10.1080/00107514.2018.1576338
42. Shuman ES, Barry JF, DeMille D (2010) Laser cooling of a diatomic molecule. Nature 467(7317):820. https://doi.org/10.1038/nature09443
43. Loudon R (1973) The quantum theory of light. Clarendon Press, Oxford
44. Scully MO, Zubairy MS (2002) Quantum optics. Cambridge University Press, Cambridge
45. Ungar PJ, Weiss DS, Riis E, Chu S (1989) Optical molasses and multilevel atoms: theory. J Opt Soc Am B 6(11):2058. http://josab.osa.org/abstract.cfm?URI=josab-6-11-2058
46. Lett PD, Phillips WD, Rolston SL, Tanner CE, Watts RN, Westbrook CI (1989) Optical molasses. J. Opt. Soc. Am. B 6(11):2084
47. Barry JF, McCarron DJ, Norrgard EB, Steinecker MH, DeMille D (2014) Magneto-optical trapping of a diatomic molecule. Nature 512:286 EP. https://doi.org/10.1038/nature13634
48. Norrgard EB, McCarron DJ, Steinecker MH, Tarbutt MR and DeMille D (2016) Sub-millikelvin dipolar molecules in a radio-frequency magneto-optical trap. Phys Rev Lett 116:063004. https://doi.org/10.1103/PhysRevLett.116.063004
49. Hummon MT, Yeo M, Stuhl BK, Collopy AL, Xia Y, Ye J (2013) 2d magneto-optical trapping of diatomic molecules. Phys Rev Lett 110:143001. https://doi.org/10.1103/PhysRevLett.110.143001
50. Zhelyazkova V, Cournol A, Wall TE, Matsushima A, Hudson JJ, Hinds EA, Tarbutt MR, Sauer BE (2014) Laser cooling and slowing of caf molecules. Phys Rev A 89:053416. https://doi.org/10.1103/PhysRevA.89.053416
51. Anderegg L, Augenbraun BL, Chae E, Hemmerling B, Hutzler NR, Ravi A, Collopy A, Ye J, Ketterle W, Doyle JM (2017) Radio frequency magneto-optical trapping of caf with high density. Phys Rev Lett 119:103201. https://doi.org/10.1103/PhysRevLett.119.103201
52. Williams HJ, Truppe S, Hambach M, Caldwell L, Fitch NJ, Hinds EA, Sauer BE, Tarbutt MR (2017) Characteristics of a magneto-optical trap of molecules. New J Phys 19(11):113035. https://doi.org/10.1088/1367-2630/aa8e52

53. Truppe S, Williams HJ, Hambach M, Caldwell L, Fitch NJ, Hinds EA, Sauer BE, Tarbutt MR (2017) Molecules cooled below the doppler limit. Nat Phys 13:1173. https://doi.org/10.1038/nphys4241

54. Truppe S, Williams HJ, Fitch NJ, Hambach M, Wall TE, Hinds EA, Sauer BE, Tarbutt MR (2017) An intense, cold, velocity-controlled molecular beam by frequency-chirped laser slowing. New J Phys 19(2):022001. https://doi.org/10.1088/1367-2630/aa5ca2

55. Doyle JM, Friedrich B, Kim J, Patterson D (1995) Buffer-gas loading of atoms and molecules into a magnetic trap. Phys Rev A 52(4):R2515

56. Brünken S, Kluge L, Stoffels A, Pérez-Ríos J, Schlemmer S (2017) Rotational state-dependent attachment of he atoms to cold molecular ions: an action spectroscopic scheme for rotational spectroscopy. J Mol Spectrosc 332:67

57. Krems RV, Stwalley WC, Friedrich B (eds) (2009) Cold molecules: theory, experiment, applications. CRC Press, Boca Raton

58. Chichkov BN, Momma C, Nolte S, von Alvensleben F, Tünnermann A (1996) Femtosecond, picosecond and nanosecond laser ablation of solids. Appl Phys A 63(2):109. https://doi.org/10.1007/BF01567637

59. Olander DR (2009) Laser-pulse-vaporization of refractory materials. Pure Appl Chem 62.1:123

60. Parker JK, Garland NL, Nelson HH (2002) Kinetics of the reaction Al + SF_6 in the temperature range 499–813 K. J Phys Chem A 106:307

61. Patterson D, Rasmussen J, Doyle JM (2009) Intense atomic and molecular beams via neon buffer-gas cooling. New J Phys 11(5):055018. https://doi.org/10.1088/1367-2630/11/5/055018

62. deCarvalho R, Doyle JM, Friedrich B, Guillet T, Kim J, Patterson D, Weinstein JD (1999) Buffer-gas loaded magnetic traps for atoms and molecules: a primer. Eur Phys J D Atomic Mol Opt Plasma Phys 7(3):289

63. Hutzler NR, Lu HI, Doyle JM (2012) The buffer gas beam: an intense, cold and slow source for atoms and molecules. Chem Rev 112:4803

64. Weinstein JD, deCarvalho R, Guillet T, Friedrich B, Doyle JM (1998) Magnetic trapping of calcium monohydride molecules at millikelvin temperatures. Nature 395(6698):148

65. Campbell WC, Groenenboom GC, Lu H-I, Tsikata E, Doyle JM (2008) Time-domain measurement of spontaneous vibrational decay of magnetically trapped NH. Phys Rev Lett 100:083003. https://doi.org/10.1103/PhysRevLett.100.083003

66. Barry JF, Shuman ES, DeMille D (2011) A bright, slow cryogenic molecular beam source for free radicals. Phys Chem Chem Phys 13(42):18936. https://doi.org/10.1039/C1CP20335E

67. Hasted JB (1964) Physics of atomic collisions. Butterworths advanced physics series'. Monographs on ionization and electrical discharges in gases. Butterworths, Washington, 1964.

68. Carr LD, DeMille D, Krems RV, Ye J (2009) Cold and ultracold molecules: science, technology and applications. New J Phys 11:055049

69. Vexiau R, Borsalino D, Lepers M, Orbán A, Aymar M, Dulieu O, Bouloufa-Maafa N (2017) Dynamic dipole polarizabilities of heteronuclear alkali dimers: optical response, trapping and control of ultracold molecules. Int Rev Phys Chem 36(4):709

70. Vexiau R, Bouloufa N, Aymar M, Danzl JG, Mark MJ, Nägerl HC, Dulieu O (2011) Optimal trapping wavelengths of cs2 molecules in an optical lattice. Eur Phys J D 65(1):243. https://doi.org/10.1140/epjd/e2011-20085-4

71. Deiß M, Drews B, Denschlag JH, Bouloufa-Maafa N, Vexiau R, Dulieu O (2015) Polarizability of ultracold Rb_2 molecules in the rovibrational ground state of $a^3\sigma_u^+$. New J Phys 17(6):065019. https://doi.org/10.1088/1367-2630/17/6/065019

72. Wing WH (1984) Prog Quantum Electron 8:181

73. Earnshaw S (1842) On the nature of the molecular forces which regulate the constitution of the luminiferous ether. Trans Camb Philos Soc 7:97

74. Pethick CJ, Smith H (2002) Bose-Einstein condensation in dilute gases. Cambridge Unviersity Press, Cambridge

75. Panofsky WKH, Phillips M (1983) Classical electricity and magnetism. Dover, New York
76. Jefimenko OD (1989) Electicity and magnetism. Electret Scientific Company, West Virginia
77. Pérez-Ríos J, Sanz AS (2013) How does a magnetic trap work? Am J Phys 81(11):836. https://doi.org/10.1119/1.4819167
78. Hess HF, Bell DA, Kochanski GP, Kleppner D, Greytak TJ (1984) Phys Rev Lett 52:1520
79. Ketterle W, van Druten NJ (1996) Adv At Mol Opt Phys 37:181
80. Doyle JM, Freidrich B, Kim J, Patterson D (1995) Phys Rev A 52:2515
81. Bergmann K, Nägerl H-C, Panda C, Gabrielse G, Miloglyadov E, Quack M, Seyfang G, Wichmann G, Ospelkaus S, Kuhn A, Longhi S, Szameit A, Pirro P, Hillebrands B, Zhu X-F, Zhu J, Drewsen M, Hensinger WK, Weidt S, Halfmann T, Wang H-L, Paraoanu GS, Vitanov NV, Mompart J, Busch T, Barnum TJ, Grimes DD, Field RW, Raizen MG, Narevicius E, Auzinsh M, Budker D, Pálffy A, Keitel CH (2019) Roadmap on stirap applications. J Phys B Atomic Mol Opt Phys 52(20):202001. https://doi.org/10.1088/1361-6455/ab3995
82. Will SA, Park JW, Yan ZZ, Loh H, Zwierlein MW (2016) Coherent microwave control of ultracold ^{23}Na^{40}K molecules. Phys Rev Lett 116:225306. https://doi.org/10.1103/PhysRevLett.116.225306
83. Guo M, Zhu B, Lu B, Ye X, Wang F, Vexiau R, Bouloufa-Maafa N, Quéméner G, Dulieu O, Wang D (2016) Creation of an ultracold gas of ground-state dipolar ^{23}Na^{87}Rb molecules. Phys Rev Lett 116:205303. https://doi.org/10.1103/PhysRevLett.116.205303
84. Takekoshi T, Reichsöllner L, Schindewolf A, Hutson JM, Le Sueur CR, Dulieu O, Ferlaino F, Grimm R, Nägerl H-C (2014) Ultracold dense samples of dipolar rbcs molecules in the rovibrational and hyperfine ground state. Phys Rev Lett 113:205301. https://doi.org/10.1103/PhysRevLett.113.205301
85. Ni KK, Ospelkaus S, de Miranda MHG, Pe'er A, Neyenhuis B, Zirbel JJ, Kotochigova S, Julienne PS, Jin DS, Ye J (2008) A high phase-space-density gas of polar molecules. Science 322(5899):231. https://doi.org/10.1126/science.1163861
86. Ospelkaus S, Ni K-K, Quéméner G, Neyenhuis B, Wang D, de Miranda MHG, Bohn JL, Ye J, Jin DS (2010) Controlling the hyperfine state of rovibronic ground-state polar molecules. Phys Rev Lett 104:030402. https://doi.org/10.1103/PhysRevLett.104.030402
87. Yang H, Zhang D-C, Liu L, Liu Y-X, Nan J, Zhao B, Pan J-W (2019) Observation of magnetically tunable feshbach resonances in ultracold ^{23}na^{40}k + ^{40}k collisions. Science 363(6424):261. https://doi.org/10.1126/science.aau5322. http://science.sciencemag.org/content/363/6424/261.abstract
88. Collopy AL, Ding S, Wu Y, Finneran IA, Anderegg L, Augenbraun BL, Doyle JM, Ye J (2018) 3d magneto-optical trap of yttrium monoxide. Phys Rev Lett 121:213201. https://doi.org/10.1103/PhysRevLett.121.213201
89. Jones KM, Tiesinga E, Lett PD, Julienne PS (2006) Ultracold photoassociation spectroscopy: long-range molecules and atomic scattering. Rev Mod Phys 78:483. https://doi.org/10.1103/RevModPhys.78.483
90. Viteau M, Chotia A, Allegrini M, Bouloufa N, Dulieu O, Comparat D, Pillet P (2008) Optical pumping and vibrational cooling of molecules. Science 321(5886):232
91. Danzl JG, Haller E, Gustavsson M, Mark MJ, Hart R, Bouloufa N, Dulieu O, Ritsch H, Nägerl H-C (2008) Quantum gas of deeply bound ground state molecules. Science 321(5892):1062. https://doi.org/10.1126/science.1159909. https://science.sciencemag.org/content/321/5892/1062.full.pdf, https://science.sciencemag.org/content/321/5892/1062
92. Danzl JG, Mark MJ, Haller E, Gustavsson M, Hart R, Adegunde J, Hutson JM, Nägerl HC (2010) An ultracold high-density sample of rovibronic ground-state molecules in an optical lattice. Nat Phys 6:265
93. Deiglmayr J, Grochola A, Repp M, Mörtlbauer K, Glück C, Lange J, Dulieu O, Wester R, Weidemüller M (2008) Formation of ultracold polar molecules in the rovibrational ground state. Phys Rev Lett 101(13):133004.
94. Yan M, DeSalvo BJ, Huang Y, Naidon P, Killian TC (2013) Rabi oscillations between atomic and molecular condensates driven with coherent one-color photoassociation. Phys Rev Lett 111:150402. https://doi.org/10.1103/PhysRevLett.111.150402

95. Hamley CD, Bookjans EM, Behin-Aein G, Ahmadi P, Chapman MS (2009) Photoassociation spectroscopy of a spin-1 bose-einstein condensate. Phys Rev A 79:023401. https://doi.org/10.1103/PhysRevA.79.023401

96. Kobayashi J, Izumi Y, Enomoto K, Kumakura M, Takahashi Y (2009) Spinor molecule in atomic bose–einstein condensate. Appl Phys B 95(1):37 https://doi.org/10.1007/s00340-008-3307-9

97. Blasing DB, Pérez-Ríos J, Yan Y, Dutta S, Li C-H, Zhou Q, Chen YP (2018) Observation of quantum interference and coherent control in a photochemical reaction. Phys Rev Lett 121:073202. https://doi.org/10.1103/PhysRevLett.121.073202

98. Landau LD, Lifshitz EM (1958) Quantum mechanics. Butterworth-Heinemann

99. Köhler T, Góral K, Julienne PS (2006) Production of cold molecules via magnetically tunable feshbach resonances. Rev Mod Phys 78:1311. Oxford, UK. https://doi.org/10.1103/RevModPhys.78.1311

100. Timmermans E, Tommasini P, Hussein M, Kerman A (1999) Feshbach resonances in atomic bose–einstein condensates. Phys Rep 315(1):199. https://doi.org/10.1016/S0370-1573(99)00025-3. http://www.sciencedirect.com/science/article/pii/S0370157399000253

101. Cubizolles J, Bourdel T, Kokkelmans SJ, Shlyapnikov GV, Salomon C (2003) Production of long-lived ultracold li_2 molecules from a fermi gas. Phys Rev Lett 91:240401. https://doi.org/10.1103/PhysRevLett.91.240401

102. Jochim S, Bartenstein M, Altmeyer A, Hendl G, Chin C, Denschlag JH, Grimm R (2003) Pure gas of optically trapped molecules created from fermionic atoms. Phys Rev Lett 91:240402. https://doi.org/10.1103/PhysRevLett.91.240402

103. Strecker KE, Partridge GB, Hulet RG (2003) Conversion of an atomic fermi gas to a long-lived molecular bose gas. Phys Rev Lett 91:080406. https://doi.org/10.1103/PhysRevLett.91.080406

104. Dürr S, Volz T, Marte A, Rempe G (2004) Observation of molecules produced from a bose-einstein condensate. Phys Rev Lett 92:020406

105. Herbig J, Kraemer T, Mark M, Weber T, Chin C, Nägerl H-C, Grimm R (2003) Preparation of a pure molecular quantum gas. Science 301(5639):1510. https://doi.org/10.1126/science.1088876. http://science.sciencemag.org/content/301/5639/1510.abstract

106. Wang F, He X, Li X, Zhu B, Chen J, Wang D (2015) Formation of ultracold narb feshbach molecules. New J Phys 17(3):035003. https://doi.org/10.1088/1367-2630/17/3/035003

107. Heo M-S, Wang TT, Christensen CA, Rvachov TM, Cotta DA, Choi J-H, Lee Y-R, Ketterle W (2012) Formation of ultracold fermionic nali feshbach molecules. Phys Rev A 86:021602. https://doi.org/10.1103/PhysRevA.86.021602

Ultracold Molecular Collisions

5

5.1 Introduction

It is well-known that molecules show electronic, vibrational, and rotational degrees of freedom. In principle, by taking into account the role of these degrees of freedom, it is possible to accurately describe chemical reactions at room temperature. However, at ultracold temperatures is necessary to go one step further and include some subtle effects on the molecular Hamiltonian. In particular, it is necessary to include the fine and hyperfine splitting of the molecular states and the interaction among them [1, 2]. The interactions emerge as a consequence of diverse couplings between the nuclear spin, molecular rotation, and electron spin [1, 2].

5.2 Scattering Molecular Theory: Rigid Rotor–Rigid Rotor Scattering

5.2.1 Distinguishable Molecule–Molecule Scattering

The scattering theory for the collision of two rigid rotors was developed by Takayanagi in the 1960s [3] and further developed and applied by Green in his pioneering work on the study of rotational inelastic H_2-H_2 collisions [4]. The rigid rotor–rigid rotor collision is a four-body problem; thus, the Hamiltonian includes four kinetic energy operators associated with the motion of each of the nuclei ($\mathbf{r}_a$, $\mathbf{r}_b$, $\mathbf{r}_c$ and $\mathbf{r}_d$, as shown in Fig. 5.1). With an adequate choice of coordinates, the kinetic energy operator becomes separable, and hence the global translational motion (the kinetic energy associated with the center of mass motion) is decoupled from the rest of the degrees of freedom. The suitable coordinates are the Jacobi coordinates, which are defined by the vector joining the center of mass (CM) of each molecule $\mathbf{R}$, and the vectors that describe the orientation of the molecules $\hat{\mathbf{r}}_1$

© Springer Nature Switzerland AG 2020
J. Pérez Ríos, *An Introduction to Cold and Ultracold Chemistry*,
https://doi.org/10.1007/978-3-030-55936-6_5

Fig. 5.1 Jacobi coordinates
for a rigid rotor–rigid rotor
collision $(\hat{\mathbf{r}}_1, \hat{\mathbf{r}}_2, \mathbf{R})$ in a space
fixed frame of reference. The
vector position of each of the
nuclei is indicated for clarity

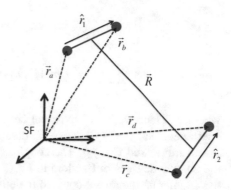

and $\hat{\mathbf{r}}_2$.[1] All these coordinates are defined with respect to a system of references
fixed on space, known as the space fixed (SF) frame, as shown in Fig. 5.1. In Jacobi
coordinates, the Hamiltonian describing the collision between two rigid rotors with
rotational constant B_0, in atomic units $(\hbar = 1)$, is given by [3, 4].

$$H = -\frac{1}{2\mu R}\frac{\partial^2}{\partial R^2}R + \frac{\hat{l}^2}{2\mu R^2} + B_0(\hat{\mathbf{j}}_1^2 + \hat{\mathbf{j}}_2^2) + V(\hat{\mathbf{r}}_1, \hat{\mathbf{r}}_2, \mathbf{R}), \tag{5.1}$$

where μ is the reduced mass of the molecule–molecule system, $V(\hat{\mathbf{r}}_1, \hat{\mathbf{r}}_2, \mathbf{R})$ stands
for the potential energy surface (PES), $\hat{\mathbf{j}}_1$, $\hat{\mathbf{j}}_2$, and $\hat{\mathbf{l}}$ are the angular momentum
operators associated with $\hat{\mathbf{r}}_1$, $\hat{\mathbf{r}}_2$ and $\hat{\mathbf{R}}$ respectively.

In the absence of external fields, the total angular momentum $\hat{\mathbf{J}} = \hat{\mathbf{j}}_1 + \hat{\mathbf{j}}_2 + \hat{\mathbf{l}}$
and its projection on the quantization axis, M, are good quantum numbers.[2] These
operators commute with the Hamiltonian (5.1); thus, it is convenient to use the
following *ansatz* to solve the time-independent Schrödinger equation

$$\Psi_\gamma^{JM} = \sum_{\gamma'} \frac{g_{\gamma'\gamma}^{JM}(R)}{R} I_{\gamma'}^{JM}(\hat{\mathbf{r}}_1, \hat{\mathbf{r}}_2, \hat{\mathbf{R}}), \tag{5.2}$$

where we separate the radial dependence on the collision coordinate $\mathbf{R}$ and the
angular degrees of freedom of the molecules $I_\gamma^{JM}(\hat{\mathbf{r}}_1, \hat{\mathbf{r}}_2, \hat{\mathbf{R}})$. Throughout this
section, we employ the following notation: $\gamma = (j_1, j_2, j_{12}, l)$, which refers to
the collision in the incoming channel, as explained in detail below. The quantum
numbers without prime generally refer to the initial states of the molecules, whereas
those primed are for their final state. The angular functions $I_{\gamma'}^{JM}(\hat{\mathbf{r}}_1, \hat{\mathbf{r}}_2, \hat{\mathbf{R}})$ come
from a first coupling between $\hat{\mathbf{j}}_1$ and $\hat{\mathbf{j}}_2$ to give $\hat{\mathbf{j}}_{12} = \hat{\mathbf{j}}_1 + \hat{\mathbf{j}}_2$ and the coupling
of this with the orbital angular momentum $\hat{\mathbf{l}}$ to yield the total angular momentum
$\hat{\mathbf{J}} = \hat{\mathbf{j}}_{12} + \hat{\mathbf{l}}$. In particular, we have

---

[1]Please, note that the $\hat{\mathbf{r}}$ stands for the unit vector of the vector position $\mathbf{r}$.

[2]In other words, the total angular momentum of the system and its projection on a given axis is
conserved.

$$I_{\gamma'}^{JM}(\hat{\mathbf{r}}_1, \hat{\mathbf{r}}_2, \hat{\mathbf{R}}) = \sum_{m'_{12}n'} (j'_{12}m'_{12}l'n'|JM) \times \tag{5.3}$$

$$\times \sum_{m'_1 m'_2} (j'_1 m'_1 j'_2 m'_2 | j'_{12} m'_{12}) Y_{j'_1}^{m'_1}(\hat{\mathbf{r}}_1) Y_{j'_2}^{m'_2}(\hat{\mathbf{r}}_2), Y_{l'}^{n'}(\hat{\mathbf{R}}).$$

where Y_j^m represents the spherical harmonics, $(.|.)$ stands for the Clebsh-Gordan coefficient, and $\gamma = (j_1, j_2, j_{12}, l)$ describes each of the triads allowed for a given J and for a fixed (j_1, j_2). The coupling scheme selected for the representation of the angular degrees of freedom in Eq. (5.3) is the so-called coupled basis, because the angular momenta are coupled to define the total angular momentum J. In fact, this basis set is especially suited for problems where the angular momentum is conserved. On the other hand, when an external field is applied, we will see that it is better to work in a basis set where no coupling between the different angular momenta is accounted for, which is called decoupled basis.

Plugging Eq. (5.3) into the time-independent Schrödinger equation $H\Psi_\gamma^{JM} = E\Psi_\gamma^{JM}$, multiplying by the complex conjugate of the angular wave function, $I_{\gamma''}^{*JM}(\hat{\mathbf{r}}_1, \hat{\mathbf{r}}_2, \hat{\mathbf{R}})$, and integrating over the angular degrees of freedom $\hat{\mathbf{r}}_1$, $\hat{\mathbf{r}}_2$ and $\hat{\mathbf{R}}$ yield the following system of coupled differential equations

$$\left[\frac{d^2}{dR^2} - \frac{l'(l'+1)}{2\mu R^2} + k_{\gamma'}^2 \right] g_{\gamma'\gamma}^{JM}(R) = 2\mu \sum_{\gamma''} V_{\gamma'\gamma''}^J(R) g_{\gamma''\gamma}^{JM}(R), \tag{5.4}$$

which are known as the coupled channel equations and they are the target equations to solve for a quantum mechanical scattering problem. In Eq. (5.4), we find

$$k_{\gamma'}^2 \equiv k_{j'_1 j'_2}^2 = 2\mu \left[E - B_e j'_1 (j'_1 + 1) - B_e j'_2 (j'_2 + 1) \right], \tag{5.5}$$

where $B_e \left(j'_1(j'_1 + 1) + j'_2(j'_2 + 1) \right)$ is the energy of the molecules in the asymptotic region (where the interaction potential is negligible), which is the rotational energy of the molecules. Therefore, $k_{\gamma'}$ is the wave vector associated with the kinetic energy of a given pair of rotational states of the colliding molecules. The potential matrix elements,

$$V_{\gamma'\gamma''}^J(R) = \int d\hat{\mathbf{r}}_1 d\hat{\mathbf{r}}_2 d\hat{\mathbf{R}} I_{\gamma'}^{*JM}(\hat{\mathbf{r}}_1, \hat{\mathbf{r}}_2, \hat{\mathbf{R}}) V(\hat{\mathbf{r}}_1, \hat{\mathbf{r}}_2, \mathbf{R}) I_{\gamma''}^{JM}(\hat{\mathbf{r}}_1, \hat{\mathbf{r}}_2, \hat{\mathbf{R}}), \tag{5.6}$$

are better evaluated by expanding the PES of the molecule–molecule interaction into an appropriate basis set for the angular degrees of freedom as [5, 6]

$$V(\hat{\mathbf{r}}_1, \hat{\mathbf{r}}_2, \mathbf{R}) = \sum_{\lambda_1 \lambda_2 \lambda} f^{\lambda_1, \lambda_2, \lambda}(R) \sum_{m_{\lambda_1} m_{\lambda_2} m_\lambda} A_{m_{\lambda_1} m_{\lambda_2} m_\lambda}^{\lambda_1 \lambda_2 \lambda} \left(\hat{\mathbf{r}}_1, \hat{\mathbf{r}}_2, \hat{\mathbf{R}} \right). \tag{5.7}$$

In general $A^{\lambda_1\lambda_2\lambda}_{m_{\lambda_1}m_{\lambda_2}m_\lambda}\left(\hat{\mathbf{r}}_1,\hat{\mathbf{r}}_2,\hat{\mathbf{R}}\right)$ is a superposition of spherical harmonics on the different angular degrees of freedom. In Eq. (5.7), $f^{000}(R)$ does not have any dependence on the rotational degrees of freedom of the molecules. Thus, scattering events in this potential term will be elastic, since no change in the internal state is possible. That is the reason why this term is called the spherical component of the potential. In the same vein, terms in which $\lambda_1 \neq 0$ or $\lambda_2 \neq 0$ (and hence $\lambda \neq 0$) are related to state-changing collisions and are known as anisotropic components of the potential. For diatomic molecule–molecule interactions it is preferable to use bipolar spherical harmonics as the basis set for the angular degrees of freedom as [5–7]

$$V(\hat{\mathbf{r}}_1,\hat{\mathbf{r}}_2,\mathbf{R}) = \sum_{\lambda_1\lambda_2\lambda} f^{\lambda_1,\lambda_2,\lambda}(R) \sum_m \begin{pmatrix} \lambda_1 & \lambda_2 & \lambda \\ m & -m & 0 \end{pmatrix} Y^m_{\lambda_1}(\hat{\mathbf{r}}_1) Y^{-m}_{\lambda_2}(\hat{\mathbf{r}}_2), \qquad (5.8)$$

where $(:::)$ stands for the 3-j symbol. Finally, plugging Eq. (5.8) into Eq. (5.6) leads to a more simple expression for the evaluation of the potential matrix elements as [4]

$$V^J_{\gamma'\gamma''}(R) = \sum_{\lambda_1\lambda_2\lambda} f^{\lambda_1,\lambda_2,\lambda}(R)(-1)^{J+j'_1+j'_2+j''_{12}}[\lambda][\lambda_1][\lambda_2][j''_1][j''_2][j'_1][j'_2]$$

$$[l''][l'][j''_{12}][j'_{12}]\begin{pmatrix} \lambda & l'' & l' \\ 0 & 0 & 0 \end{pmatrix}\begin{pmatrix} \lambda_1 & j''_1 & j'_1 \\ 0 & 0 & 0 \end{pmatrix}\begin{pmatrix} \lambda_2 & l''_2 & l'_2 \\ 0 & 0 & 0 \end{pmatrix}$$

$$\begin{Bmatrix} l'' & l' & \lambda \\ j'_{12} & j''_{12} & J \end{Bmatrix} \begin{Bmatrix} j''_{12} & j''_2 & j''_1 \\ j'_{12} & j'_2 & j'_1 \\ \lambda & \lambda_2 & \lambda_1 \end{Bmatrix}, \qquad (5.9)$$

where $[n] = \sqrt{2n+1}$, $(:::)$, $\{:::\}$ and $\{\overset{:}{:::}\}$ are the $3-j$, $6-j$ and $9-j$ respectively [8–10].

Equation (5.4) establishes the equation that the radial component of the scattering wave function must fulfil. Moreover, $g^{JM}_{\gamma'\gamma}(R)$ has to satisfy particular boundary conditions to ensure that the solution contains the proper physics. In particular, we impose regularity at the origin, $g^{JM}_{\gamma'\gamma} \to 0$ when $R \to 0$, and the following asymptotic behavior [3]

$$g^{JM}_{\gamma'\gamma}(R \to \infty) \to \frac{e^{-\imath\left(Rk_{\gamma'}-l'\pi/2\right)}}{\sqrt{k_\gamma}}\delta_{\gamma'\gamma} + \frac{e^{\imath\left(Rk_{\gamma'}-l'\pi/2\right)}}{\sqrt{k_{\gamma'}}}S^{JM}_{\gamma',\gamma}. \qquad (5.10)$$

Where $S^{JM}_{\gamma',\gamma}$ is the scattering matrix for a given J and M quantum number and for the $\gamma \to \gamma'$ channel. It is appropriate to introduce the transition matrix, which relates to the scattering matrix as

$$S^{JM}_{\gamma',\gamma} = \delta_{\gamma'\gamma} - 2\imath T^{JM}_{\gamma',\gamma}. \qquad (5.11)$$

Finally, the asymptotic expression for the radial wave function can be recast, in terms of the transition matrix, as

$$g_{\gamma'\gamma}^{JM}(R \to \infty) \to \frac{\sin\left(k_\gamma R - l\pi/2\right)}{\sqrt{k_\gamma}}\delta_{\gamma'\gamma} - \frac{e^{\imath\left(Rk_{\gamma'} - l'\pi/2\right)}}{\sqrt{k_{\gamma'}}}T_{\gamma',\gamma}^{JM}. \quad (5.12)$$

The coupled differential equations in Eq. (5.4), with the pertinent potential matrix elements through Eq. (5.9), are solved numerically in a given range of the collision coordinate. Next, comparing its solution at large values of the collision coordinate, i.e., in the asymptotic region, with Eq. (5.12) it is possible to obtain the transition matrix. A more detailed explanation is given in Sect. 5.5. However, a question remains: how can we calculate different scattering observables from the transition matrix?

The answer to this question relies on comparing the asymptotic form, Eq. (5.12), with the *physical* scattering wave function.[3] As we have seen in Sect. 2.2.1, the scattering wave function can be expressed as an incoming plane wave plus a spherically divergent outgoing wave, which in the case of molecule–molecule collision reads as

$$\Psi_\alpha(R \to \infty) \to e^{\imath \mathbf{k}_{j_1 j_2} \cdot \mathbf{R}} Y_{j_1}^{m_1}(\hat{\mathbf{r}}_1) Y_{j_2}^{m_2}(\hat{\mathbf{r}}_2) + \sum_{\alpha'} f_{\alpha'\alpha}(\hat{\mathbf{R}}) \frac{e^{\imath k_{j_1' j_2'} R}}{R} Y_{j_1'}^{m_1'}(\hat{\mathbf{r}}_1) Y_{j_2'}^{m_2'}(\hat{\mathbf{r}}_2),$$

$$(5.13)$$

where $\alpha = (j_1, m_1, j_2, m_2)$ refers to the initial state of the molecules. j_1 and j_2 are the initial rotational states of the molecules and m_1 y m_2 their projection on the z axis of the SF frame. The scattering wave function in Eq. (5.13) is similar to the one described in Sect. 2.2.1. However, in this case it is necessary to account for the state- and orientation-dependent scattering amplitude $f_{\alpha'\alpha}(\hat{\mathbf{R}})$, as well as the state-dependent wave vector. The scattering wave function, Eq. (5.13), is described on a basis on which the angular momenta are decoupled. Nevertheless, the scattering wave function can be rendered in terms of the coupled basis as [3, 11]

$$\Psi_{j_1 m_1 j_2 m_2} = \sum_{JMj_{12}l} A_{m_1 m_2}^{JMj_{12}l} \Psi_{j_1 j_2 j_{12} l}^{JM}. \quad (5.14)$$

Keeping in mind Eq. (5.2) and the asymptotic form of $g_{\gamma'\gamma}^{JM}(R)$ (the radial part of the solution of the Schrödinger equation) given in Eq. (5.12), Eq. 5.14 yields

---

[3]The term physical here denotes the fact that the scattering wave function can be always be decomposed as an incoming plane wave plus a divergent spherical wave times the amplitude of scattering.

$$\Psi_{j_1m_1j_2m_2}(R \to \infty) = \sum_{JMj_{12}l} A_{m_1m_2}^{JMj_{12}l} \frac{\sin\left(k_\gamma R - l\pi/2\right)}{R\sqrt{k_\gamma}} I_\gamma^{JM} - \tag{5.15}$$

$$\sum_{JMj_{12}l} A_{m_1m_2}^{JMj_{12}l} \sum_{\gamma'} \frac{e^{\iota\left(k_{\gamma'}R - l'\pi/2\right)}}{R\sqrt{k_{\gamma'}}} T_{\gamma'\gamma}^{JM} I_{\gamma'}^{JM}.$$

Next, comparing the first term in Eqs. (5.15) and (5.13)

$$\sum_{JMj_{12}l} A_{m_1m_2}^{JMj_{12}l} \frac{\sin\left(k_\gamma R - l\pi/2\right)}{R\sqrt{k_\gamma}} I_\gamma^{JM} = e^{\iota\mathbf{k}_{j_1j_2}\cdot\mathbf{R}} Y_{j_1}^{m_1}(\hat{\mathbf{r}}_1) Y_{j_2}^{m_2}(\hat{\mathbf{r}}_2), \tag{5.16}$$

and using the plane wave expansion in spherical harmonics [12, 13],

$$e^{\iota\mathbf{k}_{j_1j_2}\cdot\mathbf{R}} = \sum_{ln} 4\pi\iota^l \frac{\sin\left(k_{j_1j_2}R - \frac{l\pi}{2}\right)}{k_{j_1j_2}R} Y_n^{*l}(\hat{\mathbf{k}}_{j_1j_2}), Y_n^l(\hat{\mathbf{R}}), \tag{5.17}$$

one finds

$$\sum_{JMj_{12}l} A_{m_1m_2}^{JMj_{12}l} \frac{\sin\left(k_\gamma R - l\pi/2\right)}{R\sqrt{k_\gamma}} I_\gamma^{JM}$$

$$= \sum_{ln} \frac{4\pi\iota^l}{k_\gamma R} \sin\left(k_\gamma R - l\pi/2\right) Y_{j_1}^{m_1}(\hat{\mathbf{r}}_1) Y_{j_2}^{m_2}(\hat{\mathbf{r}}_2) Y_l^n(\hat{\mathbf{R}}) Y_l^{*n}(\hat{\mathbf{k}}_\gamma). \tag{5.18}$$

At this point it is useful to apply the orthonormal properties of the Clebsch–Gordan coefficients to invert Eq. (5.3) yielding

$$Y_{j_1}^{m_1}(\hat{\mathbf{r}}_1) Y_{j_2}^{m_2}(\hat{\mathbf{r}}_2) Y_l^n(\hat{\mathbf{R}}) = \sum_{JMj_{12}} (j_1m_1j_2m_2|j_{12}m_{12})(j_{12}m_{12}ln|JM) I_\gamma^{JM}(\hat{\mathbf{r}}_1, \hat{\mathbf{r}}_2, \hat{\mathbf{R}}), \tag{5.19}$$

which allows the decoupled basis to be expressed in terms of the coupled basis. Thus, Eq. (5.18) leads to

$$A_{m_1m_2}^{JMj_{12}l} = \frac{4\pi\iota^l}{\sqrt{k_\gamma}} (j_1m_1j_2m_2|j_{12}m_{12})(j_{12}m_{12}ln|JM) Y_l^n(0,0), \tag{5.20}$$

where, without loss of generality, we choose the orientation of the incoming wave vector $\mathbf{k}_\gamma$ along the z axis of the SF frame.

Once $A_{m_1m_2}^{JMj_{12}l}$ is known, we have all we need to calculate the scattering amplitude. This is done by comparing the second term in Eqs. (5.13) and (5.15), yielding

$$\sum_{\alpha'} f_{\alpha'\alpha}(\hat{\mathbf{R}}) \frac{e^{ik_{j_1'j_2'}R}}{R} Y_{j_1'}^{m_1'}(\hat{\mathbf{r}}_1) Y_{j_2'}^{m_2'}(\hat{\mathbf{r}}_2) = \quad (5.21)$$

$$-\sum_{JMj_{12}l} A_{m_1m_2}^{JMj_{12}l} \sum_{\gamma'} \frac{e^{i\left(k_{\gamma'}R - l'\pi/2\right)}}{R\sqrt{k_{\gamma'}}} T_{\gamma',\gamma}^{JM} I_{\gamma'}^{JM} =$$

$$= -\sum_{JMj_{12}l} A_{m_1m_2}^{JMj_{12}l} \sum_{j_1'j_2'j_{12}'l'} \frac{e^{(ik_{\gamma'}R - l'\pi/2)}}{R\sqrt{k_{\gamma'}}} T_{\gamma',\gamma}^{JM} \sum_{m_{12}'n'} (j_{12}'m_{12}'l'n'|JM)$$

$$\sum_{m_1'm_2'} (j_1'm_1'j_2'm_2'|j_{12}'m_{12}') Y_{j_1'}^{m_1'}(\hat{\mathbf{r}}_1) Y_{j_2'}^{m_2'}(\hat{\mathbf{r}}_2) Y_{l'}^{n'}(\hat{\mathbf{R}}),$$

and finally leading to

$$f_{\alpha'\alpha}(\hat{\mathbf{R}}) = -\frac{4\pi}{\sqrt{k_\gamma k_{\gamma'}}} \sum_{JMj_{12}j_{12}'} \sum_{ll'} \sum_{n'm_{12}'} i^{l-l'} (j_1m_1j_2m_2|j_{12}m_{12})(j_{12}m_{12}ln|JM) \quad (5.22)$$

$$(j_{12}'m_{12}'l'n'|JM)(j_1'm_1'j_2'm_2'|j_{12}'m_{12}') T_{\gamma',\gamma}^{JM} Y_l^n(0,0) Y_{l'}^{n'}(\hat{\mathbf{R}}).$$

The differential cross section given in Eq. (2.7) may be generalized to account for the different channels for the molecule–molecule collision as

$$\frac{d\sigma}{d\Omega_R}(\hat{\mathbf{R}})(\alpha \to \alpha') = \frac{k_{\gamma'}}{k_\gamma} |f_{\alpha'\alpha}(\hat{\mathbf{R}})|^2. \quad (5.23)$$

In order to define the cross section for a given process it is important to think about the detection scheme. For instance, in general, detectors do not discriminate molecules by their angular momentum projection. Furthermore, since no external field is applied, no deviations from the spin-weighted population are expected. Thus, detectors detect molecules in the same angular momentum state independently of their angular momentum projection. In this situation, the differential cross section has to be summed over the final states and averaged over initial states, unless the incoming molecules are polarized[4] [12, 14]. As a consequence, the cross section reads as

$$\sigma(j_1j_2 \to j_1'j_2') = \frac{1}{(2j_1+1)(2j_2+1)} \sum_{m_1m_2m_1'm_2'} \int \frac{d\sigma}{d\Omega_R}(\hat{\mathbf{R}})(\alpha \to \alpha')d\Omega_R, \quad (5.24)$$

where it is noticed that the cross section only depends on the rotational quantum states of the molecules but not on its projection. Substituting the obtained scattering amplitude from Eq. (5.22) into Eq. (5.24), taking into account that

---

[4]Polarized molecules show a preferable angular momentum projection.

$$Y_l^n(0,0) = \sqrt{\frac{2l+1}{4\pi}} \delta_{n,0}, \tag{5.25}$$

and the orthonormality properties of the Clebsch–Gordan coefficients [8–10], the cross section reads as [4, 14, 15]:

$$\sigma^d(j_1 j_2 \to j_1' j_2') = \frac{4\pi}{k_{j_1 j_2}^2 (2j_1+1)(2j_2+1)} \sum_{Jj_{12}j_{12}'l'l} (2J+1)|T_{j_1'j_2'j_{12}'l' \ j_1 j_2 j_{12} l}^{JM}|^2. \tag{5.26}$$

The cross section can also be recast in terms of the scattering matrix by means of Eq. (5.11) as

$$\sigma^d(j_1 j_2 \to j_1' j_2') = \frac{\pi}{k_{j_1 j_2}^2 (2j_1+1)(2j_2+1)} \sum_{Jj_{12}j_{12}'l'l} (2J+1)|\delta_{j_1'j_2'j_{12}'l' \ j_1 j_2 j_{12} l}$$

$$- S_{j_1'j_2'j_{12}'l' \ j_1 j_2 j_{12} l}^{JM}|^2. \tag{5.27}$$

The superscript d has been added to explicitly show that this expression for the cross section is only valid for distinguishable molecules. Finally, it is worth emphasizing that the transition matrix, $T_{j_1'j_2'j_{12}'l' \ j_1 j_2 j_{12} l'}^{JM}$, and the scattering matrix, $S_{j_1'j_2'j_{12}'l' \ j_1 j_2 j_{12} l'}^{JM}$, do not depend on M, and they are a function of the collision energy.

5.2.2 Identical Molecule–Molecule Scattering

In collisions between identical particles it is necessary to account for their indistinguishability. For instance, for $O_2 - O_2$ collisions, we can define the following two-particle state $O_2(1) - O_2(2)$, but this is equivalent to $O_2(2) - O_2(1)$, because the two oxygen molecules are identical. Therefore, the scattering wave function cannot change, with the exception of a given defined scalar, under the exchange of molecules. Formally, the Hamiltonian in Eq. (5.1) commutes with the exchange operator $P_{AB}(\mathbf{r}_1, \mathbf{r}_2, \mathbf{R}) \to (\mathbf{r}_2, \mathbf{r}_1, -\mathbf{R})$. Thus, the scattering wave function must be an eigenstate of the exchange operator with $+1$ (even) or -1 (odd) eigenvalues. The cross section for an even scattering wave is denoted as σ^+, whereas it is labeled as σ^- for an odd scattering wave function [5]. Assuming that the nuclear spin and electronic state of the molecules do not change during the collision, the cross section is given by

$$\sigma = W^s \sigma^+ + W^a \sigma^-, \tag{5.28}$$

where $W^{s,a}$ represents the weights related to the nuclear spin for the symmetric (s) and antisymmetric (a) wave functions [5].

The incoming wave part of the physical scattering wave function must be an eigenstate of the exchange operator too. Thus, following Gioumousis and Curtiss [16], it is found that

$$\Psi^{\zeta,in}_{j_1 m_1 j_2 m_2}(\mathbf{k}_{j_1 j_2}, \hat{\mathbf{r}}_1, \hat{\mathbf{r}}_2, \mathbf{R})$$

$$= \frac{1}{\sqrt{2}} \left(e^{i\mathbf{k}_{j_1 j_2}\cdot\mathbf{R}} Y^{m_1}_{j_1}(\hat{\mathbf{r}}_1) Y^{m_2}_{j_2}(\hat{\mathbf{r}}_2) + \zeta e^{-i\mathbf{k}_{j_1 j_2}\cdot\mathbf{R}} Y^{m_2}_{j_2}(\hat{\mathbf{r}}_1) Y^{m_1}_{j_1}(\hat{\mathbf{r}}_2) \right), \qquad (5.29)$$

where ζ stands for the symmetry with respect to the exchange operator. It is worth emphasizing that $\Psi^{\zeta,in}_{j_1 m_1 j_2 m_2}(\mathbf{k}_{j_1 j_2}, \hat{\mathbf{r}}_1, \hat{\mathbf{r}}_2, \mathbf{R})$ is a complete orthonormal set, with the constraint that the vectors $\mathbf{k}_{j_1 j_2}$ must be defined on a half-sphere. Although, there is no constraint over the values of (j_1, m_1, j_2, m_2), in particular, defining $\bar{\alpha} = (j_2, m_2, j_1, m_1)$, $\Psi^{\zeta,in}_{\alpha}(\mathbf{k}_{j_1 j_2}, \hat{\mathbf{r}}_1, \hat{\mathbf{r}}_2, \mathbf{R})$ and $\Psi^{\zeta,in}_{\bar{\alpha}}(\mathbf{k}_{j_1 j_2}, \hat{\mathbf{r}}_1, \hat{\mathbf{r}}_2, \mathbf{R})$ are different wave functions, but $\Psi^{\zeta,in}_{\alpha}(\mathbf{k}_{j_1 j_2}, \hat{\mathbf{r}}_1, \hat{\mathbf{r}}_2, \mathbf{R})$ and $\Psi^{\zeta,in}_{\bar{\alpha}}(-\mathbf{k}_{j_1 j_2}, \hat{\mathbf{r}}_1, \hat{\mathbf{r}}_2, \mathbf{R})$ represent the same wave function [15].

The outgoing wave from the physical scattering wave function must be properly symmetrized too, and in analogy with Eq. (5.29), it can be expressed as

$$\Psi^{\zeta,out}_{\alpha}(R \to \infty) \to$$

$$\sum_{\alpha'} \frac{e^{ik_{j_1' j_2'}R}}{\sqrt{2}R} \left(f_{\alpha'\alpha}(\hat{\mathbf{R}}) Y^{m_1'}_{j_1'}(\hat{\mathbf{r}}_1) Y^{m_2'}_{j_2'}(\hat{\mathbf{r}}_2) + \zeta f_{\alpha'\alpha}(-\hat{\mathbf{R}}) Y^{m_2'}_{j_2'}(\hat{\mathbf{r}}_1) Y^{m_1'}_{j_1'}(\hat{\mathbf{r}}_2) \right), \qquad (5.30)$$

and keeping in mind

$$\sum_{j_1' m_1' j_2' m_2'} f_{\alpha'\alpha}(-\hat{\mathbf{R}}) Y^{m_2'}_{j_2'}(\hat{\mathbf{r}}_1) Y^{m_1'}_{j_1'}(\hat{\mathbf{r}}_2) = \sum_{j_2' m_2' j_1' m_1'} f_{\bar{\alpha}'\alpha}(-\hat{\mathbf{R}}) Y^{m_1'}_{j_1'}(\hat{\mathbf{r}}_1) Y^{m_2'}_{j_2'}(\hat{\mathbf{r}}_2),$$

$$(5.31)$$

Eq. (5.30) yields [3]

$$\Psi^{\zeta,out}_{\alpha}(R \to \infty) \to \sum_{\alpha'} \frac{e^{ik_{j_1' j_2'}R}}{\sqrt{2}R} f^{\zeta}_{\alpha'\alpha}(\hat{\mathbf{R}}) Y^{m_1'}_{j_1'}(\hat{\mathbf{r}}_1) Y^{m_2'}_{j_2'}(\hat{\mathbf{r}}_2). \qquad (5.32)$$

In Eq. 5.32 $f^{\zeta}_{\alpha'\alpha}(\hat{\mathbf{R}})$ represents the symmetrized scattering amplitude, which is related to the one for distinguishable particles Eq. (5.22) by

$$f^{\zeta}_{\alpha'\alpha}(\hat{\mathbf{R}}) = f_{\alpha'\alpha}(\hat{\mathbf{R}}) + \zeta f_{\bar{\alpha}'\alpha}(-\hat{\mathbf{R}}). \qquad (5.33)$$

The scattering wave function from Eq. (5.2) has to be properly symmetrized with respect to the exchange operator as

$$\Psi_\gamma^{JM,\zeta} = \sum_{\gamma'} \frac{g_{\gamma'\gamma}^{JM}(R)}{R} I_{\gamma'}^{JM,\zeta}(\hat{\mathbf{r}}_1, \hat{\mathbf{r}}_2, \hat{\mathbf{R}}), \tag{5.34}$$

where the symmetrization only affects the angular part of the scattering wave function I_γ^{JM}:

$$I_{\gamma'}^{JM,\zeta} = \frac{I_{\gamma'}^{JM} + \zeta(-1)^{j_1'+j_2'-j_{12}'+l'} I_{\bar{\gamma}'}^{JM}}{\sqrt{2\left(1 + \zeta(-1)^{-j_{12}'+l'} \delta_{j_1'j_2'}\right)}}, \tag{5.35}$$

where $\bar{\gamma}' = (j_2', j_1', j_{12}', l')$. However, in this case an ordering in needed for (j_1', j_2'), for instance $j_1' \geq j_2'$. When $j_1' = j_2'$ the values j_{12}' y l', must fulfil $(-1)^{-j_{12}'+l'} = \zeta$, because $(-1)^{-j_{12}'+l'} = -\zeta$ is prohibited by the denominator in Eq. (5.35). Next, the time-independent Schrödinger equation through the wave function in Eq. (5.34) yields a set of coupled differential equations, and after imposing proper boundary conditions, the transition matrix for identical particles $T_{\gamma'\gamma}^{JM,\zeta}$ is rendered. Finally, the transition matrix for identical molecules can be expressed in terms of the one for distinct particles as

$$T^{JM,\zeta} = \frac{T_{\gamma'\gamma}^{JM} + \zeta(-1)^{j_1'+j_2'-j_{12}'+l'} T_{\bar{\gamma}'\gamma}^{JM}}{\sqrt{1 + \zeta(-1)^{l'-j_{12}'}\delta_{j_1'j_2'}}\sqrt{1 + \zeta(-1)^{l-j_{12}}\delta_{j_1j_2}}}. \tag{5.36}$$

Following the same logic as in Sect. 5.2.1 for the definition of the cross section in terms of the scattering amplitude, the cross section for the collision of two identical molecules as [3]:

$$\sigma_{(j_1j_2 \to j_1'j_2')}^{\zeta}(k_{j_1j_2}) = \frac{1}{(2j_1+1)(2j_2+1)} \sum_{m_1m_2m_1'm_2'} \int \frac{d\sigma}{d\Omega_R}(\alpha \to \alpha')d\Omega_R = \tag{5.37}$$

$$= \frac{1}{(2j_1+1)(2j_2+1)} \sum_{m_1m_2m_1'm_2'} \int \frac{k_{j_1'j_2'}}{k_{j_1j_2}} |f_{\alpha'\alpha}^\zeta(\hat{\mathbf{R}})|^2 d\Omega_R =$$

$$= \frac{4\pi(1+\delta_{j_1j_2})(1+\delta_{j_1'j_2'})}{k_{j_1j_2}^2(2j_1+1)(2j_2+1)} \sum_{Jj_{12}j_{12}'l'l}^{'} (2J+1)|T_{j_1'j_2'j_{12}'l'\ j_1j_2j_{12}l}^{JM,\zeta}|^2,$$

where the prime in the summation implies that the values of j_{12}, l, j_{12}', l' are constrained for $j_1 = j_2$ and/or $j_1' = j_2'$, as specified above.

5.2.3   The Role of Indistinguishability in Molecule–Molecule Collisions

A comparison between the expressions for the cross section for distinguishable and indistinguishable particles helps us to understand what indistinguishability of particles really means. In particular, the differential cross section for the indistinguishable, following Eqs. (5.37) and (5.33), reads as

$$\frac{d\sigma^{\zeta}_{\alpha \to \alpha'}}{d\Omega} = \frac{k_{j'_1 j'_2}}{k_{j_1 j_2}} |f^{\zeta}_{\alpha'\alpha}(\hat{\mathbf{R}})|^2 = \qquad (5.38)$$

$$= \frac{k_{j'_1 j'_2}}{k_{j_1 j_2}} \left(|f_{\alpha'\alpha}(\hat{\mathbf{R}})|^2 + |f_{\bar{\alpha}'\alpha}(-\hat{\mathbf{R}})|^2 + 2\zeta \, Re \left(f^*_{\alpha'\alpha}(\hat{\mathbf{R}}) f_{\bar{\alpha}'\alpha}(-\hat{\mathbf{R}}) \right) \right).$$

Noting that $\alpha = (j_1, m_1, j_2, m_2)$ and $\bar{\alpha} = (j_2, m_2, j_1, m_1)$ in Eq. (5.38), it is possible to interpret such an equation as a result of two different effects. The first is purely *classical*: the detector detects a particle with j'_1 in the direction $\hat{\mathbf{R}}$, although in one case it does it as a projectile and in a second, as a target, and both are indistinguishable for the device. The second effect is purely quantum mechanical and appears as a consequence of the interference between trajectories associated with the two classical processes, and is given by the last term of the right of Eq. (5.38).

The indistinguishability effects appearing in the scattering amplitude are translated into the cross section, by plugging Eq. (5.38) into Eq. (5.37), leading to

$$\sigma^{\zeta}_{(j_1 j_2 \to j'_1 j'_2)} = \sigma^{d}_{(j_1 j_2 \to j'_1 j'_2)} + \sigma^{d}_{(j_1 j_2 \to j'_2 j'_1)} + \zeta \sigma^{\text{int}}, \qquad (5.39)$$

where the interference reads as

$$\sigma^{\text{int}} = \frac{8\pi}{k^2_{j_1 j_2}(2j_1 + 1)(2j_2 + 1)} \sum_{J j_{12} j'_{12} l l'} (2J + 1)(-1)^{j'_1 + j'_2 - j'_{12} + l'}$$

$$Re \left(T^{JM*}_{j'_1 j'_2 j'_{12} l' \; j_1 j_2 j_{12} l} T^{JM}_{j'_2 j'_1 j'_{12} l' \; j_1 j_2 j_{12} l} \right). \qquad (5.40)$$

5.3   Molecular Collisions in an External Field

In the previous section we have introduced the cross section for molecule–molecule collision, assuming that the molecules behave as rigid rotors. The expression relies on the fact that the angular momentum, $\mathbf{J}$, and its projection on the quantization axis, are good quantum numbers. However, this scenario is no longer valid when the collision occurs in the presence of an external field. The external field breaks the spherical symmetry of the problem, thus defining a privileged direction (the direction itself $\hat{\mathbf{z}}$ of the external field). Under these premises, the wave function has to be expanded in the decoupled basis. Furthermore, the direction of the quantization axis is determined by the direction of the external field rather than by an axis parallel to the incoming part of the physical scattering wave function [17], Eq. (5.13). The present formulation may be applicable to the case of molecular collisions in the absence of external fields, as we have already pointed out in Sect. 5.2.1.

This section is structured as the previous one, i.e., first, the collision between distinct molecules is discussed and then the cross section when the molecules are identical is explored.

5.3.1 Distinguishable Molecules

In the presence of external fields it is better to express the scattering wave function in terms of the decoupled basis, i.e., the product of the eigenfunctions for the internal degrees of freedom of the molecules. In this case, assuming that molecules behave as rigid rotors, the scattering wave function in the decoupled basis set reads as

$$\Psi_{\alpha ln} = \sum_{\alpha' l' n'} \frac{g(R)_{\alpha' l' n', \alpha ln}}{R} Y_{j_1'}^{m_1'}(\hat{\mathbf{r}}_1) Y_{j_2'}^{m_2'}(\hat{\mathbf{r}}_2) Y_{l'}^{n'}(\hat{\mathbf{R}}), \tag{5.41}$$

where it is worth remembering that $\alpha = (j_1, m_1, j_2, m_2)$. Plugging Eq. (5.41) in the time-independent Schrödinger $H\Psi_{\alpha ln} = E\Psi_{\alpha ln}$, and using the Hamiltonian in Eq. (5.1), one obtains the set of coupled equations for $g(R)_{\alpha' l' n', \alpha ln}$. However, in this case, the potential matrix elements would have a different expression. $g(R)_{\alpha' l' n', \alpha ln}$ must be regular at the origin and it must fulfil the following asymptotic form

$$g(R)_{\alpha' l' n', \alpha ln}(R \to \infty) \to \frac{\sin\left(k_{j_1 j_2} R - l\pi/2\right)}{\sqrt{k_{j_1 j_2}}} \delta_{\alpha\alpha'} \delta_{ll'} \delta_{nn'} -$$

$$- \frac{e^{i\left(k_{j_1' j_2'} R - l'\pi/2\right)}}{\sqrt{k_{j_1' j_2'}}} T_{\alpha' l' n', \alpha ln}. \tag{5.42}$$

The physical scattering wave function can be expressed in terms of the decoupled basis set as (in analogy to Eq. (5.14)

$$\Psi_\alpha = \sum_{ln} A_\alpha^{ln} \Psi_{\alpha ln}, \tag{5.43}$$

and keeping in mind Eqs. (5.41) and (5.42), Eq. (5.43) reads as

$$\Psi_\alpha = \sum_{ln} A_\alpha^{ln} \frac{\sin\left(k_{j_1 j_2} R - l\pi/2\right)}{\sqrt{k_{j_1 j_2}} R} Y_{j_1}^{m_1}(\hat{\mathbf{r}}_1) Y_{j_2}^{m_2}(\hat{\mathbf{r}}_2) Y_l^n(\hat{\mathbf{R}}) -$$

$$\sum_{ln\alpha'l'n'} A_\alpha^{ln} \frac{e^{i\left(k_{j_1'j_2'}R - l'\pi/2\right)}}{\sqrt{k_{j_1'j_2'}}R} T_{\alpha'l'n',\alpha ln} Y_{j_1'}^{m_1'}(\hat{\mathbf{r}}_1) Y_{j_2'}^{m_2'}(\hat{\mathbf{r}}_2) Y_{l'}^{n'}(\hat{\mathbf{R}}). \tag{5.44}$$

The physical wave function also reads as

$$\Psi_\alpha = e^{i\mathbf{k}_{j_1 j_2}\cdot\mathbf{R}} Y_{j_1}^{m_1}(\hat{\mathbf{r}}_1) Y_{j_2}^{m_2}(\hat{\mathbf{r}}_2) + \sum_{\alpha'} f(\hat{\mathbf{k}}_{j_1 j_2}, \hat{\mathbf{R}})_{\alpha',\alpha} \frac{e^{ik_{j_1'j_2'}R}}{R} Y_{j_1'}^{m_1'}(\hat{\mathbf{r}}_1) Y_{j_2'}^{m_2'}(\hat{\mathbf{r}}_2), \tag{5.45}$$

where it is easy to identify the incoming plane wave part and the outgoing divergent wave term. Indeed, Eq. (5.45) looks very similar at Eq. (5.13), although the scattering amplitude depends on the orientation between the incident flux $\mathbf{k}_{j_1 j_2}$ with respect to the axis of the external field. This dependence emerges as a consequence of having a preferred symmetry axis on the problem owing to the presence of the external field.

Comparing the first terms of Eqs. (5.44) and (5.45) and assuming a plane wave expansion in spherical harmonics, Eq. (5.17), one finds

$$A_{j_1 m_1 j_2 m_2}^{ln} = \frac{4\pi i^l}{\sqrt{k_{j_1 j_2}}} Y_l^{*n}(\hat{\mathbf{k}}_{j_1 j_2}). \tag{5.46}$$

Next, using Eq. (5.46) in the second term of Eq. (5.44) and comparing it with the second term in Eq. (5.45) renders the scattering amplitude as

$$f(\hat{\mathbf{k}}_{j_1 j_2}, \hat{\mathbf{R}})_{\alpha',\alpha} = -\sum_{ll'nn'} \frac{4\pi i^{l-l'}}{\sqrt{k_{j_1 j_2} k_{j_1' j_2'}}R} Y_l^{*n}(\hat{\mathbf{k}}_{j_1 j_2}) Y_{l'}^{n'}(\hat{\mathbf{R}}) T_{\alpha'l'n',\alpha ln}. \tag{5.47}$$

The differential cross section given by Eq. (5.23) can be expressed as

$$\frac{d\sigma}{d\Omega_k d\Omega_R} = \frac{k_{j_1' j_2'}}{k_{j_1 j_2}} |f(\hat{\mathbf{k}}_{j_1 j_2}, \hat{\mathbf{R}})|^2. \tag{5.48}$$

Integrating Eq. (5.48) over all final orientations of the scattered molecules, and averaging over the initial orientation of the incoming flux leads to the cross section as [18]

$$\sigma(j_1 m_1 j_2 m_2 \rightarrow j_1' m_1' j_2' m_2')$$

$$= \frac{1}{4\pi} \int\int \frac{k_{j_1' j_2'}}{k_{j_1 j_2}} |f(\hat{\mathbf{k}}_{j_1 j_2}, \hat{\mathbf{R}})_{j_1' m_1' j_2' m_2' l'n', j_1 m_1 j_2 m_2 ln}|^2 d\Omega_k d\Omega_R. \tag{5.49}$$

Here, it is important to notice that the cross section depends on the rotational quantum number j_i and also on its projection m_i. This situation is different from the free-field case, where the cross section only depends on j_i. When an external field is present, the cross section must depend on the projection of the angular momentum of the molecules on the quantization axis, since that is the only good quantum number. Finally, taking into account Eq. (5.47) in Eq. (5.49) yields

$$\sigma(j_1 m_1 j_2 m_2 \to j_1' m_1' j_2' m_2') = \frac{4\pi}{k_{j_1 j_2}^2} \sum_{ll'nn'} |T_{j_1' m_1' j_2' m_2' l'n', j_1 m_1 j_2 m_2 ln}|^2. \tag{5.50}$$

5.3.2 Indistinguishable Molecules

As we have seen in Sect. 5.2.2, the physical scattering wave function and the scattering wave function must be eigenfunctions of the exchange operator owing to the indistinguishability of the colliding molecules. Therefore, Eq. (5.33) must be satisfied and Eq. (5.49) leads to

$$\sigma^\zeta(j_1 m_1 j_2 m_2 \to j_1' m_1' j_2' m_2') = \frac{1}{4\pi} \int \int \frac{k_{j_1' j_2'}}{k_{j_1 j_2}} |f(\hat{\mathbf{k}}_{j_1 j_2}, \hat{\mathbf{R}})_{\alpha,\alpha}^\zeta|^2 d\Omega_k d\Omega_R, \tag{5.51}$$

which is the cross section for the collision of two identical molecules in the presence of an external field. The scattering amplitude in Eq. (5.51) is given by

$$f^\zeta(\hat{\mathbf{k}}_{j_1 j_2}, \hat{\mathbf{R}})_{\alpha',\alpha} = f(\hat{\mathbf{k}}_{j_1 j_2}, \hat{\mathbf{R}})_{\alpha',\alpha} + \zeta f(\hat{\mathbf{k}}_{j_1 j_2}, -\hat{\mathbf{R}})_{\bar{\alpha}',\alpha} =$$

$$- \sum_{ll'nn'} \frac{4\pi i^{l-l'}}{\sqrt{k_{j_1 j_2} k_{j_1' j_2'}}} Y_l^{*n}(\hat{\mathbf{k}}_{j_1 j_2}) Y_{l'}^{n'}(\hat{\mathbf{R}}) T_{\alpha'l'n',\alpha ln} -$$

$$\zeta \sum_{ll'nn'} \frac{4\pi i^{l-l'}}{\sqrt{k_{j_1 j_2} k_{j_1' j_2'}}} Y_l^{*n}(\hat{\mathbf{k}}_{j_1 j_2}) Y_{l'}^{n'}(-\hat{\mathbf{R}}) T_{\bar{\alpha}'l'n',\alpha ln} =$$

$$- \sum_{ll'nn'} \frac{4\pi i^{l-l'}}{\sqrt{k_{j_1 j_2} k_{j_1' j_2'}}} Y_l^{*n}(\hat{\mathbf{k}}_{j_1 j_2}) Y_{l'}^{n'}(\hat{\mathbf{R}}) \left(T_{\alpha'l'n',\alpha ln} + \zeta(-1)^{l'} T_{\bar{\alpha}'l'n',\alpha ln} \right), \tag{5.52}$$

where Eq. (5.47) has been employed for the scattering amplitude for distinct molecules. Equation (5.52) establishes how the symmetrized scattering amplitude relates to the transition matrix for distinct particles. However, it is still necessary to know how the transition matrix for the collision of identical particles is defined in terms of the one for distinct particles. To this end, it is better to begin with the definition of the symmetrized angular part of the scattering wave function, which is given by

$$I_{\alpha l n}^{\zeta}(\hat{\mathbf{r}}_1, \hat{\mathbf{r}}_2, \hat{\mathbf{R}}) = \frac{Y_{j_1}^{m_1}(\hat{\mathbf{r}}_1) Y_{j_2}^{m_2}(\hat{\mathbf{r}}_2) Y_l^n(\hat{\mathbf{R}}) + \zeta(-1)^l Y_{j_2}^{m_2}(\hat{\mathbf{r}}_1) Y_{j_1}^{m_1}(\hat{\mathbf{r}}_2) Y_l^n(\hat{\mathbf{R}})}{\sqrt{2(1 + \zeta(-1)^l \delta_{j_1 j_2} \delta_{m_1 m_2})}}.$$

(5.53)

It is possible to evaluate the transition matrix elements on this basis yielding

$$T_{j_1' m_1' j_2' m_2' l' n', j_1 m_1 j_2 m_2 l n}^{\zeta} = \langle I_{\alpha' l' n'}^{\zeta}(\hat{\mathbf{r}}_1, \hat{\mathbf{r}}_2, \hat{\mathbf{R}}) | T | I_{\alpha l n}^{\zeta}(\hat{\mathbf{r}}_1, \hat{\mathbf{r}}_2, \hat{\mathbf{R}}) \rangle =$$

$$\frac{\langle \alpha' l' n' | T | \alpha l n \rangle + \zeta(-1)^l \langle \alpha' l' n' | T | \bar{\alpha} l n \rangle + \zeta(-1)^{l'} \langle \bar{\alpha}' l' n' | T | \alpha l n \rangle + \langle \bar{\alpha}' l' n' | T | \bar{\alpha} l n \rangle}{\sqrt{2(1 + \zeta(-1)^l \delta_{j_1 j_2} \delta_{m_1 m_2})} \sqrt{2(1 + \zeta(-1)^{l'} \delta_{j_1' j_2'} \delta_{m_1' m_2'})}}.$$

(5.54)

Taking into account that $\langle \bar{\alpha}' l' n' | T | \bar{\alpha} l n \rangle = \langle \alpha' l' n' | T | \alpha l n \rangle$ and

$$\zeta(-1)^l \langle \alpha' l' n' | T | \bar{\alpha} l n \rangle = \zeta(-1)^{l'} \langle \bar{\alpha}' l' n' | T | \alpha l n \rangle,$$

(5.55)

Eq. (5.54) yields

$$T_{\alpha' l' n', \alpha l n}^{\zeta} = \frac{T_{\alpha' l' n', \alpha l n} + \zeta(-1)^{l'} T_{\bar{\alpha}' l' n', \alpha l n}}{\sqrt{(1 + \zeta(-1)^l \delta_{j_1 j_2} \delta_{m_1 m_2})} \sqrt{(1 + \zeta(-1)^{l'} \delta_{j_1' j_2'} \delta_{m_1' m_2'})}}.$$

(5.56)

Finally, plugging Eq. (5.56) into Eq. (5.52) it is possible to express the scattering amplitude in terms of the symmetrized transition matrix as

$$f^{\zeta}(\hat{\mathbf{k}}_{j_1 j_2}, \hat{\mathbf{R}})_{\alpha' \alpha} =$$

$$- \sum_{l l' n n'} \frac{4\pi i^{l-l'}}{\sqrt{k_{j_1 j_2} k_{j_1' j_2'}}} Y_l^{*n}(\hat{\mathbf{k}}_{j_1 j_2}) Y_{l'}^{n'}(\hat{\mathbf{R}}) T_{\alpha_1' l' n', \alpha l n}^{\zeta} \sqrt{(1 + \zeta(-1)^l \delta_{j_1 j_2} \delta_{m_1 m_2})}$$

$$\sqrt{(1 + \zeta(-1)^{l'} \delta_{j_1' j_2'} \delta_{m_1' m_2'})}.$$

(5.57)

Therefore, the cross section given by Eq. (5.51) reads as

$$\sigma_{(j_1 m_1 j_2 m_2 \to j_1' m_1' j_2' m_2')}^{\zeta}(k_{j_1 j_2}) = \frac{4\pi(1 + \delta_{j_1 j_2} \delta_{m_1 m_2})(1 + \delta_{j_1' j_2'} \delta_{m_1' m_2'})}{k_{j_1 j_2}^2} \sum_{l l' n n'}' |T_{\alpha' l' n', \alpha l n}^{\zeta}|^2,$$

(5.58)

where the summation is restricted to the values of l and l' to avoid doubling counting of the states when the final and initial states are identical, as in the Sect. 5.2.2.

It is worth emphasizing that the cross section for identical molecule–molecule collisions shows the same functional form, independently of the basis to describe the angular dependence of the scattering wave function. In fact, one notices that

Eq. (5.58) contains the same symmetrization factors as in Eq. (5.37). However, in the decoupled basis the symmetrization also affects the magnetic quantum number m of each of the molecules.

5.4   Ultracold Collisions of Oxygen Molecules in the Presence of an External Magnetic Field

In previous sections, we have introduced the quantum mechanical scattering theory with and without external fields. Nevertheless, we have not seen how this machinery works in a real case scenario. For this reason, we chose to present some results regarding the collision of oxygen molecules at ultracold temperatures in the presence of an external magnetic field. Oxygen is a paramagnetic molecule, showing a spin–rotation coupling that leads to intriguing relaxation mechanisms [19, 20]. Moreover, oxygen has been proposed as a possible candidate for evaporative cooling [21], thus envisioning the possibility of having a dense ultracold gas of oxygen molecules.

5.4.1   Zeeman Hamiltonian for $^3\Sigma$ Molecules

O_2 has a $^3\Sigma$ ground electronic state; thus, the interaction Hamiltonian is better described within the Hund's case b [22]. Therefore, the spin–orbit interaction is negligible and the electronic spin is coupled to the molecular rotation to yield the total angular momentum. However, in the presence of an external field, it is convenient to work in the decoupled basis (see Sect. 5.3), which in this case reads as $|N M_N S M_S\rangle$. Assuming that O_2 is in its vibrational ground state, the Zeeman Hamiltonian reads as [23]

$$\hat{H} = B_e \hat{N}^2 + 2\mu_B \vec{B} \cdot \hat{S} + \gamma_{SR} \hat{N} \cdot \hat{S} + \frac{2}{3}\lambda_{SS}\sqrt{\frac{24\pi}{5}} \sum_q Y_{2q}^*(\hat{r})[\hat{S} \otimes \hat{S}]_q^{(2)}, \quad (5.59)$$

where $\hat{N}^2$ is the angular momentum of rotation (in previous Sections it has been identified by $\hat{j}^2$), B_e is the rotational constant, μ_B is the Bohr magneton, $\vec{B}$ is the external magnetic field, and $\hat{S}$ is the electronic spin. In Eq. (5.59) appear two extra terms apart from the rotational and Zeeman terms. The first term is the so-called spin–rotation, $\gamma_{SR} \hat{N} \cdot \hat{S}$ ($\gamma_{SR} \sim 1$ mK), which is a consequence of a coupling between the rotation of the molecule and the electronic spin. The last term is the spin–spin interaction ($\lambda_{SS} \sim 1$ K), which arises through the coupling of the magnetic dipolar moments of the electrons dispersed into the molecule. This last term only appears for electronic states whose spin is ≥ 1. These two terms, of relativistic origin, are expressed in their effective form (see Ref. [1] for more detail), and t γ_{SR} and λ_{SS} can be calculated from *ab initio* quantum chemistry methods or determined from accurate spectroscopy data for the molecule.

The rotational and Zeeman parts of the Hamiltonian, Eq. (5.59), are diagonal in the decoupled basis, and the matrix elements read as

$$
\langle N M_N S M_S | B_e \hat{N}^2 + 2\mu_B B S_z | N' M_{N'} S' M_{S'} \rangle =
$$

$$
\delta_{NN'} \delta_{SS'} \delta_{M_N M_{N'}} \delta_{M_S M_{S'}} \left(B_e N(N+1) + 2\mu_B B M_S \right), \tag{5.60}
$$

where we have assumed that the z axis of the SF frame is parallel to the external magnetic field, $\vec{B}$.

The spin–rotation term in the decoupled basis is better obtained by representing the scalar product between the two angular momenta in terms of the angular momentum ladder operators as

$$
\gamma_{SR} \hat{N} \cdot \hat{S} = \gamma_{SR} \left(N_z S_z + \frac{1}{2} (N_+ S_- + N_- S_+) \right). \tag{5.61}
$$

And taking into account how the angular momentum ladder operators act over the decoupled basis [9, 10] leads to

$$
\langle N M_N S M_S | \gamma_{SR} \hat{N} \cdot \hat{S} | N' M_{N'} S' M_{S'} \rangle = \delta_{NN'} \delta_{SS'} \Big[\delta_{M_N M_{N'}} \delta_{M_S M_{S'}} \gamma_{SR} +
$$

$$
+ \delta_{M_N M_{N'} \pm 1} \delta_{M_S M_{S'} \mp 1} \frac{\gamma_{SR}}{2} \left(N(N+1) - M_{N'}(M_{N'} \pm 1) \right)^{1/2} \times
$$

$$
\times \left(S(S+1) - M_{S'}(M_{S'} \mp 1) \right)^{1/2} \Big], \tag{5.62}
$$

where it is noticed that the spin–rotation interaction couples states in such a way that $M_S + M_N$ is conserved.

The evaluation of the spin–spin term is somewhat more involved since it contains vector couplings, giving rise to tensors of order 2. It is desirable to separate the part that depends on the spin from the part that depends only on the molecular rotation:

$$
\langle N M_N S M_S | \frac{2}{3} \lambda_{SS} \sqrt{\frac{24\pi}{5}} \sum_q (-1)^q Y_{2-q}(\hat{r}) [\hat{S} \otimes \hat{S}]_q^{(2)} | N' M_{N'} S' M_{S'} \rangle =
$$

$$
= \frac{2}{3} \lambda_{SS} \sqrt{\frac{24\pi}{5}} \sum_q (-1)^q \langle N M_N | Y_{2-q} | N' M_{N'} \rangle \times \langle S M_S | [\hat{S} \otimes \hat{S}]_q^{(2)} | S' M_{S'} \rangle, \tag{5.63}
$$

where the term involving rotational states reads as

$$
\langle N M_N | Y_{2-q} | N' M_{N'} \rangle = (-1)^{-M_N} \sqrt{(2N+1)(2N'+1)} \sqrt{\frac{5}{4\pi}} \times
$$

$$
\begin{pmatrix} N & 2 & N' \\ -M_N & -q & M_{N'} \end{pmatrix} \begin{pmatrix} N & 2 & N' \\ 0 & 0 & 0 \end{pmatrix}. \tag{5.64}
$$

On the other hand, the term that couples the spin of the two open shell electrons of oxygen is evaluated by means of the Wigner–Eckart theorem yielding [1, 9, 10]

$$\langle SM_S||[\hat{S}\otimes\hat{S}]_q^{(2)}|S'M_{S'}\rangle = (-1)^{S-M_S}\begin{pmatrix} S & 2 & S' \\ -M_S & q & M_{S'}\end{pmatrix}\langle S||[\hat{S}\otimes\hat{S}]^{(2)}||S'\rangle, \quad (5.65)$$

where the reduced matrix element is [1, 9, 10]

$$\langle S||[\hat{S}\otimes\hat{S}]^{(2)}||S'\rangle = \sqrt{5}S(S+1)(2S+1)\begin{Bmatrix} 1 & 1 & 2 \\ S & S & S\end{Bmatrix}. \quad (5.66)$$

Taking into account that O_2 is $^3\Sigma$ molecule in Eq. (5.66), and that $q = M_{N'} - M_N$, otherwise Eq. (5.64) is null, the matrix elements of the spin-spin interaction Hamiltonian read as

$$\langle NM_N SM_S|\frac{2}{3}\lambda_{SS}\sqrt{\frac{24\pi}{5}}\sum_q(-1)^q Y_{2-q}(\hat{r})[\hat{S}\otimes\hat{S}]_q^{(2)}|N'M_{N'}S'M_{S'}\rangle =$$

$$= \frac{2\sqrt{30}}{3}\lambda_{SS}(-1)^{S-M_S-M_{N'}}\begin{pmatrix} N & 2 & N' \\ -M_N & M_N - M_{N'} & M_{N'}\end{pmatrix}$$

$$\sqrt{2N+1}\sqrt{2N'+1}\begin{pmatrix} N & 2 & N' \\ 0 & 0 & 0\end{pmatrix}\begin{pmatrix} S & 2 & S' \\ -M_S & M_{N'} - M_N & M_{S'}\end{pmatrix}. \quad (5.67)$$

Here, it is noted that the spin–spin interaction term couples states with distinct values of M_S and M_N, although $M_S + M_N$ remains constant.

In Fig. 5.2 the eigenvalues of Eq. (5.59) for $^{16}O_2$ and $^{17}O_2$ are shown as a function of the magnitude of an external magnetic field. The first peculiarity that may strike the reader is that only odd rotational quantum numbers appear for $^{16}O_2$, and only even for $^{17}O_2$. The explanation lies in the inherent nature of the atoms forming the molecules. The nucleus of ^{17}O is a fermion with spin $I = 5/2$, then the $^{17}O_2$ molecule shows nuclear spin $I = 0, \ldots, 5$, and the total wave function (electronic + nuclear + rotational) must be antisymmetric with respect to the exchange of the nuclei [1, 2]. Assuming that $^{17}O_2$ can be prepared in a $M_I = I = 5$ state, and taking into account that the electronic wave function is anti-symmetric regarding the exchange of atoms, imply that N can only be an even number. In the case of $^{16}O_2$, ^{16}O has a null nuclear spin; therefore, it is a boson and the total wave function must be symmetric against exchange of atoms. However, its electronic wave function is anti-symmetric with respect to the exchange of particles; therefore, N has to be odd to guarantee the proper symmetrization of the wave function.

In Fig. 5.2 the levels associated with the lower rotational state are depicted as blue or red solid lines. The blue lines correlate with low-field seeker states (lfs), i.e., states that increase their energy as the magnitude of the magnetic field increases. The red lines correspond to high-field seeker (hfs) states, in which their energy decreases as the magnitude of the magnetic field increases. The reader may go to Sect. 4.3 for more detailed information about these states.

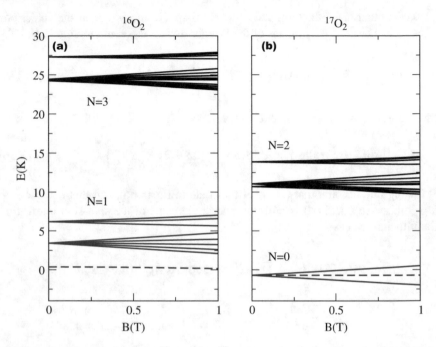

Fig. 5.2 Zeeman energy levels for $^{16}O_2$ (**a**) and $^{17}O_2$ (**b**) as a function of the magnetic field (in T). The high-field seeker states are shown in red whereas the high-field seekers are presented in blue. The spectroscopic constants employed for $^{16}O_2$ are $\gamma_{SR} = -0.0089\,cm^{-1}$, $\lambda_{SS} = 1.985\,cm^{-1}$ and $B_e = 1.438\,cm^{-1}$ [23]. For $^{17}O_2$ we have taken $\gamma_{SR} = -0.00396\,cm^{-1}$, $\lambda_{SS} = 1.985\,cm-1$ and $B_e = 1.353\,cm^{-1}$ [24]

Low-field seeker states, are the only ones that can be trapped in a magnetic field, as shown in Sect. 4.7. Therefore, it is crucial to ensure that the molecules remain in one of the trappable states during their time in the trap. However, the lfs states are not the lowest energy state for the collision, and hence inelastic collisions are possible. Indeed, the majority of the effort in the community is to find a system in which the elastic collisions are a factor of 100 or more (see Sect. 4.8) frequent than inelastic ones. Here, we explore whether O_2 is one of these molecules.

5.4.2   Spin Relaxation and Zeeman Suppression

Spin flip collisions (also called spin exchange collisions) are responsible for molecular and atomic losses in magnetic traps, as explained in Sect. 4.7, and hence, they determine the chances of success of evaporative cooling techniques. Therefore, understanding the physics behind spin flip collisions is mandatory to find the best molecular and atomic candidates for evaporative cooling. In particular, spin flip collisions between $^3\Sigma$ molecules occur via the so-called spin relaxation mechanism [17, 19], and it is explained below.

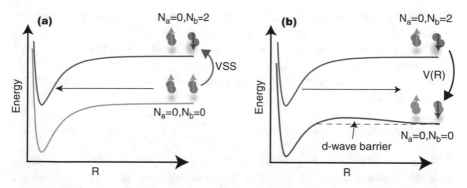

Fig. 5.3 Schematic representation of the spin relaxation mechanism. In panel (**a**), the spin–spin interaction, V_{SS}, changes M_{S_b} at the same time that N_b and M_{N_b} change. In panel (**b**), the PES induces the transition between states with the same spin projection but different rotational quantum numbers. The red dotted line denotes the asymptotic value for wave-s ($l = 0$), showing that the output channel must be in wave-d ($l = 2$) (see text for details)

Let us assume the collision between two $^{17}O_2$ molecules (a and b). These molecules are in the lowest energy lfs state in a given magnetic field; thus, the two-molecule state reads as

$$|N_a = 0, M_{N_a} = 0, M_{S_a} = 1, N_b = 0, M_{N_b} = 0, M_{S_b} = 1\rangle. \qquad (5.68)$$

The lowest energy lfs in $^{17}O_2$ is not the lowest energy state, and hence several inelastic channels are open. In particular, the state (5.68) after a collision may end up in the state

$$|N_a = 0, M_{N_a} = 0, M_{S_a} = 1, N_b = 0, M_{N_b} = 0, M_{S_b} = -1\rangle, \qquad (5.69)$$

where one of the molecules has changed its spin projection after the collision, which leads to an exothermic process leading to the loss of both atoms from the trap. How does this transition occur? The answer is the spin relaxation mechanism, which is depicted in Fig. 5.3. In the spin relaxation mechanism, two factors come into play. First, the spin–spin interaction term of the molecular Hamiltonian (5.59) mixes the states $|N_b = 0, M_{N_b} = 0, M_{S_b} = 1\rangle$ and $|N_b = 2, M_{N_b} = 2, M_{S_b} = -1\rangle$ (see Eq. (5.67)). Second, the intermolecular potential allows the coupling between the component $|N_b = 2, M_{N_b} = 2, M_{S_b} = -1\rangle$ of the initial state and a final state with components $|N_b = 0, M_{N_b} = 0, M_{S_b} = -1\rangle$, thus arriving at the state shown in Eq. (5.69). Therefore, spin relaxation depends on the strength of the spin–spin coupling $\lambda_S S$ constant and on the anisotropy of the underlying PES for the molecule–molecule interaction. Indeed, Krems and Dalgarno showed that the magnetic field affects the couplings of the potential for interaction with the input and output channel of the collision, and will therefore play a crucial role in obtaining effective inelastic sections. In addition, it can be shown that the spin relaxation process is more efficient when the ratio λ_{SS}/B_e, is larger [17].

Let us assume the spin relaxation mechanism at ultracold temperatures; thus, the entrance channel will be in a s-wave regime ($l = 0$ and $m_l = 0$). In the presence of an external magnetic field, the projection of the total angular momentum is conserved M, i.e.,

$$M_{N_a} + M_{S_a} + M_{N_b} + M_{S_b} + m_l = M_{N_a'} + M_{S_a'} + M_{N_b'} + M_{S_b'} + m_{l'}. \tag{5.70}$$

Therefore, if the state $|N_a = 0, M_{N_a} = 0, M_{S_a} = 1, N_b = 0, M_{N_b} = 0, M_{S_b} = 1\rangle$ is the entrance channel with $m_l = 0$, the conservation of M leads to that $m_l' \geq 2$ for the state $|N_a = 0, M_{N_a} = 0, M_{S_a} = 1, N_b = 0, M_{N_b} = 0, M_{S_b} = -1\rangle$ as the final state. In other words, the outgoing channel is in a d-wave. This brings very interesting physics into play: the outgoing channel shows a barrier. Thus, if the kinetic energy of the outgoing channel is not enough to overcome the barrier, spin relaxation will be highly suppressed. In particular, by tuning the magnetic field, it is possible to make the energy difference between the entrance channel and the state $|N_a = 0, M_{N_a} = 0, M_{S_a} = 1, N_b = 0, M_{N_b} = 0, M_{S_b} = -1\rangle$ smaller than the height of the d-wave barrier, thus leading to the suppression of the spin relaxation (which in this case can only take place by means of tunneling through the barrier). This magnetic assisted suppression is called Zeeman suppression.

Zeeman suppression is the method in which the inelastic collisions are screened by tuning the external magnetic field. The energy difference between the states $|N_a = 0, M_{N_a} = 0, M_{S_a} = 1, N_b = 0, M_{N_b} = 0, M_{S_b} = 1\rangle$ and $|N_a = 0, M_{N_a} = 0, M_{S_a} = 1, N_b = 0, M_{N_b} = 0, M_{S_b} = -1\rangle$ can be estimated as $2\mu_B B$. At a long-range distance the molecule–molecule potential is given by

$$V_l(R) = \frac{l(l+1)}{2\mu R^2} - \frac{C_6}{R^6}, \tag{5.71}$$

where C_6 is the long-range coefficient (or van der Waals coefficient) for the molecule–molecule collision. Next, solving for R^*

$$\left. \frac{dV_l(R)}{dR} \right|_{R=R^*} = 0, \tag{5.72}$$

renders the height of the d-wave barrier, $V_l(R^*)$, which in this case is $3/(\sqrt{C_6}\mu^{3/2})$. Finally, comparing the energy difference between the entrance and outgoing channels with the height of the barrier yields

$$B \leq \frac{2}{3\mu_B \mu^{3/2}\sqrt{C_6}}. \tag{5.73}$$

This represents the region where Zeeman suppression becomes effective and hence the possibility of trapping $^3\Sigma$ molecules may work.

5.4.3   Molecular Scattering Theory in Action

The Hamiltonian for the interaction between two $^3\Sigma$ molecules in the presence of an external field reads as (in Jacobi coordinates)

$$\hat{H} = -\frac{1}{2\mu R}\frac{\partial^2}{\partial R^2}R + \frac{\hat{l}^2}{2\mu R^2} + V(\mathbf{R}, \hat{\mathbf{r}}_1, \hat{\mathbf{r}}_2) + \hat{H}_1 + \hat{H}_2, \tag{5.74}$$

where, $\hat{l}$ is the orbital angular moment of collision, μ is the reduced mass, and $V(\mathbf{R}, \hat{\mathbf{r}}_1, \hat{\mathbf{r}}_2)$ is the PES for fixed intramolecular separations (rigid rotor approximation, i.e., $r_1 = r_2 = r_e$). The Hamiltonian (5.74) is equivalent to the Hamiltonian in (5.1); however, in this case the molecular Hamiltonian is the Zeeman Hamiltonian given by Eq. (5.59). In (5.74) we have not taken into account the hyperfine structure since it is expected that its effects are negligible under typical experimental conditions [25]. In the same vein, the magnetic dipole–dipole interaction is neglected, following Avdeenkov and Bohn, who demonstrated that their influence on the effective section was $\lesssim\%1$ [25].

In the presence of an external magnetic field, only the projection of the total angular momentum on the quantization axis is a good quantum number, as we have seen in Sect. 5.3. To calculate the scattering observables for the collision at hand we work in the decoupled basis, as in Sect. 5.3. Although, we need to include the spin degrees of freedom (we also change to the nomenclature of the rotation quantum number, $j_\alpha \to N_\alpha$, following Sect. 5.4.1), which in the decoupled basis leads to

$$|\tau_1 \tau_2 l m_l\rangle = |\tau_1\rangle |\tau_2\rangle |l m_l\rangle, \tag{5.75}$$

where $|\tau_\alpha\rangle = |N_\alpha M_{N_\alpha} S_\alpha M_{S_\alpha}\rangle$. Since the collision is between identical molecules, the scattering wave function must be symmetrized based on the nuclear spin of the atoms, and hence the angular part of the scattering wave function reads as

$$I_{\tau_1 \tau_2}^{\eta \epsilon M}(\hat{r}_1, \hat{r}_2, \hat{R}) = \frac{1}{\left(2\left(1 + \delta_{\tau_1, \tau_2}\right)\right)^{1/2}}\left(|\tau_1 \tau_2\rangle + \eta\epsilon |\tau_2 \tau_1\rangle\right)|l m_l\rangle. \tag{5.76}$$

The basis of Eq. (5.76) is formed by normalized functions and constitutes a well-ordered set with $\tau_a \geq \tau_b$. These functions are eigenfunction of the molecule exchange operator with eigenvalue τ. The same functions are eigenfunctions of the space inversion operator E^* with eigenvalue $\epsilon = (-1)^{N_a + N_b + l}$. In addition, the superscript M refers to the projection of the total angular momentum on the z axis of the SF reference system, a magnitude that is conserved during the collision. Thus, the scattering wave function may be expressed as

$$\Psi^{\eta\epsilon M} = \frac{1}{R} \sum_{\tau_a \geq \tau_b l m_l} u^{\eta\epsilon M}_{\tau_a \tau_b l m_l}(R) I^{\eta\epsilon M}_{\tau_a \tau_b l m_l}(\hat{r}_a, \hat{r}_b, \hat{R}). \tag{5.77}$$

Next, plugging Eq. (5.77) into the time-independent Schrödinger and using the same approach as the one explained for the derivation of Eq. (5.4), the following set of coupled differential equations appear

$$\left[\frac{1}{2\mu}\frac{d^2}{dR^2} - \frac{l(l+1)}{2\mu R^2} + E\right] u^{\eta\epsilon M}_{\tau_1 \tau_2 l m_l}(R) =$$

$$\sum_{\tau_1' \geq \tau_2' l' m_l'} \langle I^{\eta\epsilon M}_{\tau_1 \tau_2 l m_l} |(V + \hat{H}_1 + \hat{H}_2)| I^{\eta\epsilon M}_{\tau_1' \tau_2' l' m_l'}\rangle u^{\eta\epsilon M}_{\tau_1' \tau_2' l' m_l'}(R), \tag{5.78}$$

where E represents total collision energy. It is worth mentioning that the molecular asymptotic Hamiltonian $\hat{H}_1 + \hat{H}_2$ is not diagonal in the basis (5.76), owing to the spin–spin and spin–rotation interaction terms that couple states with different N, M_N and M_S quantum numbers. The potential matrix elements are given by

$$\langle I^{M\eta\epsilon}_{\tau_1 \tau_2 l m_l} |V| I^{M\eta\epsilon}_{\tau_1' \tau_2' l' m_l'}\rangle = \frac{1}{[(1 + \delta_{\tau_1, \tau_2})(1 + \delta_{\tau_1', \tau_2'})]^{1/2}} \times$$

$$\times \left[\langle \tau_1 \tau_2 l m_l |V| \tau_1' \tau_2' l' m_l'\rangle + \eta\epsilon \langle \tau_1 \tau_2 l m_l |V| \tau_2' \tau_1' l' m_l'\rangle\right]. \tag{5.79}$$

The interaction between two molecules $^3\Sigma$ shows several spin multiplicities that arise from the composition of the electronic spin of both molecules $S = S_a + S_b = 0, 1, 2$. The intermolecular potential can be expressed in the total spin basis as [26].

$$V(\mathbf{R}, \hat{\mathbf{r}}_1, \hat{\mathbf{r}}_2) = \sum_{S=0}^{2} \sum_{M_S=-S}^{S} V_S(\mathbf{R}, \hat{\mathbf{r}}_1, \hat{\mathbf{r}}_2)|SM_S\rangle\langle SM_S|, \tag{5.80}$$

where M_S projection of the total spin, $M_S = M_{S_a} + M_{S_b}$, and the three multiplicities, singlet ($S = 0$), triplet ($S = 1$), and quintuplet ($S = 2$) are included.[5] Next, plugging Eq. (5.80) into Eq. (5.79) yields

$$\langle N_a M_{N_a} S_a M_{S_a} N_b M_{N_b} S_b M_{S_b} l m_l |V| N_a' M_{N_a'} S_a M_{S_a'} N_b' M_{N_b'} S_b M_{S_b'} l' m_{l'}\rangle$$

$$= \delta_{M_{S_a}+M_{S_b},M_{S_a'}+M_{S_b'}} \sum_{S=0}^{2} \begin{pmatrix} S_a & S_b & S \\ M_{S_a} & M_{S_b} & -M_{S_a} - M_{S_b} \end{pmatrix} \begin{pmatrix} S_a & S_b & S \\ M_{S_a'} & M_{S_b'} & -M_{S_a'} - M_{S_b'} \end{pmatrix}$$

$$\times (2S+1) \langle N_a M_{N_a} N_b M_{N_b} l m_l | V_S(\vec{R}, \hat{r}_a, \hat{r}_b) | N_a' M_{N_a'} N_b' M_{N_b'} l' m_{l'} \rangle.$$

$$(5.81)$$

The potential matrix elements can be readily evaluated by means of the expansion of the PES in bipolar spherical harmonics, as shown in Eq. (5.8) yielding

$$\langle N_a M_{N_a} S_a M_{S_a} N_b M_{N_b} S_b M_{S_b} l m_l | V | N_a' M_{N_a'} S_a' M_{S_a'} N_b' M_{N_b'} S_b' M_{S_b'} l' m_{l'} \rangle =$$

$$= \sum_{S=0}^{2} \sum_{\lambda_a,\lambda_b,\lambda} f_S^{\lambda_a,\lambda_b,\lambda}(R) \left[(2N_a+1)(2N_a'+1)(2N_b+1)(2N_b'+1) \right]^{1/2} \times$$

$$\times \left[(2l+1)(2l'+1)(2\lambda_a+1)(2\lambda_b+1)(2\lambda+1) \right]^{1/2} (-1)^{M_{N_a}+M_{N_b}+m_l} \times$$

$$\times (2S+1) \begin{pmatrix} S_a & S_b & S \\ M_{S_a} & M_{S_b} & -M_{S_a} - M_{S_b} \end{pmatrix} \begin{pmatrix} S_a & S_b & S \\ M_{S_a'} & M_{S_b'} & -M_{S_a'} - M_{S_b'} \end{pmatrix}$$

$$\begin{pmatrix} \lambda_a & \lambda_b & \lambda \\ M_{N_a} - M_{N_a'} & M_{N_b} - M_{N_b'} & m_l - m_l' \end{pmatrix} \begin{pmatrix} N_a & \lambda_a & N'a \\ -M_{N_a} & M_{N_a} - M_{N_a'} & M_{N_a'} \end{pmatrix}$$

$$\begin{pmatrix} N_a & \lambda_a & N_a' \\ 0 & 0 & 0 \end{pmatrix} \begin{pmatrix} N_b & \lambda_b & N'b \\ -M_{N_b} & M_{N_b} - M_{N_b'} & M_{N_b'} \end{pmatrix}$$

$$\begin{pmatrix} N_b & \lambda_b & N_b' \\ 0 & 0 & 0 \end{pmatrix} \begin{pmatrix} l & \lambda & l' \\ -m_l & m_l - m_{l'} & m_{l'} \end{pmatrix} \begin{pmatrix} l & \lambda & l' \\ 0 & 0 & 0 \end{pmatrix}. \quad (5.82)$$

Once potential matrix elements are known, one is ready to solve the system of coupled differential equations Eq. (5.78) by means of the numerical method that is best suited for the proposal of the study. Here, we chose the coupled channel method, in which the scattering wave function (5.77) is *propagated* from the short-range region until the long-range region where the proper asymptotic conditions, Eq. (5.56), are imposed. However, at the last point of propagation, we have to make a change of basis in the propagator, going from the totally decoupled basis to the basis that diagonalizes $\hat{H}_1 + \hat{H}_2$, which we label as $|\zeta_1, \zeta_2 l m_l \rangle$ (the Zeeman levels of the molecules), i.e.,

$$\left[\hat{H}_1 + \hat{H}_2 \right] \psi_{\zeta_1 \zeta_2 l m_l}^{M\eta\epsilon} = (\epsilon_{\xi_1} + \varepsilon_{\zeta_2}) \psi_{\zeta_1 \zeta_2 l m_l}^{M\eta\epsilon}. \quad (5.83)$$

Finally, after performing the cited changes of basis one gets the cross section as

$$\sigma^{M\eta\epsilon}_{\zeta_1\zeta_2 \to \zeta_1'\zeta_2'} = \frac{\pi\left(1+\delta_{\zeta_1,\zeta_2}\right)}{k^2_{\zeta_1\zeta_2}} \sum_{lm_l l'm_{l'}} |T^{M\eta\epsilon}_{\zeta_1\zeta_2 lm_l;\zeta_1'\zeta_2'l'm_{l'}}|^2, \tag{5.84}$$

where the indexes on the summation are restricted, following the recipe of Sect. 5.3.2, to avoid counting the same state twice.

5.4.3.1 Labeling of the States and Intermolecular Potentials

The spin–spin interaction and spin–rotation interaction mix states with different spins and rotational quantum numbers; therefore, it is proper to use $|NM_N SM_S\rangle$ to label the different states for the scattering process. Indeed, we have already pointed out that the cross section given by Eq. (5.84) depends on ζ_1 and ζ_2, which are one of the states that diagonalize the Zeeman Hamiltonian. Figure 5.4 shows the energy for different asymptotic scattering states, $|\zeta_1, \zeta_2\rangle$, as a function of the magnitude of the external magnetic field. In particular, these states correlate with the lower rotational energy level in $^{17}O_2$. The states $|3, 3\rangle$ and $|1, 1\rangle$ are the most relevant for ultracold temperatures. The first because it is the most prominent lfs state and the second because is the lowest energy state; thus, it does not show any inelastic channel.

From this part of the section and on, we focus on collision of oxygen molecules through two PES: the *ab initio* PES of Bartolomei et al. [31] and the empirical PES by Aquilanti et al. [27], and we will refer to it as Perugia PES.

Figure 5.5 shows the spatial dependence of the matrix elements of the potential $\langle 3, 3|V|3, 3\rangle$, $\langle 3, 3|V|3, 1\rangle$ and $\langle 3, 1|V|3, 1\rangle$, where $|3, 1\rangle$ corresponds to a (final)

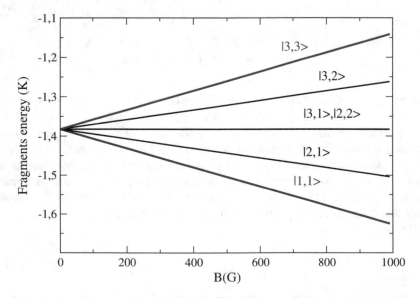

Fig. 5.4 Asymptotic energy levels for $^{17}O_2(\zeta_a)^{17}O_2(\zeta_b)$ depending on the magnetic field. In general, we consider that the initial state of the molecules is the state *low-field seeker* $|\zeta_1, \zeta_2\rangle = |3, 3\rangle$. (Figure adapted with permission from Ref. [30] Copyright (2020) (AIP Publishing))

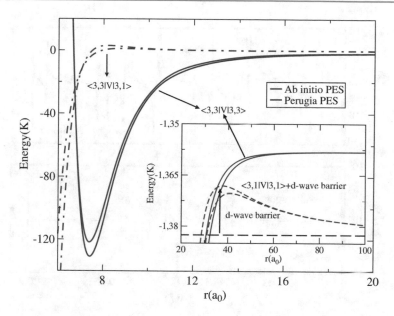

Fig. 5.5 Comparison of the Perugia PES and *ab initio* potential matrix elements for the fragments $|3, 3\rangle$ and $|3, 1\rangle$ for a given magnetic field $B=100$ G ($1G=10^{-4}$ T). The inset shows the long-range behavior, where the d-wave barrier is highlighted. Note that orbital angular momentum for the entrance ($|3, 3\rangle$) and outgoing ($|3, 1\rangle$) channels are 0 and 2 respectively. (Figure reproduced with permission from Ref. [30] Copyright (2020) (AIP Publishing))

state in which one of the molecules has undergone a spin flip process in a fixed magnetic field. These states are connected by the spin relaxation mechanism explained in Sect. 5.4.2. The analysis of the efficiency of the spin relaxation from both interaction potentials may elucidate part of the role of the intermolecular potential in the spin relaxation. In Fig. 5.5 we observe that the matrix elements associated with both intermolecular potentials are quite similar. Although, a quantitative difference appears in the d-wave barrier for both intermolecular potentials. The heights of the barriers differ by a few mK, owing to the different long-range behavior of both PES. The long-range coefficient C_6^{0006} defines the height of the d-wave barrier, as shown in Sect. 5.4.2, and also plays a crucial role on other magnitudes relevant for ultracold physics, such as the van der Waals energy E_{vdW}, and van der Waals length R_{vdW}, introduced in Sect. 1.2.2.

---

[6]This is equivalent to C_6; however, the superscript is omitted when dealigning for spherical symmetric potentials.

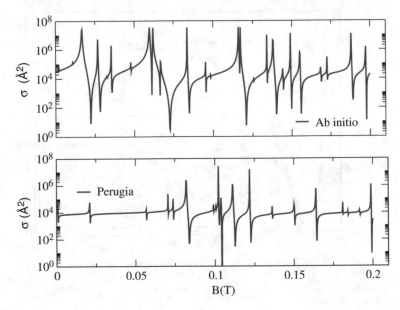

Fig. 5.6 Elastic cross section as a function of the magnetic field for the high-field state seeker $|1, 1\rangle$. The collision kinetic energy is μK

5.4.4   Fano–Feshbach Resonances in Ultracold Molecule–Molecule Collisions

Fano–Feshbach resonances are the key ingredient in controlling and tuning atom–atom interactions and, in this case, molecule–molecule interactions, as shown in Sect. 2.5.2. The presence of Fano–Feshbach resonances is better noticed in the elastic collision from the lowest energetic scattering channel: $|1, 1\rangle$ state (a hfs), and denoted as the solid red line in Fig. 5.4. The calculations for the *ab initio* PES and the Perugia PES have been performed at the same collision energy $E_k = 1\,\mu$K. In this calculation $N_a = N_b = 0, 2, 4$, partial waves $l = 0, 2, 4$, and $M = -2$, only s-wave collisions have been employed. The results are presented in Fig. 5.6, where we observe a large number of Fano–Feshbach resonances. In particular, we find a density of resonances ~ 100/T for both intermolecular potentials (an order of magnitude greater than those observed in atoms). In addition, since this is similar for both potentials, it is deduced that the number of bound states supported by both potentials must be very similar. However, the width and position of the Fano–Feshbach resonances depend strongly on the intermolecular, as expected, since they strongly depend on bound states' position and the coupling between the incoming and closed channels.

The number of Fano–Feshbach resonances depends on the number of bound states supported by the four-body interaction O$_2$-O$_2$. Therefore, the number of Fano–Feshbach resonances should be very similar for the $|1, 1\rangle$ and $|3, 3\rangle$ states,

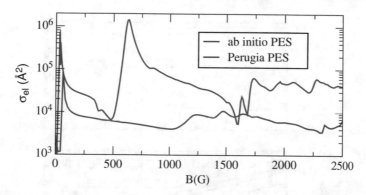

Fig. 5.7 Elastic cross section for the lfs state $|3, 3\rangle$ as a function of the magnetic field. Results for the *ab initio* PES and Perugia PES are shown. (Figure adapted with permission from Ref. [30] Copyright (2020) (AIP Publishing))

owing to the small energy difference between these states in comparison with the depth of the O_2-O_2 potential $\sim$200 K [31]. However, the elastic cross section for the lfs state $|3, 3\rangle$, displayed in Fig. 5.7, shows a smaller number of Fano–Feshbach resonances in comparison with the one observed in Fig. 5.6. This behavior is due to the presence of inelastic collisions (absent for the $|1, 1\rangle$) that modify the line shape of the resonances [30, 32].

5.4.5 Cross Section as a Function of the Collision Energy: The Wigner Threshold Laws

The cross section as a function of the magnetic field has revealed the presence of Fano–Feshbach resonances at ultracold temperatures, which is very informative for understanding the intermolecular interaction of the system at hand. But, so far, we have not seen how ultracold physics emerges in a collision, in particular, the Wigner threshold laws (see Sects. 2.4.2 and 2.2.2). To observe ultracold phenomena it is necessary to study the cross section as a function of the collision energy. In particular, we represent different scattering observables for the lfs $|3, 3\rangle$ as a function of the collision energy: the elastic cross section, the total inelastic cross section given by

$$\sigma_{\text{loss}} = \sigma_{3,3\to3,1} + \sigma_{3,3\to2,2} + \sigma_{3,3\to2,1} + \sigma_{3,3\to1,1}, \tag{5.85}$$

and the factor

$$\gamma = \frac{\sigma_{\text{el}}}{\sigma_{\text{loss}}}. \tag{5.86}$$

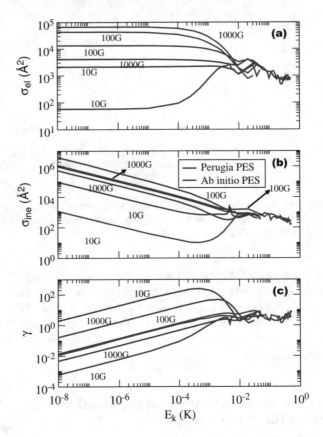

Fig. 5.8 Scattering observables as a function of the collision energy for $^{17}O_2$-$^{17}O_2$ collisions: (**a**) elastic cross section; (**b**) total inelastic cross section; and (**c**) the γ factor relevant for evaporative cooling. (Figure taken with permission from Ref. [30] Copyright (2020) (AIP Publishing))

This is the factor that defines the efficiency of evaporative cooling. The results for such scattering observables as a function of the collision energy for different values of the magnetic field, and using the *ab initio* and Perugia intermolecular potentials are shown in Fig. 5.8.

In the upper panel of the Figure it is observed that the elastic cross section tends to a constant value, depending on the magnetic field applied, as the collision kinetic energy decreases. This is the Wigner threshold law introduced in Sect. 2.2.2. In the same vein, in panel (b) of that Figure it is noted that the total inelastic cross section increases as the kinetic energy decreases as $\sim E_k^{-1/2} \sim k^{-1}$, confirming Wigner's threshold law for the inelastic cross section (see Sect. 2.4.2). Comparing the results for the Perugia and *ab initio* intermolecular potentials, which, although different, show very similar values for the relevant parameters at ultracold temperatures, leads to drastic deviations in the cross section for the same magnetic field in the ultracold regime, $E_k < E_{vdW}$. Nevertheless, for $E_k > E_{vdW}$ the cross section becomes insensitive to the details of the intermolecular potential. Therefore, we conclude that the cross section at ultracold temperatures strongly depends on the details of the intermolecular potential, as one may expect by looking at the relevant energy range for the ultracold regime. Finally, the results for γ strongly depend on the details of intermolecular potential. However, since the accuracy of any of the potentials presented in this chapter is lower than E_{vdw}, it is not possible to make any quantitative claim. Indeed, this is the general scenario in ultracold chemistry: it is possible to know the long-range interaction very precisely but not the short-range interaction.

5.4.6 Do Short-Range Interactions Play a Role in Ultracold Collisions?

At ultracold energies, it is generally assumed that the relevant physics is captured by the long-range of the interparticle interaction. However, scattering and ultimately chemistry, happens predominantly at short-range distances, where the coupling is enough to induce the expected transition from reactants to product states. Therefore, it is interesting to see what is the role of short-range interactions in ultracold collisions.

To investigate the relevance of short-range interactions versus long-range interactions, in Fig. 5.9 we show the results of the elastic and total inelastic cross sections for O_2-O_2, where in one of the calculations the long-range anisotropy is suppressed. In particular, we impose an exponential decay for all terms of the potential from a distance $R>19$ a_0, except for the isotropic term ($\lambda_1 = \lambda_2 = \lambda = 0$). In Fig. 5.9 hardly any variation in the structure of the elastic and inelastic cross section as a function of the long-range anisotropy is observed; therefore, the structure of the resonances is dominated by the short range of the potential. This shows that for O_2-O_2 collisions at ultracold temperatures the short-range physics is crucial in order to have a clear picture of the ultracold processes within this system. Indeed, the same is true for almost any scattering process at ultracold temperatures in which some reactive channel is open. However, owing to the lack of accurate potentials, it is preferable to assume some kind of loss probability for the reactants at some short-range region. Indeed, this is the core idea behind universality models for ultracold chemical reactions [33].

5.5 Computational Techniques

In this last section of the chapter, we introduce one of the best well-known and useful techniques for solving the Schrödinger equation for scattering states: Numerov's method. Numerov's method is a finite-difference method of solving the time-independent Schrödinger equation. In particular, it is based on the discretization of the unknown function (in our case $g_{\gamma'\gamma}^{JM}(R)$) on a grid of the independent variable (R, in our case). Moreover, it assumes that the unknown function and its first

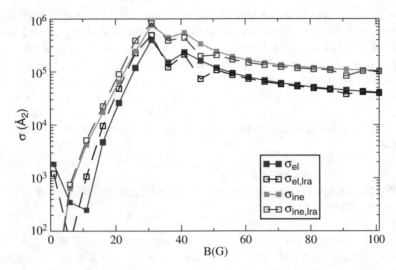

Fig. 5.9 The effect of long-range anisotropy in O_2-O_2 ultracold collisions: elastic and total inelastic cross section for the lfs state, $|3, 3\rangle$, as a function of the magnetic field and constant collision energy of $10\,\mu K$. The solid lines refer to the results for the *ab initio* PES, elastic cross section (red) and total inelastic cross section (orange). The dashed lines stand for the calculations employing the *ab initio* PES, where the long-range anisotropy has been suppressed: elastic collisions ($\sigma_{el,lra}$, in black) and total inelastic cross section ($\sigma_{ine,lra}$, in blue). (Figure adapted with permission from Ref. [30] Copyright (2020) (AIP Publishing))

derivative are continuous functions in the whole range of the dependent variable R. Therefore, it is possible to Taylor expand the unknown function along the grid points on R. Next, keeping in mind the differential equation, Eq. (5.4), which dictates the evolution of the unknown function, it can be shown that exist a way to link the value of the unknown function in two consecutive grid points. The relation of the values of the function in consecutive points is given by the propagator, and allows the value of the unknown function in the whole range of R to be found after knowing its value at one point.

5.5.1 Single-Channel Scattering

Let us start with the single-channel scattering problem defined by Eq. (2.14). This equation can be further simplified by $\psi_l(R) = \phi(R)/R$, leading to

$$\phi''(R) = \frac{2\mu}{\hbar^2}\left(V(R) + \frac{l(l+1)\hbar^2}{2\mu R^2} - E\right)\phi(R), \qquad (5.87)$$

where E is the collision energy and $\phi''(R) \equiv \frac{d^2\phi(R)}{dR^2}$. Equation (5.87) can be recast as

$$\phi''(R) = \frac{2\mu}{\hbar^2} \left(V^l(R) - E \right) \phi(R), \tag{5.88}$$

where the superscript l emphasizes that the potential includes the centrifugal barrier according to the quantum number l. As described above, we first take the independent variable, the collision coordinate R, in a grid given by $R_i = R_{\text{ini}} + (i - 1)\Delta$ for $i = 1 \ldots N$, where Δ is the step size, N is the number of grid points, and R_{ini} is the smaller value of the collision coordinate. In this grid, it is possible to define the values of $\phi(R_i) \equiv \phi_i$, and hence Eq. (5.88) yields

$$\phi_i'' = \frac{2\mu}{\hbar^2} \left(V_i^l - E \right) \phi_i. \tag{5.89}$$

Assuming that ϕ and its first derivative are continuous functions, it is legitimate to Taylor expand such a function around any point of the grid. In particular, its Taylor expansion on the points $n \pm 1$ reads as

$$\phi_{n\pm1} = \phi_n \pm \Delta\phi_n' + \frac{\Delta^2}{2}\phi_n'' \pm \frac{\Delta^3}{6}\phi_n''' + \frac{\Delta^4}{24}\phi_n^{(iv)} + O(\Delta^5), \tag{5.90}$$

which are added together yielding

$$\phi_{n+1} + \phi_{n-1} = 2\phi_n + \Delta^2\phi_n'' + \frac{\Delta^4}{12}\phi_n^{(iv)} + O(\Delta^5). \tag{5.91}$$

Now, performing a Taylor expansion of the second derivative of $\phi(R)$ in the points $n \pm 1$

$$\phi_{n\pm1}'' = \phi_n'' \pm \Delta\phi_n''' + \frac{\Delta^2}{2}\phi_n^{(iv)} \pm \frac{\Delta^3}{6}\phi_n^{(v)} + \frac{\Delta^4}{24}\phi_n^{(vi)} + O(\Delta^5), \tag{5.92}$$

and performing $\phi_{n+1} + \phi_{n-1} - \frac{\Delta^2}{12}(\phi_{n+1}'' + \phi_{n-1}'')$ by means of Eqs. (5.91) and (5.92) we find

$$\phi_{n+1} + \phi_{n-1} - \frac{\Delta^2}{12}(\phi_{n+1}'' + \phi_{n-1}'') = 2\phi_n + \frac{10\Delta^2}{12}\phi_n'' + O(\Delta^6). \tag{5.93}$$

This equation can be further simplified taking into account Eq. (5.89) and reads as

$$\gamma_{n+1}\phi_{n+1} + \beta_n\phi_n + \alpha_{n-1}\phi_{n-1} = 0, \tag{5.94}$$

where

$$\gamma_{n+1} = 1 - \frac{\Delta^2}{12} \frac{2\mu}{\hbar^2} \left(V_{n+1}^l - E \right)$$

$$\beta_n = -2 - \frac{10\Delta^2}{12} \frac{2\mu}{\hbar^2} \left(V_n^l - E \right)$$

$$\alpha_{n-1} = 1 - \frac{\Delta^2}{12} \frac{2\mu}{\hbar^2} \left(V_{n-1}^l - E \right). \tag{5.95}$$

It is worth noting that the error is proportional to Δ^6; therefore, it is preferable to use small step sizes to minimize the errors during the propagation of the wave function. We assume that a function $\hat{R}$ exists, called propagator, that fulfils

$$\phi_i = \hat{R}_i \phi_{i+1}. \tag{5.96}$$

Making use of the condition $\phi(0) = 0$, thus $\phi_1 = 0$,[7] one finds $\hat{R}_1 = 0$ since $\phi_1 = 0$. For the calculation of the next iteration we employ Eq. (5.94) with $n = 2$

$$\gamma_3\phi_3 + \beta_2\phi_2 + \alpha_1\phi_1 = \gamma_3\phi_3 + \beta_2\phi_2, \tag{5.97}$$

and keeping in mind that $\phi_2 = \hat{R}_2\phi_3$ one finds

$$\hat{R}_2 = -\frac{\gamma_3}{\beta_2}. \tag{5.98}$$

Making $\phi_{n-1} = \hat{R}_{n-1}\phi_n$ in (5.94) yields

$$\gamma_{n+1}\phi_{n+1} + \beta_n\phi_n + \alpha_{n-1}\hat{R}_{n-1}\phi_n = 0, \tag{5.99}$$

which is compared with Eq. (5.96), leads to the general recurrence relation

$$\hat{R}_n = - \left(\beta_n + \alpha_{n-1}\hat{R}_{n-1} \right)^{-1} \gamma_{n+1}. \tag{5.100}$$

We have seen in Sect. 2.2.1 that asymptotically the wave function is given by (Fig. 5.10)

$$\phi_l(R \to \infty) \approx A_l(k) \sin\left(kR + \delta_l(k) \right); \tag{5.101}$$

therefore, the wave function in the $N - 1$ and N points is given by

$$\phi_N = A_l(k) \sin\left(kR_N + \delta_l \right)$$

---

[7]This condition is general for scattering problems. One needs to choose the initial propagation point deeply in the classical forbidden region to ensure that the wave function is null at that point.

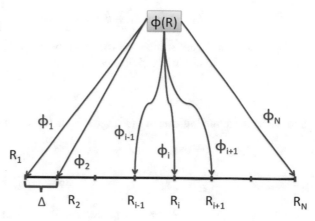

Fig. 5.10 Discretization of the collision coordinate R and the wave function ϕ

$$\phi_{N-1} = A_l(k) \sin(kR_{N-1} + \delta_l). \tag{5.102}$$

The propagator derived in Eq. (5.100) correlates these wave functions leading to

$$\hat{R}_{N-1} = \frac{A_l(k) \sin(kR_{N-1} + \delta_l)}{A_l(k) \sin(kR_N + \delta_l)}, \tag{5.103}$$

which can be further simplified, yielding

$$\tan \delta_l(k) = \frac{\sin(kR_{N-1}) - \hat{R}_{N-1} \sin(kR_N)}{\hat{R}_{N-1} \cos(kR_N) - \cos(kR_{N-1})}. \tag{5.104}$$

This equation leads to the computation of the phase shift for any potential $V(R)$, and with this, one is ready to calculate the elastic cross section through Eq. (2.30).

5.5.2 Multi-channel Scattering: Elastic and Inelastic Channels

Let us start by expressing Eq. (5.4) as in the form of Eq. (5.88)

$$\bar{g}'' = \frac{2\mu}{\hbar^2} \left(\bar{V}(R)^l - E\bar{I} \right) \bar{g}, \tag{5.105}$$

where the bar above a variable indicates that it is an array. The term $\bar{V}^l$ is the matrix of potential matrix elements between different states plus the centrifugal term in Eq. (5.4).

As in the single-channel case, we discretize the collision coordinate. At the same time, let us assume that the wave function and its first derivative are continuous in all the collisional distances. Thus, it is possible to Taylor expand $\bar{g}$ and its second derivative, analogously to Eqs. (5.90) and (5.92), and they are combined as in Eq. (5.93), and making use of Eq. (5.105) we find that

$$\bar{\gamma}_{n+1}\bar{g}_{n+1} + \bar{\beta}_n\bar{g}_n + \bar{\alpha}_{n-1}\bar{g}_{n-1} = 0. \tag{5.106}$$

This equation is very similar to Eq. (5.94) for the single-channel problem. At this point it is useful to apply Eq. (5.96), which in the multi-channel case reads as

$$\bar{g}_i = \hat{\bar{R}}_i\bar{g}_{i+1}, \tag{5.107}$$

and using the same logic as in the single-channel problem we find

$$\hat{\bar{R}}_1 = 0$$

$$\hat{\bar{R}}_2 = -\bar{\beta}_2^{-1}\bar{\gamma}_3$$

$$\hat{\bar{R}}_i = -\left(\bar{\alpha}_{i-1}\hat{\bar{R}}_{i-1} + \bar{\beta}_i\right)^{-1}\bar{\gamma}_{i+1}. \tag{5.108}$$

Finally, applying the proper asymptotic conditions to $\bar{g}$ (a generalization of the ones in Eq. (5.101)) and keeping in mind that $\bar{g}_{N-1} = \hat{\bar{R}}_{N-1}\bar{g}_N$ leads to

$$\bar{K} = \left(\hat{\bar{R}}_{N-1}\cos\bar{k}R_N - \cos\bar{k}R_{N-1}\right)^{-1}\left(\sin\bar{k}R_{N-1} - \hat{\bar{R}}_{N-1}\sin\bar{k}R_N\right) \tag{5.109}$$

Indeed, knowing the reactance matrix is equivalent to knowing the transfer matrix and therefore the cross section is obtained.

References

1. Brown J, Carrington A (2003) Rotational spectroscopy of diatomic molecules. Cambridge University Press, Cambridge
2. Bernath PF (2005) Spectra of atoms and molecules. Oxford University Press, Oxford
3. Takayanagi K (1965) The production of rotational and vibrational transitions in encounters between molecules. Academic, pp 149–194. https://doi.org/10.1016/S0065-2199(08)60282-1, http://www.sciencedirect.com/science/article/pii/S0065219908602821
4. Green S (1975) J Chem Phys 62(6):2271. https://doi.org/10.1063/1.430752
5. Zarur G, Rabitz H (1974) J Chem Phys 60:2057
6. Englot GF, Rabitz H (1974) Dimensionality control of coupled scattering equations using partitioning techniques: the case of two molecules. Phys Rev A 10:2187. https://doi.org/10.1103/PhysRevA.10.2187
7. Barreto PRP, Cruz ACPS, Barreto RLP, Palazzetti F, Albernaz AF, Lombardi A, Maciel GS, Aquilanti V (2017) The spherical-harmonics representation for the interaction between diatomic molecules: the general case and applications to COCO and COHF. J Mol Spectrosc 337:163. https://doi.org/10.1016/j.jms.2017.05.009, http://www.sciencedirect.com/science/article/pii/S0022285217300905
8. Varshalovich D, Moskalev A, Khersonskii V (1988) Quantum theory of angular momentum: irreducible tensors, spherical harmonics, vector coupling coefficients, 3nj symbols. World Scientific. https://books.google.de/books?id=hQx_QgAACAAJ
9. Edmonds AR (1974) Angular momentum in quantum mechanics. Princeton University Press, Princeton
10. Zare RN (1988) Angular momentum. Wiley, New York

11. Zhang JZH (1999) Theory and applications of quantum molecular dynamics. World Scientific, Singapore
12. Landau LD, Lifshitz EM (1958) Quantum mechanics. Butterworth-Heinemann
13. Abramowitz M, Stegun IA (1972) Handbook of mathematical functions. Dover, New York
14. Miller WH (1976) Dynamics of molecular collisions. Plenum Press, New York
15. Pérez-Ríos J, Bartolomei M, Campos-Martínez J, Hernández MI, Hernández-Lamoneda R (2009) J Phys Chem A 113:14952
16. Gioumousis G, Curtiss CF (1958) J Chem Phys 29:996
17. Krems RV, Dalgarno A (2004) "Quantum-mechanical theory of atom-molecule and molecular collisions in a magnetic field: spin depolarization. J Chem Phys 120(5):2296. https://doi.org/10.1063/1.1636691
18. Tscherbul TV, Suleimanov YV, Aquilanti V, Krems RV (2009) New J Phys 11(5):055021
19. Volpi A, Bohn JL (2002) Phys Rev A 65:052712
20. Tscherbul TV, Suleimanov YV, Aquilanti V, Krems R (2009) New J Phys 11:055021
21. Freidrich B, deCarvalho R, Kim J, Patterson D, Weinstein JD, Doyle JM (1998) J Chem Soc Faraday Trans 94:1783
22. Herzberg G (1950) Spectra of diatomic molecules. Van Nostrand, New York
23. Mizushima M (1975) The theory of rotating diatomic molecules. Wiley, New York
24. Gazzoli G, Espositi CD (1985) Chem Phys Lett 113:501
25. Avdeenkov AV, Bohn JL (2001) Phys Rev A 64:052703. https://doi.org/10.1103/PhysRevA.64.052703
26. Teisinga E, Verhaar BJ, Stoof HTC (1993) Phys Rev A 47:4114
27. Aquilanti V, Ascenzi D, Bartolomei M, Cappelletti D, Cavalli S, de Castro Vìtores M, Pirani F (1999) J Am Chem Soc 121(46):10794
28. Wormer PES, van der Avoird A (1984) J Chem Phys 81(4):1929. https://doi.org/10.1063/1.447867
29. Bussery B, Wormer PES (1993) J Chem Phys 99:1230
30. Pérez-Ríos J, Campos-Martínez J, Hernández MI (2011) J Chem Phys 134(12):124310. https://doi.org/10.1063/1.3573968
31. Bartolomei M, Carmona-Novillo E, Hernández MI, Campos-Martínez J, Hernández-Lamoneda R (2010) J Chem Phys 133(12):124311
32. Hutson JM (2007) New J Phys 9:152
33. Idziaszek Z, Julienne PS (2010) Phys Rev Lett 104(11):113202. https://doi.org/10.1103/PhysRevLett.104.113202

Three-Body Collisions at Ultracold Temperatures: An Effective Field Theory Approach

6

The present book focusses on the study of cold and ultracold chemical reactions, which are mainly a consequence of the encounter of two or three bodies (atoms, molecules, ions, or Rydbergs) at short-range distances, although they are highly influenced by the long-range behavior of interparticle interaction. These reactions, like most of the chemical reactions in nature, can be viewed as few-body precesses. However, when one thinks about few-body processes or few-body physics, it is customary to think of nuclear systems rather than chemical systems. Nevertheless, the tools for treating few-body nuclear systems are totally extensible to chemical reactions, and this will be the topic of the present chapter. In particular, we present how some of the tools and ideas from few-body physics in nuclear physics (and even high-energy physics) are readily applicable to ultracold collisions, leading to a new and interesting perspective of the problem related to the universality of chemical reactions.

6.1   The Renormalization Group

In every field of physics, the nature of observable phenomena depends on the energy scale at which a given system is studied. In other words, the physics that it is explored changes as a function of the energy scale that is being probed. Indeed, the relation between the physics and the energy scale is more profound and fundamental, something intrinsic to the laws of physics and hence universal. The renormalization group method is aimed at studying the most difficult problems in physics involving a large number of degrees of freedom such as relativistic quantum field theory or critical phenomena. In this section, we follow the seminal and impressive work of Wilson and Kogut [1], a work that we consider a masterpiece of modern physics.

Let us assume a physical system A composed of spins separated by a distance l_0, which is the smallest length scale of the system at hand, and hence it represents the

© Springer Nature Switzerland AG 2020
J. Pérez Ríos, *An Introduction to Cold and Ultracold Chemistry*,
https://doi.org/10.1007/978-3-030-55936-6_6

highest energy scale Λ_0.[1] At the given energy scale, the system A can be described by the Hamiltonian H_0. At this level, the system is intractable owing to the presence of the many degrees of freedom (large number of spins). Furthermore, if our aim is to study low-energy physics, it seems difficult to apply this approach; however, the renormalization group can be applied to this scenario. In particular, we integrate out some degrees of freedom by increasing the length scale of the system, for instance $l_1 = 2l_0$, with the corresponding change in the Hamiltonian $H_1 = \tau(H_0)$, and energy scale $\Lambda_1 \propto \Lambda_0/2$, where τ is a general transformation that depends on the system at hand. Here, Λ_0 will be referred to as an ultraviolet cutoff, which is related to the smaller length scale of the system at hand, l_0. The transformation τ can be applied iteratively: $H_2 = \tau(H_1)$, $H_3 = \tau(H_2)$ and so on and so forth. After n iterations the length scale of the system is $l_n = 2^n l_0$ and the subsequent energy scale is given by $\Lambda_n \propto \Lambda_0 2^{-n}$. Therefore, by means of the application of the transformation τ iteratively it is possible to predict the Hamiltonian for a given energy scale Λ. A schematic representation of the renormalization group approach is shown in panel (a) of Fig. 6.1. The most exciting physics occurs when $n \to \infty$; in such a case, the simplest possible solution is a fixed point, i.e., $\tau(H_n) = H_n$, which is better to recast as $\tau(H^*) = H^*$ in order to avoid any confusion with the iterative process. The solution of this equation is totally independent of the choice of the initial Hamiltonian H_0; therefore, it represents an universal behavior of the system. In particular, in a fixed point the theory is scale-free, and can thus be viewed as a conformal theory. Indeed, the universality classes in phase transitions and critical phenomena are directly related to the nature of the fixed points of the renormalization group [1]. On the other hand, as $n \to \infty$, other different solutions are raised, such as limit cycle and ergodic or turbulent behavior. Each of those is related to different universality properties of the system at hand. Here, special attention is given to the limit cycle, which is related to the three-body universality [2, 3].

The renormalization group is usually presented in a more general way, assuming a continuous change on the length scale l (instead of discrete one as was assumed in the previous paragraph) of the system, which leads to the key concept of the beta function [1, 4]. In this case, it is useful to write down the most general initial Hamiltonian of our system A as

$$H_0 = \sum_i g_i^0 O_i, \tag{6.1}$$

where O_i represents all kind of operators compatible with the symmetries of the system A, e.g., gauge invariance, Galilean invariance, Lorentz invariance, etc. In Eq. (6.1) g_i^0 denotes the coupling constants associated with each single operator of the Hamiltonian. The iterative application of τ into the Hamiltonian translates into

---

[1]Please, note that Λ_0 defines the momentum; therefore, the proper energy of the system will be $\propto \Lambda_0^2$.

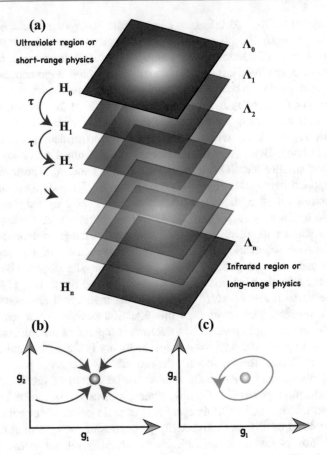

Fig. 6.1 Renormalization group approach. Panel (**a**); schematic representation of the renormalization group theory, in particular, it is stressed how the physics of the system changes with energy scale Λ_i. Each single plane is associated with a new Hamiltonian, mimicking the idea of Hamiltonian flow of the renormalization group approach of Wilson. Panels (**b**) and (**c**); two different topologies associated with the beta function $\beta(\mathbf{g})$, which is represented for a two-coupling constant case for the sake of simplicity. Panel (**b**) represents a fixed point, whereas the panel (**c**) represents a limit cycle

an evolution on the coupling constants $g_i^0 \to g_i^1 \to g_i^2$, and so on. Therefore, the renormalization group transformation can be viewed as an operation on the vectorial space of the coupling constants $\mathbf{g} = (g_1, g_2, \ldots)$. Finally, taking into account a continuous change on the length scale, it is found

$$l\frac{d\mathbf{g}}{dl} = -\beta(\mathbf{g}), \tag{6.2}$$

which can be recast in terms of the ultraviolet cutoff as

$$\Lambda \frac{d}{d\Lambda} \mathbf{g} = \beta(\mathbf{g}), \tag{6.3}$$

where the beta function $\beta(\mathbf{g})$ is in general a complicated nonlinear function of the coupling constants. The beta function contains all the information regarding the universality classes of the system at hand. In particular, fixed points appear when the condition, $\beta(\mathbf{g}) = 0$, is satisfied, which implies that in such case the coupling constants do not depend on the ultraviolet cutoff, reaching a universal behavior. An example of a fixed point is schematically presented in panel (b) of Fig. 6.1.

The behavior of a particular physical system close to a fixed point is dictated by the functional dependence of the coupling constants with the ultraviolet cutoff Λ. In particular, the behavior of the coupling constants, attached to an operator, close to a fixed point can be classified as:

- $g_i \propto \Lambda^\nu$, relevant or super-renormalizable in the context of high-energy physics. For instance, the coupling constant in a ϕ^2 scalar field theory [5].
- $g_i \propto \Lambda^0$, marginal or renormalizable, for example, the electric charge in quantum electrodynamics and in general these kind of coupling constants show a logarithmic dependence on the ultraviolet cutoff.
- $g_i \propto \Lambda^{-\nu}$, irrelevant or nonrenormalizable , e.g., interactions in mesonic chiral perturbation field theory [6].

The study of the renormalization group flow given by Eq. (6.3) may also lead to the existence of limit cycles. In such a scenario, the beta function describes a closed loop in the space of coupling constants, as it is presented in panel (c) of Fig. 6.1. This behavior resembles the stable points in the phase portrait of an harmonic oscillator, and consequently it shows a periodic dependence on the ultraviolet cutoff as [3, 7]:

$$\mathbf{g}(\Lambda) = \mathbf{g}\left(\frac{\theta_0 + 2\pi \ln\left(\frac{\Lambda}{\Lambda_0}\right)}{\ln(S_0)}\right), \tag{6.4}$$

where Λ_0 describes the initial cutoff where the flow starts, θ_0 is some initial phase attached to it, Λ is the running cutoff, and S_0 can be interpreted as an auxiliary scale. This scale determines the periodicity of the limit cycle through a discrete transformation $\Lambda = S_0^n \Lambda_0$, if n is a positive integer. This transformation clearly represents a universal geometric scaling law for the energy and length scales.

6.2   Effective Field Theory Approach for Low-Energy Two-Body Collisions

The low-energy scattering between two identical bosons can be described in terms of an effective field theory, in which the Lagrangian density is given by

$$\mathcal{L} = \phi^\dagger \left(i \frac{\partial}{\partial t} + \frac{1}{2} \nabla^2 \right) \phi - \frac{g_2}{4} (\phi^\dagger \phi)^2, \tag{6.5}$$

where here and in this section $\hbar = 1$ and $m = 1$ have been assumed. The Lagrangian density $\mathcal{L}$ is invariant under the Galilean transformations instead of the Lorentz transformations, which implies that ϕ represents a nonrelativistic scalar field.

The scattering amplitude $i\mathcal{A}$ associated with a two-body collision is obtained from the Lagrangian $\mathcal{L}$ by summing up all different amputated connected Feynman diagrams, which in this case are[2]

$$\times + \bowtie + \Join + \ldots = \otimes . \tag{6.6}$$

The following Feynman rule applies for the field-free propagator of the nonrelativistic field ϕ

$$\underline{\qquad\qquad} = \frac{i}{p_0 - \frac{\mathbf{p}^2}{2} + i\epsilon}, \tag{6.7}$$

where $\mathbf{p}$ represents the momentum of the field and p_0 stands for the energy of the field. The vertices are represented by the following Feynman diagram

$$\times = -ig_2, \tag{6.8}$$

where the two-body interaction coupling constant clearly shows its importance for fixing the interaction strength. In principle, the scattering amplitude $\mathcal{A}$ depends on the momentum and energies of the four external lines; however, in the the center of mass frame, the incoming momentum of the particles can be taken as $\mathbf{p}$ and $-\mathbf{p}$, and for the final momentum we have $\mathbf{k}$ and $-\mathbf{k}$. Furthermore, the conservation of energy enforces the scattering amplitude to be a function of the energy $E = p^2 = k^2$, i.e., $p = k$.

The scattering amplitude, $i\mathcal{A}(E)$, schematically represented through the Feynman diagrams in Eq. (6.6), is calculated by a nonperturbative treatment by applying the Feynman rules from Eqs. (6.7) and (6.8) to the infinite series associated with a single bubble yielding

$$i\mathcal{A}(E) = -ig_2 \sum_{i=0}^{\infty} \xi(E)^i, \tag{6.9}$$

---

[2] Generally, in quantum field theory fermionic fields are represented by a solid line, whereas the scalar fields are shown as dashed lines.

where $\xi(E)$ represents the amplitude associated with a one-loop diagram when one of the vertices is factorized out, and it reads as

$$\xi(E) = -\iota \frac{g_2}{2} \int \frac{dp_0}{2\pi} \int \frac{d^3p}{(2\pi)^3} \frac{\iota}{p_0 - \frac{\mathbf{p}^2}{2} + \iota\epsilon} \frac{\iota}{E - p_0 - \frac{\mathbf{p}^2}{2} + \iota\epsilon}. \tag{6.10}$$

The integral over p_0 can be performed by contour integration yielding

$$\xi(E) = \frac{g_2}{2} \int \frac{d^3p}{(2\pi)^3} \frac{1}{E - \mathbf{p}^2 + \iota\epsilon}, \tag{6.11}$$

which is a divergent integral in the variable $\mathbf{p}$. However, this problem is circumvented by imposing an ultraviolet cutoff Λ, i.e., the integral in Eq. (6.11) is evaluated for values of the momentum satisfying $p < \Lambda$ yielding

$$\xi(E) = -\frac{g_2}{4\pi^2} \left(\Lambda - \frac{\iota\pi}{2} \sqrt{E + \iota\epsilon} \right). \tag{6.12}$$

This equation is plugged into Eq. (6.9), and after performing the summation, the scattering amplitude recasts as

$$\mathcal{A}(E) = \frac{-g_2}{1 + \frac{g_2}{4\pi^2} \left(\Lambda - \frac{\iota\pi}{2} \sqrt{E + \iota\epsilon} \right)}. \tag{6.13}$$

Therefore, the scattering amplitude depends on the cutoff. However, the physics must be independent of any regularization approach (which is just a mathematical technique), and hence the scattering amplitude must be independent of such cutoff. The cutoff dependence on Eq. (6.13) is eliminated by renormalizing the coupling constant g_2. This is achieved by imposing that at zero energy, the scattering amplitude must be proportional to the two-body scattering length a, $\mathcal{A}(0) = -8\pi a$ (the 8π factor reflects the fact that we are dealing with identical bosons), i.e.,

$$\frac{g_2}{1 + \frac{g_2}{4\pi^2} \Lambda} = 8\pi a. \tag{6.14}$$

The solution of this equation for g_2 reads as

$$g_2 = \frac{8\pi a}{1 - \frac{2a\Lambda}{\pi}}, \tag{6.15}$$

and it depends on the cutoff as well as on the scttering length. Finally, substituting back g_2 into Eq. (6.13), the physical scattering amplitude reads as

$$\mathcal{A}(E) = \frac{8\pi a}{-1 + \iota a \sqrt{E + \iota\epsilon}}. \tag{6.16}$$

The physical scattering amplitude $\mathcal{A}(E)$ shows the existence of poles when $E = -1/a^2$, which corresponds to the existence of bound states, which can be viewed as divergences in the coupling constant, as shown in the caption of Fig. 6.2. Although the coupling constant is divergent, the physical observables are independent of the coupling constant (see Eq. (6.16)).

The renormalization of the exposed effective field theory is now analyzed from a renormalization group perspective. Using the dimensionless coupling constant given by [2]

$$g_2^a = \frac{\Lambda g_2}{4\pi^2}, \tag{6.17}$$

it is straightforward to show that the beta function reads as

$$\Lambda \frac{dg_2^a}{d\Lambda} = g_2^a(g_2^a + 1). \tag{6.18}$$

In the renormalization group flow equations, Eq. (6.18), the variation of the coupling constant with respect to Λ is performed for a given value of a. However, it is worth noting that Λ is an ultraviolet cutoff rather than a variable length scale of the system. Nevertheless, this is the terminology and customary approach in the field

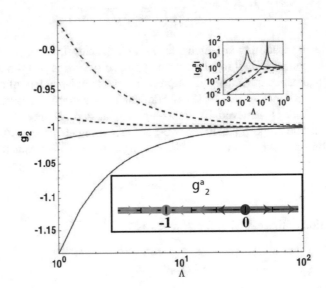

Fig. 6.2 Dimensionless renormalized coupling constant $g_2^a = \frac{\Lambda g_2}{4\pi^2}$ as a function of the ultraviolet cutoff for different values of the scattering length a: solid blue line is 10, dashed blue is -10, solid red is 100 and dashed red is -100. The scattering length as well as the cutoff is given in units of the length scale of the system, which has been set to one for the sake of simplicity. Upper inset: absolute value of the renormalized coupling constant for small cutoff values, i.e., low energies. Lower inset: renormalization group flow associated with the effective field theory $\mathcal{L}$, see text for details

of effective field theory for ultracold collisions. Equation (6.18) shows the existence of two fixed points, the first for $g_2^a = 0$ which represents a non-interacting effective field theory. The second fixed point is found for $g_2^a = -1$, which can be interpreted as an ultraviolet fixed point since $g1_2^a \to -1$ as $\Lambda \to \infty$, as shown in Fig. 6.2, for different values of the scattering length.

This renormalization group approach seems to be the standard and accepted approach to the subject; however, some considerations must be emphasized in order to be as precise as the subject matter requires:

- The fixed point for $g_2^a = 0$, is a trivial one and usually is referred to as the infrared fixed point, and all abelian quantum field theories presented it.
- The consideration of $g_2^a < 0$ is totally unphysical, since it implies that the field theory is unbounded from below. In other words, the field theory does not have a ground state. Therefore, in the field theory at hand, for $a > 0$ only for $\Lambda < \pi/2a$ is the field theory stable. However, for $a < 0$ the field theory is totally unstable and hence is not predictive.
- In virtue of the previous point the considered value of $g_2^a = -1$ as a fixed point must be ruled out, since the present field theory is only well defined and adequate for low-energy scattering.

6.3   Effective Field Theory for Three-Body Bosonic Collisions Involving Short-Ranged Interactions

The low-energy scattering among three identical bosons can be described within the framework of effective field theory. In particular, the Lagrangian density is given by

$$\mathcal{L}_{3b} = \phi^\dagger \left(\imath \frac{\partial}{\partial t} + \frac{1}{2} \nabla^2 \right) \phi - \frac{g_2}{4} (\phi^\dagger \phi)^2 - \frac{g_3}{36} (\phi^\dagger \phi)^3, \tag{6.19}$$

which is almost identical to the $\mathcal{L}$ for two-body collisions (see Eq. (6.5)), except for the last term, since it involves a term proportional to ϕ^6. This term is associated with a three-body collision and hence the coupling constant g_3 refers to the three-body coupling. In this section, as in the previous one, $\hbar = 1$ and $m = 1$ is assumed for simplicity in the notation. The Lagrangian density $\mathcal{L}_{3b}$ obeys the same invariance relations as $\mathcal{L}$. It is worth emphasizing that the Lagrangian density from Eq. (6.19) cannot be theoretically justified apart from assuming a particular model. In other words, Eq. (6.19) reflects the physics of a particular model that it is generally employed for the study of three-body collision at ultracold temperatures. This may be relevant for the reader interested in the fundamentals of effective quantum field theory.

The Feynman rules associated with $\mathcal{L}_{3b}$ are the same as those shown in Eqs. (6.7) and (6.8), but a new one is required in relation to the three-body term

$$\times = -\imath g_2,$$

$$(6.20)$$

Taking into account these Feynman rules, it is possible to calculate the scattering amplitude for three-body collisions by including different contributions: tree-level and one-loop level. However, in this case, the scattering amplitude is very hard to calculate using nonperturbative techniques similar to those employed for the two-body effective field theory. Nevertheless, the scattering amplitude can be computed by solving the Skorniakov–Ter-Martirosian (STM) integral equation [8], which is introduced below. The STM integral equation is related to a six-point Green's function $\langle 0|T \left(\phi\phi\phi\phi^\dagger\phi^\dagger\phi^\dagger \right) |0\rangle$, where T denotes the time-ordering operator, and the dependence on the spatial and temporal degrees of freedom are not explicitly shown. However, the problem of three-interacting bosons involving short-ranged forces is better studied within a different framework owing to Bedaque, Hammer, and van Klock [9–12], which is based on the introduction of a new scalar field d: the dimer field, which is a composite quantum field operator. In this case, the Lagrangian density is given by [9–12]

$$\mathcal{L}_{d-a} = \phi^\dagger \left(\imath \frac{\partial}{\partial t} + \frac{1}{2}\nabla^2 \right) \phi + \frac{g_2}{4} d^\dagger d$$

$$- \frac{g_2}{4}(d^\dagger \phi^2 + (\phi^\dagger)^2 d) - \frac{g_3}{36}(d^\dagger d\phi^\dagger \phi). \qquad (6.21)$$

This new Lagrangian density leads to new Feynman rules for the interaction vertices

$$\rightequals\!\!< = \frac{-\imath g_2}{2}$$

$$>\!\!\!- = \frac{-\imath g_2}{2}$$

$$\times = \frac{-\imath g_3}{36},$$

$$(6.22)$$

corresponding to the decay of the dimer into two particles, the composition of two particles into a dimer, and a particle-dimer event, respectively. Furthermore, $\mathcal{L}_{d-a}$ induces the following propagators for the two fields

$$\underline{\qquad} = \frac{\imath}{p_0 - \frac{\mathbf{P}^2}{2} + \imath\epsilon}$$

$$\underline{\underline{\qquad}} = \frac{4\imath}{g_2}. \tag{6.23}$$

The first term of this equation is associated with the propagator of the scalar field ϕ, which is the same as in the effective field theory for two-body collisions (see Eq. (6.7)). The second term represents the propagator of the dimer field, but it does not show any dynamics, as one would expect owing to the absence of spatial or time derivatives of such field in $\mathcal{L}_{d-a}$. However, it is necessary to keep in mind that the dimer propagator shown in Eq. (6.23) is just the *bare* propagator. The dimer propagator accounts for the space–time propagation of the dimer field; hence, it accounts for any two-body collision in which the initial and final states are two-body states. The dressed dimer propagator could be obtained by using the nonperturbative treatment introduced in the previous section. However, it is preferable to use the framework of the STM integral equation [8] to obtain the propagator.

The infinite series of Feynman diagrams contributing to the dressed dimer propagator can be expressed as a finite contribution by means of the following Feynman diagrams

$$\underline{\qquad} = \underline{\underline{\qquad}} + \underline{\underline{\oslash}}\underline{\qquad}, \tag{6.24}$$

where the dressed propagator appears on both sides of the equality. The scattering amplitude, in this case, satisfies the following integral equation

$$\imath\mathcal{D}(P_0, \mathbf{P}) = \frac{4\imath}{g_2} - \frac{\imath g_2}{2} \int \frac{dq_0}{2\pi} \frac{d^3\mathbf{q}}{(2\pi)^3} \frac{\imath}{q_0 - q^2/2 + \imath\epsilon}$$

$$\frac{\imath}{P_0 - \mathbf{P}^2/4 - q_0 - q^2/2 + \imath\epsilon} \imath\mathcal{D}(P_0, \mathbf{P}), \tag{6.25}$$

where the Feynman rules associated with $\mathcal{L}_{d-a}$ have been applied. $\mathcal{D}(P_0, P)$ denotes the scattering amplitude, which correlates with the dressed dimer propagator, for a given energy P_0 and momentum $\mathbf{P}$ of the dimer. This equation is the two-body STM integral equation. The dressed propagator does not depend on q; therefore, the integral in the second term of the right-hand side of Eq. (6.25) is identical to the integral of Eq. (6.10) with $E = P_0 - P^2/4$; and in virtue of Eq. (6.12) we find

$$\mathcal{D}(P_0, \mathbf{P}) = \frac{4}{g_2} \left(1 + \frac{g_2}{4\pi^2}\left(\Lambda - \frac{\pi}{2}\sqrt{-P_0 + \mathbf{P}^2/4 - \imath\epsilon}\right)\right)^{-1}, \tag{6.26}$$

where Λ denotes an ultraviolet cutoff employed in the regularization of the integral. Finally, taking into account the Eq. (6.15), the dressed dimer propagator is written as

$$\mathcal{D}(P_0, \mathbf{P}) = \frac{32\pi}{g_2^2} \left(\frac{1}{a} - \sqrt{-P_0 + \mathbf{P}^2/4 - \iota\epsilon} \right)^{-1}, \tag{6.27}$$

where the dependence on the cutoff is imbedded in g_2. $\mathcal{D}(P_0, P)$ has a pole at $P_0 = -1/a^2 + P^2/4$ for $a > 0$, which corresponds to a dimer of binding energy $-1/a^2$ and momentum $\mathbf{P}$.

The scattering amplitude for the three-body process at hand is calculated here using the STM integral equation approach, which in this case is diagrammatically represented by

$$\tag{6.28}$$

For the calculation of the three-body scattering amplitude and hence the STM integral equation it is preferable to work in the center of mass frame, where the incoming momentum of the atom and dimer have the same magnitude but opposite signs, and the same holds for the outgoing momenta. Let us assume that the incoming momentum for the atom is $\mathbf{p}$, whereas the outgoing momentum will be $\mathbf{k}$. If one takes the energy of the incoming atom as E_A, for instance, then the dimer shows an energy $E - E_A$, E being the collision energy. In analogy, assuming for the outgoing atom an energy $E_{A'}$, then the subsequent dimer energy will be $E - E_{A'}$. In this scenario, taking into account the four kinds of graphs contributing to the three-body process, Eq. (6.29), using the dressed dimer propagator given by Eq. (6.27) in addition to the Feynman rules regarding $\mathcal{L}_{d-a}$, the amplitude of scattering for three-body collisions reads as [2, 8–11]

$$\mathcal{A}(\mathbf{p}, \mathbf{k}, E, E_A, E_{A'}) =$$

$$- \left[\frac{g_2^2/4}{E - E_A - E_{A'} - (\mathbf{p}+\mathbf{k})^2/2 + \iota\epsilon} + \frac{g_3}{36} \right]$$

$$+ \int \frac{dq_0 d^3\mathbf{q}}{(2\pi)^4} \left[\frac{g_2^2/4}{E - E_A - E_{A'} - (\mathbf{p}+\mathbf{q})^2/2 + \iota\epsilon} + \frac{g_3}{36} \right]$$

$$\times \frac{\iota}{q_0^2 - \mathbf{q}^2/2 + \iota\epsilon} \mathcal{D}(E, \mathbf{q}) \mathcal{A}(\mathbf{q}, \mathbf{k}, E, E_A, E_{A'}),$$

$$\tag{6.29}$$

which is the TSM integral equation. This equation can be expressed in a more convenient way by integrating over q_0 and setting the initial and final energies of the atoms on shell, i.e., $E_A = \mathbf{p}^2/2$ and $E_{A'} = \mathbf{k}^2/2$. By doing so, we find

$$A(\mathbf{p}, \mathbf{k}, E, \mathbf{p}^2/2, \mathbf{k}^2/2) =$$

$$-\left[\frac{g_2^2/4}{E - E_A - E_{A'} - (\mathbf{p} + \mathbf{k})/2 + \iota\epsilon} + \frac{g_3}{36}\right]$$

$$-\int \frac{d^3\mathbf{q}}{(2\pi)^3}\left[\frac{g_2^2/4}{E - E_A - E_{A'} - (\mathbf{p} + \mathbf{q})/2 + \iota\epsilon} + \frac{g_3}{36}\right]$$

$$\times \mathcal{D}(E, \mathbf{q})A(\mathbf{q}, \mathbf{k}, E, \mathbf{p}^2/2, \mathbf{k}^2/2).$$

(6.30)

The STM integral equation for three identical bosons, as is shown in Eq. (6.30), involves seven independent variables plus the effect of an extra variable, $\vec{q}$, associated with the exchange of a virtual atom. This virtual exchange is associated with the integral in Eq. (6.30), which is divergent and requires regularization. Indeed, it has been shown that the STM equation for the problem at hand does not have a unique solution [13]. However, it has an unique solution if the three-body binding energy is fixed. Another possibility for having a unique solution is by tuning the ultraviolet cutoff Λ in order to reproduce three-body scattering data [14], and hence Λ is taken as a parameter. This a priori *hand-made* renormalization procedure can be justified when the three-body STM equation is renormalized, as shown below.

The three-body STM equation can be further simplified if we restrict to the case of total orbital angular momentum $L = 0$. This can be done by averaging the scattering amplitude over the cosine of the angle between the incoming momentum of the atom and its outgoing momentum, *i.e.*, $\cos\theta = \mathbf{p} \cdot \mathbf{k}/pk$, and the result will be denoted as $A_s(p, k, E)$, which is given by

$$A_s(p, k, E) = \frac{Z_D}{2}\int_0^\pi d(\cos\theta)A(\mathbf{p}, \mathbf{k}, E, \mathbf{p}^2/2, \mathbf{k}^2/2), \qquad (6.31)$$

where the factor $Z_D/2$ has been included for convenience. Z_D stands for the residue of the dressed dimer propagator Eq. (6.27) at his pole, i.e., $Z_D = 64\pi/(ag_2^2)$. The evaluation of the integral in Eq. (6.31) leads to

$$A_s(p, k, E) = \frac{Z_D g_2^2}{4}\left[\frac{1}{2pk}\ln\left(\frac{p^2 + pk + k^2 - E - \iota\epsilon}{p^2 - pk + k^2 - E - \iota\epsilon}\right)\right.$$

$$\left. + \frac{H(\Lambda)}{\Lambda^2}\right] + \frac{4}{\pi}\int_0^\Lambda q^2 dq\left[\frac{1}{2pq}\ln\left(\frac{p^2 + pq + q^2 - E - \iota\epsilon}{p^2 - pq + q^2 - E - \iota\epsilon}\right)\right.$$

$$\left. + \frac{H(\Lambda)}{\Lambda^2}\right]\frac{A_s(q, k, E)}{-1/a + \sqrt{-E + 3q^2/4 - \iota\epsilon}},$$

(6.32)

where the explicit dependence of the dressed dimer propagator $\mathcal{D}(E, \mathbf{q})$ has been taken into account, and

$$\frac{g_3}{9g_2^2} = -\frac{H(\Lambda)}{\Lambda^2} \tag{6.33}$$

has been applied, where Λ stands for the ultraviolet cutoff. The function $H(\Lambda)$ is a dimensionless function such that $H(\Lambda)/\Lambda^2$ has to be well-defined for $\Lambda \to \infty$ [11]. This has been the standard approach to the subject and it is the one that we present here for renormalizing the three-body STM equation in the next section. In particular, it is shown that $H(\Lambda)$ is the crucial function for explaining the universality in the Efimov effect from a renormalization group approach.

6.4 Universality on the Three Boson Problem: Efimov States from a Renormalization Group Approach

It has been shown that the effective field theory associated with collision of three identical bosons leads to a scattering amplitude that depends on the ultraviolet cutoff Λ (see Eq. (6.32)). However, the scattering amplitude that ultimately gives rise to physical observables, such as the cross section, cannot depend on any arbitrary parameter; otherwise, the theory at hand will not stand as a reasonable theoretical model for the three identical boson problem. Therefore, we must find a way to get rid of the Λ dependence on the scattering amplitude.

The STM equation for three identical bosons involves a term proportional to $\mathcal{H}(\Lambda)$, which clearly induces a new energy scale on the problem. Assuming that $\mathcal{H}(\Lambda)/\Lambda^2 = 0$ when $\Lambda \to \infty$ and taking the limit $p \to \infty$ leads to a power law behavior for the scattering amplitude as a function of p. In particular, we find $\mathcal{A}_s(q, k, E) \approx p^{s-1}$; therefore, the STM equation is expressed as [13, 15–17]

$$p^{s-1} = \frac{4}{\pi\sqrt{3}p} \int_0^\infty \ln\left(\frac{p^2 + pq + q^2}{p^2 - pq + q^2}\right) q^{s-1} dq, \tag{6.34}$$

where the inhomogeneous term in Eq. (6.32) is neglected, as well as terms going as a^{-2} and E in comparison with q^2. Equation (6.34) is solved by means of the change of variables $q = wp$, leading to a Mellin transform type integral and its evaluation leads to the following equation in s:

$$1 = \frac{8}{s\sqrt{3}} \frac{\sin(\pi s/6)}{\cos(\pi s/2)}. \tag{6.35}$$

This equation has an infinite number of solutions owing to its periodic nature; however, the solution with the lowest $|s|$ leads to a pure imaginary solution ιs_0 with $s_0 = 1.0062$. Then, the most general solution for the scattering amplitude for $p \to \infty$ is

$$A_s(q, k, E) \approx \alpha p^{\iota s_0 - 1} + \beta p^{-\iota s_0 - 1}, \tag{6.36}$$

where α and β are constants. One of the constants in Eq. (6.36) is fixed by means of the inhomogeneous term in Eq. (6.32), whereas the second constant must be determined from the role of the three-body term $\mathcal{H}(\Lambda)$. Assuming $q \sim \Lambda$, $\Lambda \to \infty$ and plugging Eq. (6.36) into Eq. (6.32) yields

$$\frac{8}{\pi \sqrt{3}} \int_0^\infty q \left[\frac{1}{q^2} + \frac{\mathcal{H}(\Lambda)}{\Lambda^2} \right] \left(\alpha q^{\iota s_0 - 1} + \beta q^{-\iota s_0 - 1} \right) dq. \tag{6.37}$$

Then, if the previous integral vanishes we find

$$\mathcal{H}(\Lambda) = \frac{\alpha \Lambda^{\iota s_0} / (1 - \iota s_0) + \beta \Lambda^{-\iota s_0} / (1 + \iota s_0)}{\alpha \Lambda^{\iota s_0} / (1 + \iota s_0) + \beta \Lambda^{-\iota s_0} / (1 - \iota s_0)}, \tag{6.38}$$

and the scattering amplitude is independent of Λ, thus confirming the renormalization for the three identical bosons STM equation, as Hammer, Bedaque, and van Kolck showed [11, 12, 18]. The constants in Eq. (6.38) can be expressed in terms of a single constant Λ_* in virtue of the role of the inhomogeneous term in Eq. (6.32) as $\alpha = \sqrt{1 + s_0^2} \Lambda_*^{-\iota s_0} / 2$ and $\beta = \sqrt{1 + s_0^2} \Lambda_*^{\iota s_0} / 2$; therefore,

$$\mathcal{H}(\Lambda) = \frac{\cos(s_0 \ln[\Lambda / \Lambda_*] + \arctan s_0)}{\cos(s_0 \ln[\Lambda / \Lambda_*] - \arctan s_0)}. \tag{6.39}$$

The presence of the logarithm of the cutoff translates into the existence of a geometric scaling that is associated with Efimov states.

The three-body coupling constant g_3 is related to $\mathcal{H}(\Lambda)$ through Eq. (6.33), as has been established previously in the derivation of the STM equation (see Eq. (6.32)). However, it is convenient to introduce a dimensionless three-body coupling constant as

$$g_3^a = \frac{\Lambda^4 g_3}{144 \pi^4} = \frac{-9 \Lambda^4 \mathcal{H}(\Lambda) g_2^2}{144 \pi^4}, \tag{6.40}$$

which in virtue of Eq. (6.16) reads as

$$g_3^a = - \left(\frac{a \Lambda}{a \Lambda - \pi/2} \right)^2 \mathcal{H}(\Lambda). \tag{6.41}$$

The results for g_3^a as a function of Λ are shown in Fig. 6.4. Each of the divergences of g_3^a represents the existence of an Efimov state, and, as observed in Fig. 6.4, they follow a geometric scaling, as it is natural in the Efimov scenario [3, 19–21]. This *periodicity* in the divergences of g_3^a with respect to Λ correlates with the existence of a liming cycle in the renormalization group flow [3, 22]. Besides, by means of Eq. (6.39) it is possible to calculate the divergences of $\mathcal{H}(\Lambda)$; hence, for g_3^a,[3] which turns out to occur at

$$\Lambda_n = e^{\pi n/s_0} \left(e^{(s_0^{-1}\pi/2 + s_0^{-1}\arctan s_0)} \Lambda_* \right). \tag{6.42}$$

This equation reveals the observed geometric scaling for g_3^a. The three-body parameter Λ_* is not predicted by the renormalization of the STM equation; indeed, it is a free parameter that has to be taken by fitting to the experimental data.

The results shown in Fig. 6.4 can be interpreted in terms of the renormalization group approach. For a fixed value of Λ_* and a, the variation of Λ defines a renormalization group (RG) trajectory. All the points of a RG trajectory describe the same physical theory for a given value of Λ_* and a. However, it is preferable to introduce the beta function defined as in Eq. (6.3) and reads as

$$\Lambda \frac{d}{d\Lambda} \begin{pmatrix} g_2^a \\ g_3^a \end{pmatrix} = \begin{pmatrix} \frac{1+s_0^2}{2} \left[(g_2^a)^2 + \left(\frac{g_3^a}{g_2^a} \right)^2 \right] + (3 - s_0^2 + 2g_2^a)g_3^a \end{pmatrix}. \tag{6.43}$$

The flow associated with this equation in the space of the coupling constants g_2^a, g_3^a is shown in Fig. 6.3. This figure displays the existence of the trivial fixed point for $g_2^a = g_3^a = 0$, i.e., a non-interacting field theory. For $g_2^a = 0$ and $g_3^a \neq 0$ the beta function is clearly singular, indicating that the present theory is not predictive for such region of the coupling constant space. Finally, for g_2^a presents an ultraviolet fix point for the two-body interaction; however, it is not possible to find a fix point for g_3^a, since Eq. (6.43) does not have real roots. Instead of an ultraviolet fixed point, what is found is an ultraviolet limit cycle. In Fig. 6.3 it is noted that for $g_3^a = -1$, the flow in g_3^a points in the same direction, which is related to the existence of divergence in g_3^a as a function of Λ. Indeed, it reflects the first divergence for $a = 1$ presented in Fig. 6.4. This flow only represents half of the limit cycle.

6.5   Concluding Remarks

In this chapter, we have presented a new and useful perspective into ultracold collisions by developing a consistent quantum effective field theory. In particular, we have shown that a proper effective quantum field theory makes it possible to explain the nature of the Efimov effect as a limit cycle in the renormalization group flow of a three-body theory. Therefore, the Efimov effect is related to the intrinsic universality class within a three-body theory. Finally, we believe that having different approaches to the same problem is beneficial for the advancement of science.

---

[3]For small Λ this is not the case, since a two-body divergence emerges.

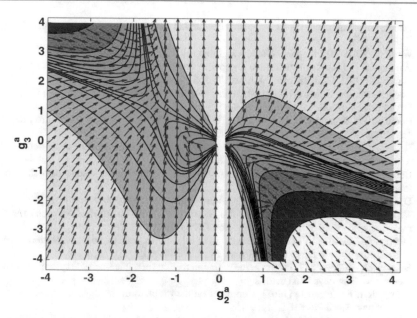

Fig. 6.3 Renormalization group flow associated with the STM equation for three identical bosons. The contours represent different slices of the $\Lambda dg_3^a/d\Lambda$ function and their values are indicated by the color bar placed at the right of the plot. The arrows represent the normalized flow associated with Eq. (6.43)

Fig. 6.4 Renormalized three-body interaction g_3^a as a function of the ultraviolet cutoff Λ for two different values of the two-body scattering length a

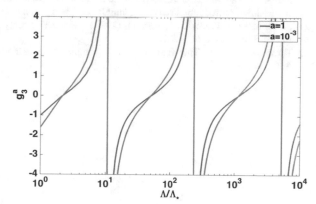

References

1. Wilson KG, Kogut J (1974) The renormalization group and the ϵ expansion. Phys Rep 12:75
2. Braaten E, Hammer HW (2006) Universality in few-body systesm with large scattering length. Phys Rep 428:259
3. Hammer HW, Platter L (2011) Efimov physics from a renormalization group perspective. Philos Trans R Soc A 369:2679
4. Wilson KG (1983) The renormalization group and critical phenomena. Rev Mod Phys 55:583

5. Peskin ME, Schoreder DV (1995) An introduction to quantum field theory. Westvie Press, Chicago
6. Weinberg S (1979) Phenomenological lagrangians. Phys A 96:327
7. Wilson KG (1971) Renormalization group and strong interactions. Phys Rev D 3:1818
8. Skorniakov GV, Ter-Martirosian KA (1957) JETP 4:648
9. Bedaque PF, van Kolck U (1998) Nucleon-deuteron scattering from and effective field theory. Phys Lett B 428:221
10. Bedaque PF, Hammer HW, van Kolck U (1998) Effective theory for neutron-deuteron scattering: energy dependence. Phys Rev C 58:R641
11. Bedaque PF, Hammer HW, van Kolck U (1999) Renormalization of the three-body system with short-range interactions. Phys Rev Lett 82:463
12. Bedaque PF, Hammer HW, van Kolck U (1999) The three-boson system with short-range interactions. Nucl Phys A 646:444
13. Danilov GS (1961) On the three-body problem with short-range forces. Sov Phys JETP 13:349
14. Kharchenko VF (1973) Solution of the Skornyakov-Ther-Martirosian equations for three nucleons with cutoff at large momenta. Sov J Phys JETP 16:173
15. Danilov GS, Lebedev VI (1963) Calculation of the doublet neutron-deuteron scattering length in the theory fo zero range forces. Sov J Phys JETP 17:1015
16. Minlos RA, Faddeev LD (1962) On the point interaction for a three-particle system in quantum mechanics. Sov Phys Dokl 6:1072
17. Minlos RA, Faddeev LD (1962) Comment on the problem of the three particles with point interactions. Sov J Phys JETP 14:1315
18. van Kolck U (1999) Effective field theory of short-range forces. Nucl Phys A 645:273
19. Efimov V (1970) Energy levels arising from resonant two-body forces in a three-body system. Phys Lett B 33(8):563. https://doi.org/10.1016/0370-2693(70)90349-7, http://www.sciencedirect.com/science/article/pii/0370269370903497
20. Efimov V, Tkachenko EG (1988) On the correlation between the triton binding energy and the neutron-deuteron doublet scattering length. Few-body Syst 4:71
21. Amorim AEA, Frederico T, Tomio L (1997) Universal aspects of efimov states and light halo nuclei. Phys Rev C 56:R2378
22. Birse MC (2011) The renormalization group and nuclear forces. Philos Trans R Soc A 369:2662

Ultracold Rydberg Atoms and Ultralong-Range Rydberg Molecules

7

Rydberg physics is considered an independent discipline within atomic, molecular, and optical physics. Most of the interest in Rydberg physics is due to the possible applications of Rydberg atoms in quantum information processing [1–4] and quantum simulation of many-body Hamiltonians [5–8]. As a consequence, most of the works and books based on Rydberg atoms assume some value for the decay rate of the atom; hence, they do not explore the nature of the different decay mechanisms [7, 8]. Such a perspective may be suitable for quantum optics, quantum information, and quantum simulations. However, if the nature of the decay of Rydberg atoms is understood, it will help to elucidate the best states and scenarios for a given application. Surprisingly enough, it turns out that the primary decay mechanism of Rydberg atoms in a high-density medium is through chemical reactions [9], the topic of this book. Indeed, we would like to go further and say that in most of the gas phase systems, chemical reactions are very often the main decoherence channels. Therefore, understanding and controlling chemical reactions is vital for the development of quantum technologies.

In this chapter, we introduce the scenario relevant to the study of chemical reactions of ultracold Rydberg atoms covered in the next chapter. Here, we introduce some basic notions about the nature and properties of Rydberg atoms. In particular, we study Rydberg–Rydberg interactions that lead to the well-known blockade effect, and introduce ultra-long range Rydberg molecules that have attracted significant attention since they were first observed in 2015 [10] and 2016 [11].

7.1 Rydberg Atoms

Any atom in a highly excited state with high principal quantum number n is defined as a Rydberg atom and fulfils the following conditions:

© Springer Nature Switzerland AG 2020
J. Pérez Ríos, *An Introduction to Cold and Ultracold Chemistry*,
https://doi.org/10.1007/978-3-030-55936-6_7

Fig. 7.1 Rydberg atom and
Rydberg states. Panel (**a**)
shows the energy level
structure of an atom, in which
Rydberg states appear before
reaching the ionization
continuum. Panel (**b**) shows
the classical trajectory of the
Rydberg electron, feeling a
positively charged ionic core

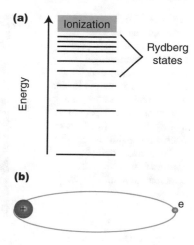

- The ionization energy of the excited state of the atom is described by the Rydberg formula $-\frac{1}{2n^{*2}}$ (in atomic units), where n^* is the *effective* principal quantum number. In other words, Rydberg atoms are hydrogen-like atoms.
- In virtue of the previous point, the trajectory of the outer electron is appropriately described through classical mechanics. That is the reason why the reader may hear that Rydberg states behave classically.

As shown in panel (a) of Fig. 7.1, every atom shows Rydberg states near the ionization continuum, and an atom in one of these states is labeled as a Rydberg atom. However, in ultracold physics, when talking about Rydberg atoms, it is assumed that a valence electron is promoted to a Rydberg state, and the motion of this electron (Rydberg electron) is classically described as shown in panel (b) of Fig. 7.1. In particular, the Rydberg electron feels the attractive force of a positively charged ionic core, and as a consequence, the electron describes an elliptical orbit around the ionic core.

However, this approach is only valid at long range, where the core electrons do not play any role. As the Rydberg electron approaches the ionic core, the core electrons no longer screen the Rydberg electron–ionic core interaction, thus altering the energy of the Rydberg electron, as shown in Fig. 7.2. As a result, the Rydberg electron feels an effective potential $V_R(r) = V_{\text{core}}(r) - 1/r$ (see Fig. 7.2), where $V_{\text{core}}(r)$ is a short-ranged potential to account for the Rydberg electron–ionic core short-range interaction. The presence of the short-range potential imposes a phase shift $(\pi \delta_{\mu_l})$ on the Rydberg electron wave function with respect to the hydrogenic one, which defines the quantum defect μ_l. Indeed, the quantum defect is a fundamental quantity used to characterize Rydberg states, since the ionization energy of a Rydberg state depends on it as

$$E = -\frac{1}{2(n - \mu_l)^2}. \tag{7.1}$$

Fig. 7.2 The physics behind quantum defects. The presence of the ionic core induces a short-ranged potential $V_{\text{core}}(r)$ that modifies the Rydberg electron wave function, leading to a phase shift with respect to the hydrogenic wave function. Such a phase shift is known as the quantum defect and appears in the effective principal quantum number $n^* = n - \mu_l$

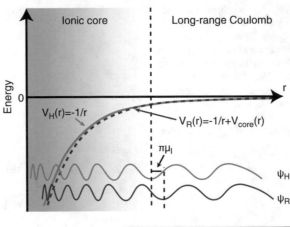

Table 7.1 Quantum defects (μ_l) of the alkali atoms taken from Ref. [17]

l	Li	Na	K	Rb	Cs
s	0.40	1.35	2.19	3.13	4.06
p	0.04	0.85	1.71	2.66	3.59
d	0.00	0.01	0.25	1.34	2.46
f	0.00	0.00	0.00	0.01	0.02
$\geq$g	0.00	0.00	0.00	0.00	0.00

Quantum defects can be obtained empirically by fitting the observed ionization energy spectra of Rydberg states to the expression given in Eq. (7.1). In addition, quantum defects can be obtained from *ab initio* calculations, in particular, using the celebrated multi-channel quantum defect theory after a *R*-matrix calculation [12] for short-range physics. We encourage the reader to look into Refs. [13–16] to find out more about this interesting and useful scattering theory since it is beyond the scope of this book. The quantum defects of the alkali atoms appear in Table 7.1, where it is noticed that for $l \geq 4$, the quantum defects are negligible, which means that those Rydberg states behave as hydrogen-like states.

7.1.1 Sommerfeld Orbits

One of the main characteristics of Rydberg states is that the motion of the Rydberg electron can be described classically. In particular, the Rydberg electron motion is explained utilizing the Sommerfeld model of the atom. The Sommerfeld model of the atom is an improvement over Bohr's atomic model since it allows the electron to move in elliptical trajectories with the positively charged atomic core placed at the focus of the ellipse, as depicted in Fig. 7.3. Within the Sommerfeld model, the trajectory of the electron is better specified in polar coordinates (r, ϕ) and conjugated momenta (p_r, p_ϕ). Therefore, the Hamiltonian of the ionic core-electron system reads as

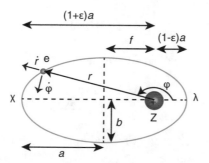

Fig. 7.3 Elliptical trajectory associated with Sommerfeld's atomic model. The electron describes an elliptical trajectory, and its position is characterized by the polar coordinates (r, ϕ) defined from the focus, where the atomic core is placed. The semi-major axis of the ellipse is a, whereas the semi-minor axis is b and the eccentricity is denoted as ϵ. Finally, the points denoted as χ, and λ stand for the aphelion and perihelion respectively

$$H = \frac{p_r^2}{2m} + \frac{p_\phi^2}{2r^2m} - \frac{Ze^2}{4\pi\epsilon_0 r}, \qquad (7.2)$$

where ϵ_0 is the electric constant of vacuum m is the electron's mass, and Z denotes the effective charge of the atomic core that the electron feels. The Hamiltonian in Eq. (7.2) is time-independent and it does not depend on ϕ. Thus, the energy and angular momentum (p_ϕ) are conserved quantities.

The connection between classical and quantum mechanics relies on the quantization of the classical action, following the ideas of Bohr and Sommerfeld. In particular, the angular degree of freedom yields

$$\oint p_\phi d\phi = hl \rightarrow p_\phi = \hbar l, \qquad (7.3)$$

where l is the quantum number associated with the angular momentum. For the radial degree of freedom, the action reads as

$$\oint p_r dr = hn_r, \qquad (7.4)$$

where n_r is the quantum number regarding the motion in the radial direction. It is preferable to express the radial action as

$$\oint p_r dr = \oint \frac{p_\phi}{r^2} \left(\frac{dr}{d\phi}\right)^2 d\phi = hn_r, \qquad (7.5)$$

where we have taken into account that $p_r = m\dot{r} = mdr/d\phi\dot{\phi}$, $p_\phi = mr^2\dot{\phi}$ and $\dot{x} \equiv dx/dt$. To solve Eq. (7.4) is convenient to introduce the equation of a general conic section in polar coordinates (r, ϕ) as

$$\frac{1}{r} = C_1 + C_2 \cos \phi, \tag{7.6}$$

which for the particular case of an ellipse yields

$$\frac{1}{r} = \frac{1 + \epsilon \cos \phi}{a \left(1 - \epsilon^2\right)}, \tag{7.7}$$

where ϵ stands for the eccentricity of the ellipse. From Eq. (7.6) it is easy to show that

$$\frac{dr}{d\phi} = r \frac{\epsilon \sin \phi}{1 + \epsilon \cos \phi}, \tag{7.8}$$

and hence the action in the radial coordinate given by Eq. (7.5) yields

$$\oint \frac{p_\phi}{r^2} \left(\frac{dr}{d\phi}\right)^2 d\phi = p_\phi \int_0^{2\pi} \frac{\epsilon^2 \sin^2 \phi}{(1 + \epsilon \cos \phi)^2} d\phi. \tag{7.9}$$

The integral in Eq. (7.9) can be solved analytically, leading to a relationship between n_r and l as

$$1 - \epsilon^2 = \left(\frac{l}{n_r + l}\right)^2 = \left(\frac{l}{n}\right)^2 = \left(\frac{b}{a}\right)^2, \tag{7.10}$$

where $n = n_r + l$ is the principal quantum number.

Equation (7.10) can be interpreted as the link between the shape of the electron's orbit and the value of the angular quantum number, l, with respect to the principal quantum number. In particular, it shows that larger values of l lead to circular electronic orbits, whereas for small values of l the electron's orbit shows a larger eccentricity, as shown in Fig. (7.4) for the case of $n = 4$. In other words, for a given principal quantum number, the classical trajectory associated with a S state of an atom is more elliptical than the one with a large angular momentum. At this point, the reader may think that something is wrong, since from the very first moment

Fig. 7.4 Sommerfeld elliptical orbits for $n = 4$

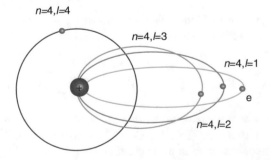

we learn quantum mechanics, we are told that S states are spherically symmetric, whereas high angular momentum states show anisotropic behavior. However, it is worth recalling that the previous statement refers to the angular degrees of freedom based on the properties of the spherical harmonics. However, if we take a look at the radial distribution function for different angular momentum states, we will find out that low angular momentum states show a more significant probability of being near the nucleus. In contrast, high angular momentum states show a negligible probability of finding the electron near the nucleus. This confirms the results based on the Sommerfeld model.

7.1.2   Properties of Rydberg Atoms

In the introduction of many papers, Rydberg atoms are defined as atoms with exaggerated properties. It is clear that, after reading this statement, the reader will have questions such as: which properties does this statement refer to? What does exaggerated mean in this context? To answer these questions and related ones that the reader may come up with, we list in Table 7.2 the main electronic, collisional, and dynamic properties of Rydberg states and how they behave as a function of the principal quantum number n.

Most of the properties of Rydberg atoms show a strong dependence on the principal quantum number, especially those related to the response to an external electric field such as the polarizability and electric field for ionization. In fact, owing to the enhancement of the polarizability and dipole moment with the principal quantum number, Rydberg–Rydberg interactions are stronger than typical atom–atom long–range interactions, which makes Rydberg atoms an excellent candidate for quantum information processing [2] and quantum simulations of many-body systems [5, 6]. Moreover, Rydberg–Rydberg and Rydberg–atom interactions may show different reaction channels owing to the strong interaction in those systems

Table 7.2 List of properties and their scaling law for Rydberg states

Property	Scaling
Energy	$= -\frac{1}{2n^{*2}}$
Energy difference	$= \frac{1}{n^{*3}}$
Size of the Rydberg orbit	$\propto n^2$
Average velocity of the electron	$\propto n^{-1}$
Probability of finding the electron at the core	$\propto n^{-3}$
Dipole moment	$\propto n^2$
Lifetime	$\propto n^3$
Polarizability	$\propto n^7$
Geometrical cross section	$\propto n^4$
Inglis–Teller limit for electric field for ionization	$\propto n^{-5}$
Classical field ionization limit	$\propto \frac{1}{16n^4}$

owing to the properties of Rydberg systems. Therefore, exaggerated properties mean that Rydberg atoms show stronger long-range interactions with ground state atoms and longer lifetimes than regular atoms and molecules in excited states.

7.2 Rydberg Blockade

The excitation of an ultracold gas of ground state atoms into Rydberg states may look like a generalization of the excitation of ground state S atoms into P state atoms. Thus, it only requires a laser light resonant with the target P state and ultracold atoms in an excited electronic state are produced. However, in the case of Rydberg atoms, the situation is not as simple as this, as one must account for the Rydberg–Rydberg interaction.

The Rydberg–Rydberg interaction between the same Rydberg state, at long range, is dominated by the induced dipole-induced dipole moment interaction that leads to the van der Waals interaction C_6/R^6 as in the case of two neutral distributions of charge separated by a large distance. However, as shown in Table 7.2 the polarizability of a Rydberg atom scales rapidly with n as n^7, leading to a gigantic van der Waals coefficient $C_6 \propto n^{11}$ [18] for long-range Rydberg–Rydberg interaction[1]. The gigantic long-range Rydberg–Rydberg interaction prevents two Rydberg atoms from exciting nearby, unless the interaction energy is equal to or less than the bandwidth of the excitation laser, Δ, which occurs at a distance r_B, as is schematically shown in panel (a) of Fig. 7.5. In panel (b) of the same figure, the Rydberg blockade radius, r_B, is illustrated through the van der Waals interaction energy for $50S + 50S$ for ^{87}Rb.

The Rydberg blockade radius (through the principal quantum number) and the density of ground-state atoms define the distinct regimes in which Rydberg–ground state atom interaction plays a role, as shown in Fig. 7.6. The yellow region is the region of interest for this chapter, as well as the next one, since having ground-state atoms within the Rydberg orbit leads to intriguing chemical reactions [9] and novel molecular binding mechanisms [19]. The white region of such a "phase diagram" corresponds to the scenario relevant for quantum information processing and quantum computing [2]. In that region, the Rydberg electron does not feel any ground-state atoms within the Rydberg orbit, and the density of Rydberg atoms controlled by r_B is high enough for the implementation of quantum gates.

---

[1]The scaling law is derived through second-order perturbation theory in which $C_6 \propto d^4/\Delta_E$, where d is the dipole moment matrix element and Δ_E is the energy difference between adjacent energy states. Then, from Table 7.2, one finds $d \propto n^2$ and $\Delta_E \propto n^{-3}$, and therefore $C_6 \propto n^{11}$, since it seems that the scaling law of the C_6 with n is affected

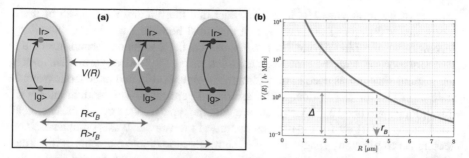

Fig. 7.5 Rydberg blockade mechanism. Panel (**a**), once a Rydberg atom is created through photon absorption, a second ground-state atom can only be promoted to the same Rydberg state if the distance with respect to the first Rydberg is larger than a threshold distance r_B called the blockade radius. (**b**) Rydberg–Rydberg long-range interaction for ^{87}Rb and $n = 50$. For a given laser bandwidth Δ the Rydberg blockade radius is determined as $r_B = (C_6/\Delta)^{1/6}$

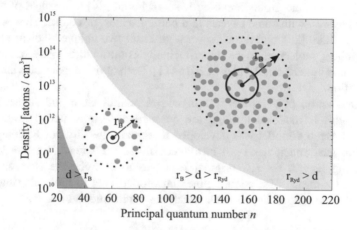

Fig. 7.6 Parameter space for Rydberg-ground state atoms showing the different regimes as a function of the principal quantum number n and density of the medium. At the lowest principal quantum number, n, and lowest densities (gray), the interparticle spacing, d, is greater than the Rydberg blockade radius, r_B, assuming an excitation bandwidth of 1 MHz. In the intermediate regime (white), the Rydberg blockade radius is larger than the interparticle spacing, but there are no background atoms overlapping with the Rydberg orbit with a radius of r_{Ryd}. In the third regime (yellow), at the highest principal quantum numbers and highest densities, many background atoms are contained within the Rydberg blockade sphere and the Rydberg orbit. (Figure reproduced with permission from Ref. [20] Copyright (2020) (IOP Publishing, Ltd))

7.3   Ultralong-Range Rydberg Molecules

Ultralong-range Rydberg molecules are exotic molecules formed by a Rydberg atom and one or more ground-state atoms. These molecules appear only in the yellow region of the "phase diagram" of Fig. 7.6, i.e., for a high density of ground-state atoms and high principal quantum numbers. The adjective exotic refers to two main aspects of the molecule. The first is the fact that the binding mechanism of these molecules is different from any type of chemical bond that we find in chemistry. In particular, the binding mechanism resides on the continuous scattering of the Rydberg electron with the ground-state atoms within the Rydberg orbit. Therefore, the ground-state atoms need to be within the Rydberg orbit, and in general, ground-state atoms within the Rydberg orbit are called perturbers. The binding energy of this unique bond is $\sim 10^{-8}$ eV to compare with the typical eV energy of the chemical bond or the ~ 0.01 eV for the binding energy of the van der Waals complexes. The second is that ultralong-range molecules show gigantic dipole moments of the order of kilo-Debye, despite the homonuclear nature of the molecule. Indeed, that is the most astonishing property of these exotic molecules: a homonuclear molecule with a massive dipole moment.

Ultralong-range Rydberg molecules were theoretically predicted by Greene et al. [21], Hamilton et al. [22] and Khuskivadze et al. [23] in the early 2000s, as a result of applying the groundbreaking ideas of Fermi [24] to the field of ultracold Rydberg physics, with some caveats and particularities, of course. They were first observed in 2009 [25], and it took several years for two unique kinds of ultralong-range Rydberg molecules to be experimentally confirmed: trilobite molecules in 2015 [10] and butterfly molecules in 2016 [11]. Both of these molecules are studied in this section.

The interaction of a Rydberg atom with a neutral atom within the classical Rydberg orbit is modeled through the quasi-free electron model developed by Fermi [24]. Such a model was developed to explain the shifts of Rydberg atoms in a buffer gas, and it has been introduced in Sect. 2.3 regarding the definition of scattering length. Fermi's original idea was to describe the electron–neutral interaction by means of a contact potential proportional to the scattering length, the so-called Fermi pseudopotential, which in atomic units is given by

$$V_{\text{ea}}(r, R) = 2\pi a[k(R)]\delta^{(3)}(r - R), \tag{7.11}$$

where r and R denote the electron and perturber position respectively. $a[k(R)]$ represents the energy-dependent electron–perturber scattering length, and $k(R) = \sqrt{2/R + 2E}$ stands for the semi-classical momentum of the Rydberg electron, where $E = -1/2n^2$ is the electron energy. Therefore, the classical turning point is $R_{\text{out}} = 2n^2$ and it is convenient to set $k(R > R_{\text{out}}) = k(R_{\text{out}})$ for evaluating the matrix elements.

The Fermi pseudopotential is adequate for treating low-energy electrons, *i.e.*, highly excited Rydberg states. However, in some systems, the electron can be momentarily trapped owing to the existence of shape resonances for the electron–ground-state atom collisions; thus, Eq. (10.5) needs to be extended to include such effects. The extension is performed following the method of Omont yielding [26]

$$V_{ea}(r, R) = 2\pi \sum_{l=0}^{\infty} (2l + 1)\delta^{(3)}(r - R) \left(\frac{-\tan \delta^l[k(R)]}{k(R)} \right) P_l \left(\frac{\overleftarrow{\nabla}_r \cdot \overrightarrow{\nabla}_r}{k^2(R)} \right),$$

(7.12)

where $\delta^l[k(R)]$ stand for the momentum-dependent l-wave phase-shift for the electron–ground-state atom interaction and $P_l(x)$ denote the Legendre polynomial of degree l and argument x. In Eq. (7.12) the backward vector symbol on the gradient implies that the gradient operator acts in the bra components of a matrix element. However, for most of the electron–ground-state atom interaction it is only necessary to include up to $l = 1$ in Eq. (7.12), leading to

$$V_{ea}(r, R) = 2\pi a[k(R)]\delta^{(3)}(r - R) - \frac{6\pi \tan \left(\delta^1[k(R)] \right)}{k(R)^3} \delta^{(3)}(r - R) \overleftarrow{\nabla}_r \cdot \overrightarrow{\nabla}_r.$$

(7.13)

Next, it is essential to account for the induced dipole–charge interaction between the ionic core and the ground-state atom within the Rydberg orbit, which reads as

$$V_{ia}(R) = -\frac{\alpha}{2R^4},$$

(7.14)

where α stands for the polarizability of the neutral atoms.[2] Therefore, the interaction energy of a ground-state atom within the Rydberg orbit is given by

$$V(r, R) = V_{ia}(R) + V_{ea}(r, R).$$

(7.15)

7.4   Potential Energy Curves for Ultralong-Range Rydberg Molecules

The potential energy curves for diatomic ultralong-range Rydberg molecules are calculated within the Born–Oppenheimer approximation as

$$H\Psi(r; R) = E(R)\Psi(r; R),$$

(7.16)

where the electronic energy of the system depends parametrically on the Rydberg–ground-state atom distance R. The electronic Hamiltonian reads as

$$H = H_0 + V(r, R),$$

(7.17)

where H_0 denotes the Hamiltonian of the Rydberg atom. The potential energy curves in Eq. (7.16) are calculated through degenerate perturbation theory, since most of the Rydberg energy states are degenerate in energy for high l, by means of the basis set $|nlm\rangle$ in which the Rydberg energy is diagonal, i.e.,

---

[2]This interaction is studied in detail in Sect. 9.4.1.

$$\langle n'l'm'|H_0|nlm\rangle = -\frac{1}{2n^{*2}}\delta_{n,n'}\delta_{l,l'}\delta_{m,m'} = -\frac{1}{2(n-\mu_l)^2}\delta_{n,n'}\delta_{l,l'}\delta_{m,m'}. \quad (7.18)$$

The basis set is defined as

$$|nlm\rangle = \psi_{nlm}(r) = \frac{u_{nl}(r)}{r}Y_l^m(\theta,\phi), \quad (7.19)$$

with

$$u_{nl}(R) = \begin{cases} \dfrac{W_{n^*,l+1/2}\left(\frac{2r}{n^*}\right)}{\sqrt{n^{*2}\Gamma(n^*+l+1)\Gamma(n^*-l)}} & l \le 3 \\ \dfrac{1}{n}\sqrt{\dfrac{\Gamma(n-l)}{\Gamma(n+l+1)}}\left(\dfrac{2r}{n}\right)^{l+1}e^{-r/n}L_{n-l-1}^{2l+1}(2r/n) & l \ge 4, \end{cases} \quad (7.20)$$

where $W_{n^*,l}(x)$ is the Whittaker function of argument x, $L_i^j(x)$ represents the associated Laguerre polynomials of argument x, and $\Gamma(x)$ is the Euler gamma function.

Assuming a single ground-state atom within the Rydberg orbit, it is better to take the z-axis in the direction of the vector joining the Rydberg core and the ground-state atom. Therefore, $\theta = 0$, and taking into account that $Y_l^m(0,\phi) = \frac{(2l+1)\delta_{m,0}}{4\pi}$, only states with $m = 0$ contribute to the energy landscape of the Rydberg atom–ground-state atom system.

Now that we have provided the proper theoretical toolkit, we are ready to delve into Rydberg physics. Let us begin by looking at the potential energy curves relevant for a Rydberg–ground-state atom interaction. In particular, we present the results for the relevant potential energy curves for 53S + 5S interaction in [87]Rb that are shown in Fig. 7.8. These curves have been obtained employing the methodology introduced in this section, i.e., perturbation theory by means of the basis shown in Eq. (7.20). For the diagonalization, we include eight different n-manifolds as well as their respective angular momentum states. In particular, two of them are above the target state and six below,[3] but taking into account the right energy ordering of the states through the quantum defects. The s-wave and p-wave energy-dependent electron-Rb phase shifts are taken from Ref. [27]. The curves in Fig. 7.7 show an oscillatory behavior that correlates with the undulations on the Rydberg electron wave function. The physics and properties of the Rydberg–ground-state atom system depend on the distance R and in which of the states it is, i.e., low-l Rydberg molecule, butterfly state, or trilobite state. All these states and their physics are explained below.

---

[3]This is a trick that leads to fairly accurate results in comparison with the Green's function formalism [22], and against accurate spectroscopy data.

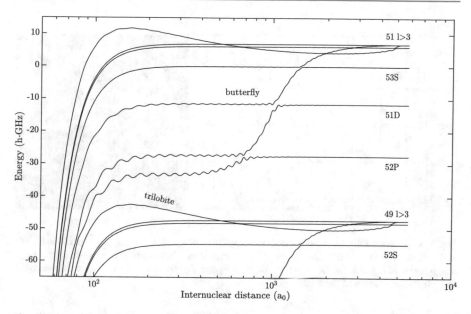

Fig. 7.7 Born–Oppenheimer potential energy curve of a $53S + 5S$ state for ^{87}Rb with their neighboring states. The energy is displayed in units of GHz, the distance in units of a_0, on a logarithmic scale. The most relevant states are labeled butterfly and trilobite, as well as the asymptotic Rydberg states

7.4.1   Low-l Ultralong-Range Rydberg Molecules

For low angular momentum Rydberg states ($l \leq 2$) the non-integer part of the quantum defects is quite significant. Hence, these low-l Rydberg states are energetically isolated. Therefore, it is possible to use the nondegenerate perturbation theory to calculate the potential energy curves associated with these states. In such a case, the potential energy curves for the Rydberg–ground-state atom interaction reads as

$$E_l^{\Sigma}(\boldsymbol{R}) = 2\pi \frac{2l+1}{4\pi} \left(a_s[k(R)] \frac{u_{nl}(R)^2}{R^2} + 3a_p^3[k(R)] \left(\frac{d(u_{nl}(R)/R)}{dR} \right)^2 \right),$$

(7.21)

which shows Σ symmetry, since it involves the s-wave term and the radial derivative of the wave function, whereas

$$E_l^{\Pi}(\boldsymbol{R}) = 6\pi a_p^3[k(R)] \left(\frac{(2l+1)l(l+1)}{8\pi} \right) \frac{u_{nl}^2(R)}{R^4},$$

(7.22)

shows a Π symmetry, since it involves angular derivatives that transform $Y_l^{|m=1|}(\theta, \phi)$ terms into $Y_l^{m=0}(\theta, \phi)$.

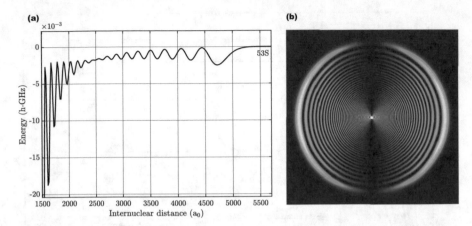

Fig. 7.8 (a) Born–Oppenheimer potential energy curve of a 53S + 5S state for ^{87}Rb. The energy is displayed in units of GHz and the distance in units of a$_0$. This figure is a zoom-in of the long-range region of Fig. 7.8. (b) Density plot for $2\pi\rho|\Psi(\rho, z, R)|$ for $R = 4300$ a$_0$. The small white dot represents the position of the ionic core

In panel (a) of Fig. 7.8 the $E_0^\Sigma(R)$ potential energy curve of a 53S + 5S state for ^{87}Rb is shown. This curve is a zoom-in of the long-range region of the panel (a) of Fig. 7.7, which looks flat in the Figure. However, in panel (a) of Fig. 7.8 a rich oscillatory structure following the electron probability is noticeable. Indeed, some of the wells support several vibrational bound states. These vibrational states are obtained by numerically solving the nuclear Schrödinger equation for the motion of the two atoms in the potential energy curve, and it reads as

$$\left(-\frac{1}{2\mu} \nabla_R^2 + E_l^{\Sigma/\Pi}(R) \right) \Psi_v(R) = E_v \Psi_v(R), \tag{7.23}$$

where ∇_R^2 stands for the Laplacian operator in R, E_v for the vibrational energy and μ is the reduced mass of the two atoms.

In panel (b) of Fig. (7.8) the probability density of the Rydberg electron (in cylindrical coordinates) when the perturber is placed at a given distance (R) is shown. The probability density shows a symmetrical behavior with respect to z, and it is very similar to the probability density of the Rydberg electron for the 53S state in Rb.

7.4.2 Trilobite Molecules

Trilobite states appear in every Rydberg–ground-state atom interaction as a consequence of the localization of the Rydberg electron near to the perturber. Such a localization occurs via the superposition of many high-l states with the same

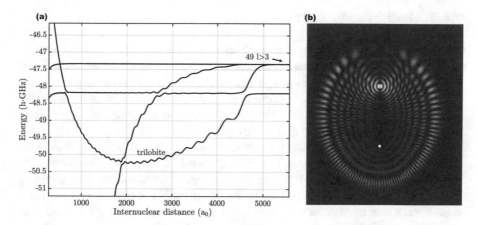

Fig. 7.9 (**a**) Born–Oppenheimer potential energy curve of a $53S + 5S$ state for ^{87}Rb with its neighboring states. The energy is displayed in units of GHz, the distance in units of a_0. This figure is a zoom-in of the long-range region of Fig. 7.8. (**b**) Density plot $2\pi\rho|\Psi(\rho, z, R)|^2$ for a trilobite state with $R = 2723$ a_0. The perturber is located underneath the double-peak structure in the vertical direction, whereas the small white dot represents the position of the ionic core

principal quantum number n as a consequence of the amalgamation of states contributing to the degenerate manifold. Figure 7.9 shows the trilobite state and trilobite wave function of a $53S + 5S$ state for ^{87}Rb with its neighboring states. In particular, panel (b) shows the density plot for the probability density of the Rydberg–ground-state atom wave function assuming that the perturber is at $R = 2723$ a_0, which is the region underneath the "twin" peaks,[4] where the electron shows its maximum probability to be found. In other words, in a trilobite state, the Rydberg electron spends most of the time near the perturber at $R \sim 1000$ a_0, leading to a giant permanent dipole moment of the molecule even though the molecule is homonuclear!

Trilobite molecules were predicted by Green et al. [21] and experimentally observed by Booth et al. [10], confirming their giant dipoles and unique nature. Since this observation, these molecules have generated more attention owing to the possibilities that these molecules may offer to quantum technologies and fundamental applications. The term trilobite molecule is employed for the bonding Rydberg–perturber when the Rydberg electron is in a high angular momentum state. Finally, we would like to explain why these states receive the name trilobite states: the density probability of such a state looks similar to a trilobite, the prehistoric creature.

---

[4]The double pick structure observed at the position of the perturber is an artefact of using cylindrical coordinates. Thus, the electron will spend most of the time near the perturber but not on two different sides of it.

7.4.3 Butterfly Molecules

The butterfly states only appear in systems where the perturber shows a p-wave shape resonance for electron-atom collisions. The presence of the resonance induces the descent of one of the curves of a given hydrogenic manifold down to a lower one. As a result, the descending curve prompts different couplings between itself and the asymptotic low-l Rydberg states. The butterfly curve is composed of high-l states with the same n as in the case of the trilobite curve.

Figure 7.10 shows the butterfly curve and wave function of a 53S + 5S state for ^{87}Rb with its neighboring states. In particular, panel (b) shows the density plot for the probability density of the Rydberg–ground-state atom wave function, assuming that the perturber is at $R = 1100$ a_0. Where the perturber is placed one observes a double-peak structure on the wave function, which indicates a strong localization of the electron in that region.[5] As a result the molecule will show a permanent dipole moment as a consequence of the positive ionic core, despite the homonuclear nature of the molecule. The term butterfly molecule is employed for Rydberg–perturber interactions where the perturber is placed on the butterfly curve. As in the case of trilobite molecules, butterfly molecules show a giant dipole moment.

Hamilton et al. [22] and Khuskivadze et al. [23] predicted the existence of butterfly molecules and Niederprüm et al. [11] observed them experimentally, confirming their giant dipole moment and showing that they readily could be excited into pendular states. Pendular states emerge as a consequence of the competition

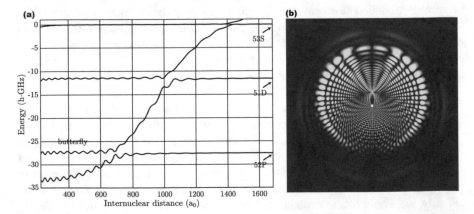

Fig. 7.10 (a) Born–Oppenheimer potential energy curve of a 53S + 5S state for ^{87}Rb with its neighboring states. The energy is displayed in units of GHz, the distance in units of a_0. This figure is a zoom-in of the long-range region of Fig. 7.8. (b) Density plot $2\pi\rho|\Psi(\rho, z, R)|^2$ for a butterfly state with $R = 1100$ a_0. The perturber is located underneath the double-peak structure in the vertical direction, whereas the small white dot represents the position of the ionic core

---

[5]The double-peak structure is an artefact of using cylindrical coordinates and hence does not have physical relevance.

between the rotational motion and alignment due to an external electric field applied to a polar molecule [28]. The observation of pendular states gives extra confidence in the determination of the size of the butterfly molecule. The term butterfly molecule comes from the fact that its wave function resembles a butterfly.

References

1. Lukin MD, Fleischhauer M, Cote R, Duan LM, Jaksch D, Cirac JI, Zoller P (2001) Dipole blockade and quantum information processing in mesoscopic atomic ensembles. Phys Rev Lett 87:037901. https://doi.org/10.1103/PhysRevLett.87.037901, https://link.aps.org/doi/10.1103/PhysRevLett.87.037901
2. Saffman M, Walker TG, Molmer K (2010) Quantum information with Rydberg atoms. Rev Mod Phys 82:2313
3. Müller M, Lesanovsky I, Weimer H, Büchler HP, Zoller P (2009) Mesoscopic Rydberg gate based on electromagnetically induced transparency. Phys Rev Lett 102:170502. https://doi.org/10.1103/PhysRevLett.102.170502, https://link.aps.org/doi/10.1103/PhysRevLett.102.170502
4. Ebert M, Kwon M, Walker TG, Saffman M (2015) Coherence and Rydberg blockade of atomic ensemble qubits. Phys Rev Lett 115:093601. https://doi.org/10.1103/PhysRevLett.115.093601, https://link.aps.org/doi/10.1103/PhysRevLett.115.093601
5. Löw R, Weimer H, Nipper J, Balewski JB, Butscher B, Büchler HP, Pfau T (2012) An experimental and theoretical guide to strongly interacting Rydberg gases. J Phys B-At Mol Opt Phys 45(11):113001
6. Balewski JB, Krupp AT, Gaj A, Hofferberth S, Löw R, Pfau T (2014) Rydberg dressing: understanding of collective many-body effects and implications for experiments. New J Phys 16(6):063012
7. Weimer H, Müller M, Büchler HP, Lesanovsky I (2011) Digital quantum simulation with Rydberg atoms. Quantum Inf Process 10(6):885. https://doi.org/10.1007/s11128-011-0303-5
8. Nguyen TL, Raimond JM, Sayrin C, Cortiñas R, Cantat-Moltrecht T, Assemat F, Dotsenko I, Gleyzes S, Haroche S, Roux G, Jolicoeur T, Brune M (2018) Towards quantum simulation with circular Rydberg atoms. Phys Rev X 8:011032. https://doi.org/10.1103/PhysRevX.8.011032, https://link.aps.org/doi/10.1103/PhysRevX.8.011032
9. Schlagmüller M, Liebisch TC, Engel F, Kleinbach KS, Böttcher F, Hermann U, Westphal KM, Gaj A, Löw R, Hofferberth S, Pfau T, Pérez-Ríos J, Greene CH (2016) Ultracold chemical reactions of a single Rydberg atom in a dense gas. Phys Rev X 6:031020. https://doi.org/10.1103/PhysRevX.6.031020, https://link.aps.org/doi/10.1103/PhysRevX.6.031020
10. Booth D, Rittenhouse ST, Yang J, Sadeghpour HR, Shaffer JP (2015) Production of trilobite Rydberg molecule dimers with kilo-Debye permanent electric dipole moments. Science 348(6230):99. https://doi.org/10.1126/science.1260722, http://science.sciencemag.org/content/348/6230/99.full.pdf, http://science.sciencemag.org/content/348/6230/99
11. Niederprüm T, Thomas O, Eichert T, Pérez-Ríos J, Greene CH, Ott H (2016) Observation of pendular butterfly Rydberg molecules. Nat Commun 7:12820
12. Burke PG (1973) The R-matrix method in atomic physics. Comput Phys Commun 6:288. https://www.sciencedirect.com/science/article/pii/0010465573900386
13. Seaton MJ (1966) Quantum defect theory II. Illustrative one-channel and two-channel problems. Proc Phys Soc 88:801
14. Greene CH, Fano U, Strinati G (1979) General form of the quantum-defect theory. Phys Rev A 19:1485
15. Greene CH, Rau ARP, Fano U (1982) General form of the quantum-defect theory. II. Phys Rev A 26:2441
16. Aymar M, Greene CH, Luc-Koenig E (1996) Multichannel Rydberg spectroscopy of complex atoms. Rev Mod Phys 68:1015

17. Burkhardt CE, Leventhal JJ (2006) The quantum defect. Springer, New York, pp 214–229. https://doi.org/10.1007/0-387-31074-6_11
18. Gallagher TF (1994) Rydberg atoms. Cambridge University Press, Cambridge. https://doi.org/10.1017/CBO9780511524530, https://www.cambridge.org/core/books/rydberg-atoms/B610BDE54694936F496F59F326C1A81B
19. Eiles MT (2019) Trilobites, butterflies, and other exotic specimens of long-range Rydberg molecules. J Phys B-At Mol Opt Phys 52(11):113001. https://doi.org/10.1088/1361-6455/ab19ca
20. Liebisch TC, Schlagmüller M, Engel F, Nguyen H, Balewski J, Lochead G, Böttcher F, Westphal KM, Kleinbach KS, Schmid T, Gaj A, Löw R, Hofferberth S, Pfau T, Pérez-Ríos J, Greene CH (2016) Controlling Rydberg atom excitations in dense background gases. J Phys B-At Mol Opt Phys 49(18):182001. https://doi.org/10.1088/0953-4075/49/18/182001
21. Greene CH, Dickinson AS, Sadeghpour HR (2000) Creation of polar and nonpolar ultra-long-range Rydberg molecules. Phys Rev Lett 85:2458. http://link.aps.org/doi/10.1103/PhysRevLett.85.2458
22. Hamilton EL, Greene CH, Sadeghpour HR (2002) Shape-resonance-induced long-range molecular Rydberg states. J Phys B-At Mol Opt Phys 35(10):L199. http://stacks.iop.org/0953-4075/35/i=10/a=102
23. Khuskivadze AA, Chibisov MI, Fabrikant II (2002) Adiabatic energy levels and electric dipole moments of Rydberg states of Rb_2 and Cs_2 dimers. Phys Rev A 66:042709
24. Fermi E (1934) Sopra lo spostamento per pressione delle righe elevate delle serie spettrali. Nouvo Cimento 11:157
25. Bendkowsky V, Butscher B, Nipper J, Shaffer JP, Low R, Pfau T (2009) Observation of ultralong-range Rydberg molecules. Nature 458(7241):1005. https://doi.org/10.1038/nature07945
26. Omont A (1977) J Phys 38:1343
27. Fabrikant II (1986) Interaction of Rydberg atoms and thermal electrons with K, Rb and Cs atoms. J Phys B-At Mol Opt Phys 19(10):1527. https://doi.org/10.1088/0022-3700/19/10/021
28. Rost JM, Griffin JC, Friedrich B, Herschbach DR (1992) Pendular states and spectra of oriented linear molecules. Phys Rev Lett 68:1299

Rydberg-Neutral Ultracold Chemical Reactions 8

The previous chapter can be viewed as a "first" approach to ultracold Rydberg atoms and ultralong-range Rydberg molecules motivated by the promising applications of these entities in quantum technologies and the quantum simulation of complex systems [1]. The approach has focused on static properties of Rydberg atoms and ultralong-range Rydberg molecules. However, nothing has been said about the dynamics of these systems and how they behave in different environments, which is the topic of the present chapter. In particular, the main ultracold chemical reactions involving Rydberg atoms and ground-state atoms are elucidated. Indeed, the Rydberg–ground-state atom chemical reactions are the primary decay mechanism of Rydberg atoms in high-density media [2], which is of capital interest for most of the applications of Rydberg atoms.

The present chapter begins with the study of elastic collisions between Rydberg and ground-state atoms, which leads to a technique to determine the density of an ultracold gas through Rydberg spectroscopy. Then, inelastic and reactive Rydberg–ground-state atom collisions are presented, leading to two main reaction mechanisms: l-mixing collisions and chemi-ionization reactions. Finally, the chapter finishes with some open questions about the ultimate nature of such reactions that are still under debate.

8.1 Rydberg-Neutral Elastic Collisions

When the Rydberg electron of an atom encounters a ground-state atom within its orbit, the electron collides elastically with the ground-state atom and its interaction is given by [3]

$$V_{ea}(\boldsymbol{r}, \boldsymbol{R}) = 2\pi a[k(R)]\delta^{(3)}(\boldsymbol{r} - \boldsymbol{R}), \tag{8.1}$$

© Springer Nature Switzerland AG 2020
J. Pérez Ríos, *An Introduction to Cold and Ultracold Chemistry*,
https://doi.org/10.1007/978-3-030-55936-6_8

where r stands for the position of the Rydberg electron, R is the position of a ground-state atom inside the Rydberg orbit and $a[k(R)]$ is the electron–ground-state atom scattering length, as seen in Sect. 2.3 and in more detail in Chap. 7. When many ground-state atoms are within the Rydberg orbit, the continuous elastic scattering of the Rydberg electron with the atoms leads to a shift on the Rydberg energy level ΔE, which can be evaluated within mean-field approximation yielding[1]

$$\Delta E = \int V_{ea}(r, R)|\Psi_{nlm}(r)|^2 \rho(R) d^3 r d^3 R = 2\pi \int a[k(R)]\rho(R)|\Psi_{nlm}(R)|^2 d^3 R.$$

(8.2)

Assuming that the density of the ultracold gas of ground-state atoms is constant within the Rydberg volume and that the electron–ground-state atom scattering length is constant, Eq. (8.2) reads as[2]

$$E = 2\pi a \bar{\rho} \int |\Psi_{nlm}(R)|^2 d^3 R = 2\pi \hbar^2 a \bar{\rho}.$$

$$2\pi a \bar{\rho}$$

(8.3)

Therefore, through the energy shift of the Rydberg spectra it is possible to infer the density of the media where the Rydberg atom is. This technique is called Rydberg spectroscopy. Indeed, Rydberg spectroscopy is a well-established technique in the Rydberg physics community to characterize the density of an ultracold gas [2, 4, 6–10]. Figure 8.1 displays the relation between different density-related magnitudes (the density, the interparticle distance, and the number of ground-state atoms inside the Rydberg orbit) and the energy shift of the Rydberg excitation of a single strontium (Sr) Rydberg atom in an ultracold gas of Sr atoms. In that Figure, it is observed that for $\Delta E = 10\,\text{MHz}$, which corresponds to a typical Bose-Einstein condensation (BEC) density of $1.3 \times 10^{14}\,\text{cm}^{-3}$, a Rydberg atom with a principal quantum number $n = 50$, 70, and 100 will contain 10, 80, and 643 ground-state atoms respectively.

Rydberg spectroscopy has been used to determine the density of different ultracold gases, including those with an electron–ground-state atom p-wave shape resonance, which ultimately leads to the emergence of butterfly states, as seen in Chap. 7. However, for these systems, the relation between density and energy shift is not as simple as the one given by Eq. (8.3) owing to the role of butterfly states as shown in Fig. 8.2, where the experimental Rydberg spectra for different principal quantum numbers in a BEC of Rb are shown. The theoretical calculations presented in the Figure are based on the quasistatic theory of the line broadening, and it is explained in the appendices. In Fig. 8.2 it is noticed that for larger values of the principal quantum number, the predictions with or without the p-wave shape resonance lead to similar predictions, which indicates that for sizeable

---

[1]Mean-field approximation is only valid when many ground-state atoms are inside the Rydberg orbit; otherwise, the discrete spectrum of dimers, trimers, tetramers, and more significant clusters should be included [4].

[2]The results of Eq. (8.3) in SI units would be $E = \frac{2\pi \hbar^2 \bar{\rho}}{m_e}$

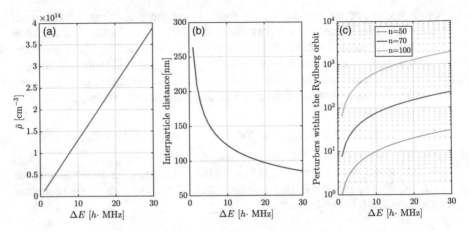

Fig. 8.1 Rydberg spectroscopy to determine the density of an ultracold gas of strontium atoms. Panel (**a**) ground-state atom density as a function of the detuning from the Rydberg excitation; panel (**b**) interparticle distance between ground-state atoms in the ultracold gas as a function of the detuning; panel (**c**) number of ground-state atoms within the Rydberg orbit for different principal quantum numbers n. The s-wave scattering length for e-Sr collisions is taken as its experimentally derived value of -12.5975 a_0 from Ref. [5]

principal quantum numbers Eq. (8.3) is more realistic. The region where the p-wave shape resonance leads to the butterfly states is more or less independent of the principal quantum number. Therefore, for higher principal quantum numbers, a more substantial fraction of the volume of the Rydberg electron orbit will be unaffected by the butterfly states. Hence, Eq. (8.3) becomes a more suitable description of the energy shift.

8.2 Rydberg-Neutral Inelastic and Reactive Collisions

When a Rydberg atom is immersed in a high-density ultracold gas, many ground-state atoms are found within the Rydberg orbit, and some of them may trigger inelastic and even reactive processes facilitated by the Rydberg electron, as displayed in Fig. 8.3. In the Figure, it is noted that the Rydberg atom may experience an inelastic process, ending up in a high angular momentum state but in a different principal quantum number. This reaction pathway is called l-mixing collision. Another possibility is that the Rydberg atom links to a ground-state atom, ejecting the Rydberg electron and leading to the formation of a molecular ion. This process is called associative ionization and it is considered to be a chemi-ionization reaction.

Rydberg atoms in high-density media at room temperature also experience inelastic processes such as l-mixing collisions. Indeed, that was a matter of exhaustive experimental study in the 1970s and 1980s [14–17], which motivated several theoretical approaches. Among them, the most prominent is the semi-classical approach of Hickman, which explicitly took into account the role of the Fermi pseudopotential, assuming a constant scattering length for the electron–

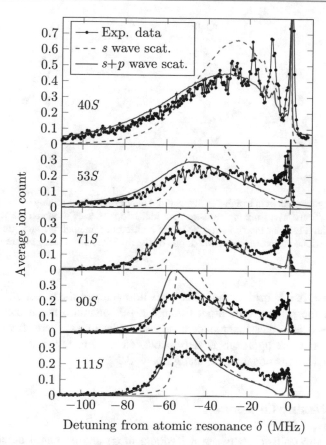

Fig. 8.2 Rydberg spectra in a Rb BEC for different principal quantum numbers. The theoretical predictions are based on the quasistatic approach for the line broadening [11–13] and using Eq. (8.1) (dashed blue line) and including the p-wave shape resonance contribution (solid red). In this figure $\delta \equiv \Delta E$. (Figure reproduced with permission from Ref. [7] Copyright (2020) (American Physical Society))

ground-state atom collision [18–20]. This approach satisfactorily describes the overall behavior of the l-mixing cross section as a function of the principal quantum number n for collisions at room temperature. The success of this approach relies on the fact that at room temperature the collision energy is much more significant than the energy difference between adjacent Rydberg states, n^{-3}. Therefore, the semi-classical framework does not apply to the case where the atoms and the Rydberg are at ultracold temperatures. Moreover, all semi-classical and even quantal approaches developed begin with the colliding partners in the asymptotic region, where the interatomic interaction is negligible with respect to the collision energy. Such a scenario is unphysical regarding the conditions of a Rydberg atom in a high-density ultracold gas, since the ground-state atoms lie within the Rydberg orbit before the collision, which is not an asymptotic state. Therefore, a new approach is required to understand the reactivity of a single Rydberg ion in a *sea* of ultracold atoms. That

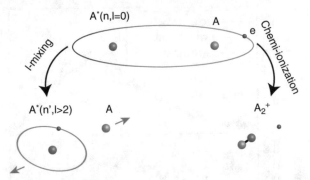

Fig. 8.3 A Rydberg atom in a high-density medium may experience l-mixing collisions or chemi-ionization reactions. The l-mixing collisions are inelastic processes in which the Rydberg atom ends up in a high angular momentum state but in a different principal quantum number. Chemi-ionization reactions are processes in which one of the final states is a charged atom or molecule

new approach is a hybrid method that takes into account the effect of the Rydberg electron on the ground-state atom through a full quantum mechanical treatment and assumes a semi-classical model at very short distances where chemi-ionization occurs [2], and the approach is reasonable. Chemi-ionization and l-mixing are explained within the novel theoretical framework.

8.2.1 L-Mixing Collisions

Every inelastic collision between a Rydberg atom and a neutral atom induces a change in the kinetic energy of the Rydberg atom associated with the energy released after the collision. Therefore, looking at the kinetic energy of Rydberg atoms is possible to reveal the nature of the inelastic process at hand. Panel (a) of Fig. 8.4 plots the observed released energy gained during a collision versus the principal quantum for Rb (nS). The change in kinetic energy correlates, for low principal quantum numbers, with the exchange energy between the excited nS Rydberg and the closest hydrogenic manifold ($n-4$), and it is given by (in atomic units)

$$\Delta E = \left(\frac{1}{2(n_1^*)^2} - \frac{1}{2(n_2^*)^2} \right) \tag{8.4}$$

$$= \left(\frac{1}{2(n - \delta_S)^2} - \frac{1}{2(n-4)^2} \right), \tag{8.5}$$

where the effective principal quantum number n^* includes the quantum defect. Therefore, the Rydberg atoms experience an inelastic collision with one of the perturber atoms in which the Rydberg state loses its energy to end up in a state with high angular momentum, i.e., an l-mixing collision.

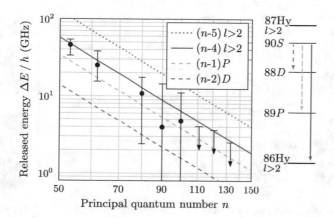

Fig. 8.4 Released energy during the l-changing collision. The black dots indicate the most likely released energy during a collision. In contrast, the bars indicate the full width at half maximum of the fitted probability distribution of the released energy. The energies observed correspond to a state change to the next-lower hydrogenic manifold $(n-4)$. The colors of the released energies are in accordance with the level scheme shown on the right, in which Hy denotes the hydrogenic manifold. (Figure adapted with permission from Ref. [2] Copyright (2020) (American Physical Society))

    The ultimate nature of l-mixing collisions resides in the Rydberg–ground-state interaction potential, which is perturbed by the ground-state atom when this is found within the Rydberg orbit, as it has been shown in detail in Chap. 7. The relevant potential energy curves (PECs) for Rb (nS) + Rb(5S) collisions have been calculated utilizing the perturbation theory approach introduced in Chap. 7. As an example, the most important PECs associated with the l-mixing collisions of Rb (53S) are shown in Fig. 8.5.

    Figure 8.5 shows that the butterfly state links different Rydberg states, and as a consequence, it is the main factor that leads to inelastic or reactive processes for Rydberg atom–ground-state atom collisions. In particular, the butterfly state induces a series of avoided crossings between different Rydberg states and itself. The branching ratio for different reaction pathways, l-mixing collisions, and chemi-ionization are determined by the nature of the cited avoided crossings and the velocity of the ground-state atom at the crossing point. Interestingly enough, the ultracold nature of the initial gas implies that the relevant kinetic energy of the ground-state atom at the crossing region is given by the energy difference between Rydberg states rather than the thermal kinetic energy of the neutrals, as it is the case for room temperature experiments [14–17]. Indeed, this is the main difference between the study of Rydberg-neutral collisions at ultracold temperatures versus the case at room temperatures. With this in mind, and assuming a Landau–Zener approach for the avoided crossings, it is possible to calculate the probability of ending up in a given state, as schematically shown in Fig. 8.5.

    l-mixing collisions relate the initial Rydberg state with the closest hydrogenic manifold. The results from Fig. 8.5 indicate that the nature of l-mixing collision is

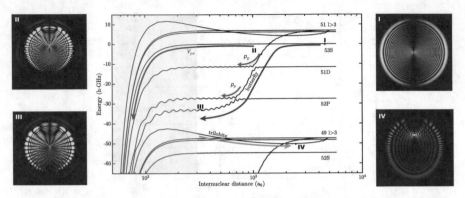

Fig. 8.5 Rydberg-ground-state atom energy landscape for l-mixing collisions. The reaction starts with a ground-state atom in the 53 S state, and as time elapses, the atom evolves with a probability p_D and p_P into the states (n-2)D and (n-1)P respectively. Also, the atom may cross diabatically the first avoided crossing owing to the butterfly state to stay in the 53S state, which drops because of the ion–atom interaction between the ionic core and the ground-state atom. However, the atom may stay in the butterfly state until reaching the strong-coupling region between the butterfly and trilobite states (shaded light green area). After the ground-state atom encounters the ionic core, it may come back toward the long-range region, and because of the butterfly–trilobite coupling, the atom may end up in the trilobite state. The wave functions for different characteristic positions of the ground-state atom are also shown at the right- and left-hand sides of the main plot

associated with the existence of an interaction between the dominant butterfly state, as the initial channel, and the trilobite state as the final channel, which correlates with the closest hydrogenic manifold. In particular, for a Rydberg-neutral distance $R \sim 300$ a_0, the butterfly and (n-1)P states approach the trilobite state coming from hydrogenic manifold leading to a coupling between these states. However, owing to the oscillatory nature of the trilobite and butterfly states, it is very involved in directly applying a Landau–Zener approach; therefore, a different treatment is required.

The dynamics at avoided crossings between two states, i and j, is described through the P-matrix, which is defined as

$$P_{ij}(R) = \langle \phi_i(R)| \frac{d}{dR} \phi_j(R) \rangle, \tag{8.6}$$

where $\phi_{i,j}(R)$ denotes the wave functions of the involved states. In particular, $\phi_i(R)$ represents the eigenstates of the Rydberg–ground-state atom interaction. In Eq. (8.6) $\langle | \rangle$ denotes the scalar product in the finite-dimensional representation of the Hilbert space associated with the basis chosen to diagonalize the Hamiltonian of Eq. (7.17) through the potential given by Eq. (7.13). The present expression of the P-matrix has been derived assuming a delta-type potential associated with the adiabatic degrees of freedom (the Rydberg electron in our case) and applying the general formalism of Clark [21]. The results of the P-matrix for the butterfly–trilobite states are shown in Fig. 8.6, where the existence of different peaks is noted.

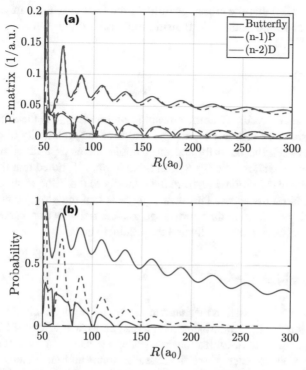

Fig. 8.6 P-matrix and probability as a function of the internuclear distance. (**a**) shows the P-matrix elements $\langle \Phi_i | \frac{\partial}{\partial R} | \Phi_j \rangle$ for the butterfly–trilobite coupling (blue), (n-1)P-trilobite (red), and (n-2)D-trilobite crossing for two different principal quantum numbers, $n = 40$ (dashed lines) and $n = 90$ (solid lines). (**b**) Probability for non-adiabatic transitions associated with the P-matrix displayed in panel (**a**), calculated as $e^{-2\pi\xi}$, $\xi = \Delta/(8vP)$ [21], where v denotes the velocity of the particle, Δ is the energy difference between the adiabatic states, and P denotes the P-matrix. (**c**) Probability of finding the Rydberg atom in different states after an ion–neutral collision: P state (blue), D-state (yellow), and in the (n-4) trilobite state (red). (Figure reproduced with permission from Ref. [2] Copyright (2020) (American Physical Society))

Each of these peaks represents a potential crossing between the two states involved. The Landau–Zener approximation may be used for treating avoided crossings as a function of a parameter (R in our case), which shows a universal feature in terms of the P-matrix [21]

$$P(R) = \frac{\alpha}{2\Delta} \frac{1}{1 + \left(\frac{\alpha R}{\Delta}\right)^2},   (8.7)$$

where Δ denotes the energy gap between the adiabatic curves, and α represents the difference in the slope of the curves at the crossing point. Thus, if the peaks of the P-matrix can be fitted utilizing a Lorentzian function, then one can conclude that the Landau–Zener approach is adequate. Indeed, for all the peaks appearing in Fig. 8.5, it has been checked that the crossings fulfilled the Landau–Zener universality class. The probability for non-adiabatic transition in terms of the P-matrix will be given by $P_{na} = e^{-2\pi\gamma}$;

$$\gamma = \frac{\Delta}{8v P_{\max}}, \tag{8.8}$$

where v denotes the velocity at the crossing point and $P_{\max}$ is the maximum value of the P-matrix at the crossing point. The results of the probability of non-adiabatic transitions between the butterfly and the trilobite states as a function of the ion–neutral distance are shown in Fig. 8.6. In this Figure, it is noted that the probability of a non-adiabatic transition, i.e., from the butterfly to the trilobite state, increases at shorter ion–neutral distances. This result suggests that the ultimate nature of the l-mixing collisions resides in the short-range ion–neutral–electron interaction, owing to a crossing between the butterfly and the trilobite states.

8.2.2   Chemi-Ionization

Chemi-ionization is a reaction mechanism in which a highly excited atom is ionized after a collision with a ground-state atom [22] and leads to two different product states. Indeed, the chemi-ionization reactions are classified as a function of the nature of the product states. If the final state contains an ionized atom then we talk about Penning ionization (PI), which is described as

$$A^* + A \rightarrow A^+ + A + e^-. \tag{8.9}$$

On the contrary, if the final state is the formation of a molecular ion we use the term associative ionization (AI), which reads as

$$A^* + A \rightarrow A_2^+ + e^-, \tag{8.10}$$

At short internuclear distances $r \lesssim 40$ a$_0$, the electronic cloud of the Rydberg ionic core and the electronic cloud of the valence electron of the neutral atom overlap, which leads to electron exchange and correlation effects. These effects, ultimately, induce a strong interaction between the ionic core and the ground-state atom. In the particular, the short-range interaction leads to two PECs, which correlate at long range with the $A^+ + A$ atomic asymptote. As shown in Fig. 8.7 for Rb*-Rb, the ground-state PEC of the molecular ion is a $^2\Sigma_g$ electronic state labeled as V_g^+, which supports a large number of bound states. These bound states are rovibrational states of the molecular ion A_2^+ (or Rb$_2^+$ for the case in Fig. 8.7). In contrast, the second PEC is a repulsive electronic state $^2\Sigma_u$, labeled as V_u^+, and in a few cases can support a few bound states.

Here, we chose to work with Rb*-Rb to illustrate the physical principle behind associative ionization. The relevant potential energy curves are presented in Fig. 8.8.

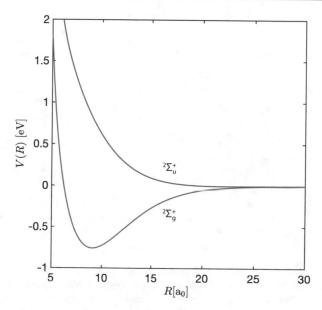

Fig. 8.7 Potential energy landscape relevant for chemi-ionization reactions. At short-range, the ion–neutral interaction for alkali atoms leads to two distinct potential energy curves. One with symmetry *gerade* and the second with *ungerade* symmetry. These calculations are for Rb-Rb$^+$ and have been carried out using *ab initio* quantum chemistry methods

We start assuming that an initial Rydberg state is populated and moving inward, and attached to this Rydberg state are the V_g^+ and V_u^+ electronic states that become close to each other in energy, since they asymptotically correlate with the same atomic asymptote Rb$^+$ + Rb. The probability of ending up in either of the two available (g, u) states as R decreases is approximately the same, $p = 1/2$ since there is no physical reason why one of the PECs should be more prominent than the other. The V_u^+ state attached to the initial Rydberg state, the thick blue solid line in Fig. 8.8, shows a crossing point at R_n with the V_g^+ potential attached to the Rb + Rb$^+$ asymptote. Thus, for $R < R_n$, V_u^+ is the lower boundary for an ionization continuum of potential energy curves, V_g^+, that correlate with the state Rb^2+e$^-$. Furthermore, R_n also represents the internuclear distance for which the energy splitting between V_g^+ and V_u^+ is equal to the Rydberg binding energy in the V_u^+ channel. Hence, for $R < R_n$, it is possible for the Rydberg electron to auto-ionize leading to the formation Rb$_2^+$ in different rovibrational states of the V_g^+ potential. The reaction probability for associative ionization is given by [23]

$$p_{AI} = p \left[1 - \exp \left(-2 \int_{R_k}^{R_n} \frac{W(R) dR}{v(R)} \right) \right], \tag{8.11}$$

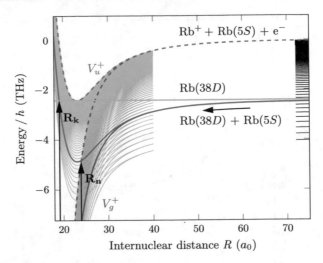

Fig. 8.8 Potential energy curves relevant to the formation of Rb_2^+ from 30D Rydberg state to the ionization threshold. For each Rydberg state, there are two PECs attached, namely, $V_u^+(R)$ showing a primarily repulsive nature. In contrast, the other, labeled $V_g^+(R)$, displays a more attractive nature that supports many bound states. The dark blue line represents the potential curve $V_u^+(R)$, which correlates with the 38D state, and the crossing point of this potential with the $V_g^{+(R)}$ PEC correlates with the onset where auto-ionization can begin into the channel $Rb_2^+ + e$ (red dashed line). This crossing point R_n represents the outer boundary of the associative ionization region. The inner boundary of this auto-ionization region ends at the inner turning point R_k, where the collision energy (solid orange line) is equal to the potential. (Figure adapted with permission from Ref. [2] Copyright (2020) (American Physical Society))

where

$$v(R) = \sqrt{\frac{2(E_k - V_{u(R)})}{\mu}}, \qquad (8.12)$$

is the radial velocity for the Rydberg-neutral collision assuming an s-wave collision. E_k denotes the collision energy, and μ is the reduced mass for the colliding partners. R_k stands for the inner classical turning point associated with the collision energy E_k, as depicted in Fig. 8.8, and $W(r)$ denotes the width of the auto-ionization resonances, which are in general proportional to $1/n^{*3}$.

The predictions based on Eq. (8.11) are shown in Fig. 8.9, where it also shows the experimental reaction probability of Rb_2^+ at different temperatures, densities, and initial states of the Rydberg atoms. It is worth noting that the theoretical results in Fig. 8.9 include the pertinent Landau–Zener probabilities at the crossing given by the energy difference between the initial Rydberg state and the butterfly state. Furthermore, the auto-ionization width has been approximated by the formula $W(R) = 0.9/n^{*3}$ in Eq. (8.11). As a result, the predictions based on Eq. (8.11) yield overall fair agreement with the experimental data. In particular, it is noted that

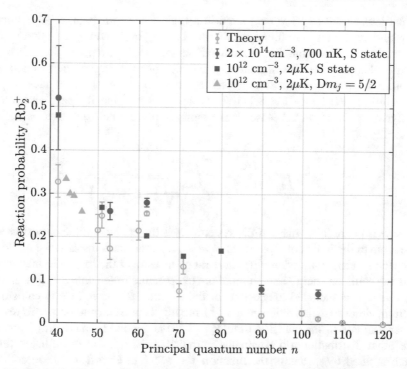

Fig. 8.9 The probability of creating deeply bound Rb_2^+ molecules as a function of the principal quantum number and under different experimental conditions. The experimental data points are calculated from the measured fraction of ionization events that produce Rb_2^+ $P_{Rb2+}(n) = N_{Rb2+}/(N_{Rb+} + N_{Rb2+})$. The error bars show the experimental uncertainty due to the finite electric field within the science chamber that drags away the molecular ions created. As a comparison, two data sets (S [4] and $D, m_j = 5/2$ [24]) from measurements of a thermal cloud added. Empty light blue circles represent the theoretical values for the probability of reaction by means of Eq. (8.11) and using the corresponding Landau–Zener probabilities. (Figure adapted with permission from Ref. [2] Copyright (2020) (American Physical Society))

the theoretical results following Eq. (8.11) show the same trend as the experimental data.

Finally, we would like to point out that the reaction probability associated with the ionization reaction can also be treated through the stochastic ionization approach, assuming the diffusion of the electron in the space of effective quantum numbers n^* [22].

8.2.3 Collision Time for Chemi-Ionization and L-Mixing Collisions: a molecular dynamics approach

The reaction time for a given process can be estimated as the time that the colliding partners take to reach the short-range region where chemistry occurs. In the particular case of a Rydberg atom in a high-density medium, a large number of ground-state atoms are found within the Rydberg orbit, and the reaction will be triggered by the first ground-state atom in reaching the short-range region. Here, one assumes that only the closest atom to the ionic core will trigger the reaction, and the atom–ionic core interaction is given by the long-range charge-induced dipole interaction $\propto R^{-4}$ (see Fig. 8.11). Thus any effect of the Rydberg electron on the dynamics is neglected. Also, it is assumed that the collision time is defined classically as

$$\tau_c = \int_{R_0}^{R_f} \frac{dR}{v(R)} = -\int_{R_0}^{R_f} \frac{dR}{\sqrt{\frac{2}{\mu}\left(E + \frac{\alpha}{2R^4}\right)}}, \tag{8.13}$$

where μ is the reduced mass of the system, R_0 is the position of the closest ground state atom to the ionic core (placed at $R = 0$), R_f represents the distance of closest approach and the point where the short-range takes over the ion–atom interaction. E_k is the collision energy in relation to the temperature of the ultracold gas, and the negative sign is included to account for the fact that $R_f < R_0$. Upon collision, a reaction, either associative ionization or l-mixing, is presumed to occur. To reach a proper estimate of the collision time, Eq. (8.13) is applied to many initial ground-state atom distributions to obtain the cumulative distribution of collision times, which is fitted by means of the function $1 - e^{-t/\tau}$ to extract the mean collision time τ as outlined in Ref. [7]. Then, the collisional loss rate is extracted from the mean collision time (Fig. 8.10).

Figure 8.11 shows a comparison between experimental data and the predictions based on the classical model of Eq. (8.13) for Sr and Rb ultracold gases. As a result, the classical model predicts reasonably well the observed rates in Sr. However, it fails to describe the collision time in Rb. At ultracold temperatures, it is reasonable to assume that $E_k \ll |\alpha/(2R_0^4)|$; hence, the collision time reads as

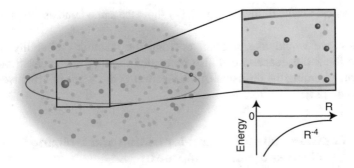

Fig. 8.10 Sketch representing the scenario for the calculation of the collision time based on a molecular dynamics approach. The reaction will be trigger by the closest ground-state atom to the ionic core that interacts through the charge-induced dipole interaction $\propto R^{-4}$

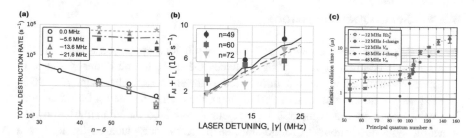

Fig. 8.11 Reaction rates and collision time in strontium and rubidium Bose–Einstein condensates. Reaction rate for inelastic and reactive processes in a Sr BEC as a function of the principal quantum number (**a**), and laser detuning or energy shift (**b**) adapted from Ref. [5]. (**c**) Collision time for inelastic and reactive processes in a Rb BEC as a function of the principal quantum numbers and for different laser detunings [2]. (Panels (**a**) and (**b**) are reproduced with permission from Ref. [5] Copyright (2020) (American Physical Society), whereas panel (**c**) is reproduced with permission from Ref. [2] Copyright (2020) (American Physical Society))

$$\tau_c \propto - \int_{R_0}^{R_f} R^2 dR \propto R_0^3. \tag{8.14}$$

The reaction rate, which is proportional to τ_c^{-1}, thus scales as R_0^{-3}. Then, taking into account that the nearest-neighbor distance for a gas with uniform density is proportional to $\rho^{-1/3}$, the reaction rate will then scale linearly with ρ. The linear dependence on the density is experimentally observed as panels (a) and (b) of Fig. 8.11 show. However, this behavior is not fulfilled in the case of Rb, as shown in panel (c) of Fig. 8.11, where it is observed that up to $n \approx 90$ Eq. (8.13) more or less reproduce the experimental trend, but it fails drastically for $n \geq 90$. As has been shown previously, in the Rb$*$–Rb interaction, the butterfly state plays a crucial role in the l-mixing collisions and associative ionization reactions. Therefore, the collision time should depend on the presence of the butterfly state and its properties, which explains why Eq. (8.13) does not describe the collision time of the processes properly under consideration in Rb.

Comparing the experimental results for the collision time for Sr and Rb with the theoretical predictions based on a molecular dynamics (MD) model, we conclude that MD simulations are not appropriate for systems having a e-ground-state atom shape resonance. The reason is that the MD simulations do not capture the totality of the physics; in particular, the role of the butterfly state is not included. However, for systems such as Sr, for which the e-ground-state atom is dominated by the s-wave scattering length, the MD model predicts the behavior of the collision time correctly as a function of the density and principal quantum number.

8.2.4 Some Open Questions

In Sect. 8.2, it has been shown that l-mixing collisions appear as a consequence of a strong interaction at $R \sim 300a_0$ between the trilobite and butterfly states. In contrast, associative ionization is dominated by the short-range behavior $R \lesssim 40\,a_0$ of the ion–ground-state atom interaction. Next, the collision time for both reactions has been measured and compared with calculations based on MD simulations. As a result, systems without e-ground-state shape resonance, such as Sr, lead to a proper description of the physics. However, it does not portray the collision time for systems that show a e-ground-state shape resonance, such as Rb. Such a discrepancy is probably because the presence of a butterfly state leads to an "acceleration" of the Rydberg electron toward the ionic core; hence, one would expect a shorter collision time in Rb than in Sr. However, the experimental findings for both systems lead to collision times of the same order of magnitude (see Fig. 8.11). How is it possible? The answer may lay in the short-range couplings between different states or the role of quantum reflection due to the steepness of the butterfly potential.

The collision of a low-energy electron with an Sr atom does not lead to any shape resonance; hence, the Sr*–Sr interaction does not show any butterfly state. So, how do l-mixing collisions appear? The answer to this question may be at $50 \lesssim R \lesssim 300$, where the charge-induced dipole interaction between the ionic core and the ground-state atom induces strong couplings between different states. Another possibility may be the existence of novel reaction pathways for l-mixing collisions.

Finally, we would like to end this chapter by emphasizing that the chemistry of Rydberg-neutral systems at ultracold temperatures is vibrant and relatively unexplored, which in our view is due to the interest of the community in its applications rather than in a deep understanding of it. We hope that the reader has seen in this chapter an opportunity to realize that ultracold and cold Rydberg atoms are more than something useful for the development of quantum technologies. Rydberg atoms were, are, and will be an essential topic of atomic, molecular, and optical physics.

References

1. Saffman M, Walker TG, Molmer K (2010) Quantum information with Rydberg atoms. Rev Mod Phys 82:2313
2. Schlagmüller M, Liebisch TC, Engel F, Kleinbach KS, Böttcher F, Hermann U, Westphal KM, Gaj A, Löw R, Hofferberth S, Pfau T, Pérez-Ríos J, Greene CH (2016) Ultracold chemical reactions of a single Rydberg atom in a dense gas. Phys Rev X 6:031020. https://link.aps.org/doi/10.1103/PhysRevX.6.031020
3. Fermi E (1934) Sopra lo spostamento per pressione delle righe elevate delle serie spettrali. Nouvo Cimento 11:157
4. Gaj A, Krupp AT, Balewski JB, Löw R, Hofferberth S, Pfau T (2014) From molecular spectra to a density shift in dense Rydberg gases. Nat Commun 5(1):4546. https://doi.org/10.1038/ncomms5546
5. Whalen JD, Camargo F, Ding R, Killian TC, Dunning FB, Pérez-Ríos J, Yoshida S, Burgdörfer J (2017) Lifetimes of ultralong-range strontium Rydberg molecules in a dense Bose-Einstein condensate. Phys Rev A 96:042702. https://link.aps.org/doi/10.1103/PhysRevA.96.042702

6. Balewski JB, Krupp AT, Gaj A, Hofferberth S, Löw R, Pfau T (2014) Rydberg dressing: understanding of collective many-body effects and implications for experiments. New J Phys 16(6):063012

7. Schlagmüller M, Leibisch TC, Nguyen H, Lochead G, Engel F, Böttcher F, Westphal KM, Kleinbach KS, Löw R, Hofferberth S, Pfau T, Pérez-Ríos J, Greene CH (2016) Probing and electron scattering resonance using Rydberg molecules within a dense and ultracold gas. Phys Rev Lett 116:053001

8. Liebisch TC, Schlagmüller M, Engel F, Nguyen H, Balewski J, Lochead G, Böttcher F, Westphal KM, Kleinbach KS, Schmid T, Gaj A, Löw R, Hofferberth S, Pfau T, Pérez-Ríos J, Greene CH (2016) Controlling Rydberg atom excitations in dense background gases. J Phys B-At Mol Opt Phys 49(18):182001. http://doi.org/10.1088/0953-4075/49/18/182001.

9. Löw R, Weimer H, Nipper J, Balewski JB, Butscher B, Büchler HP, Pfau T (2012) An experimental and theoretical guide to strongly interacting Rydberg gases. J Phys B-At Mol Opt Phys 45(11):113001

10. Li W, Pohl T, Rost JM, Rittenhouse ST, Sadeghpour HR, Nipper J, Butscher B, Balewski JB, Bendkowsky V, Löw R, Pfau T (2011) A homonuclear molecule with a permanent electric dipole moment. Science 334(6059):1110. http://doi.org/10.1126/science.1211255, http://science.sciencemag.org/content/334/6059/1110.full.pdf, http://science.sciencemag.org/content/334/6059/1110

11. Kuhn H (1934) Xciii. Pressure shift and broadening of spectral lines. Lond Edinb Philos Mag J Sci 18(122):987. https://doi.org/10.1080/14786443409462572

12. Kuhn H, Lindemann FA (1937) Pressure broadening of spectral lines and van der Waals forces i—influence of argon on the mercury resonance line. Proc R Soc A-Math Phys 158(893):212. https://doi.org/10.1098/rspa.1937.0015

13. Collins GW (1989) The fundamentals of stellar astrophysics II. W. H. Freeman and Co., New York

14. Gallagher TF, Edelstein SA, Hill RM (1975) Collisional angular momentum mixing in Rydberg states of sodium. Phys Rev Lett 35:644. https://link.aps.org/doi/10.1103/PhysRevLett.35.644

15. Gallagher TF, Edelstein SA, Hill RM (1977) Collisional angular-momentum mixing of Rydberg states of Na by He, Ne, and Ar. Phys Rev A 15:1945. https://link.aps.org/doi/10.1103/PhysRevA.15.1945

16. Gallagher TF, Cooke WE, Edelstein SA (1978) Collisional angular momentum mixing of f states of Na. Phys Rev A 17:904. https://link.aps.org/doi/10.1103/PhysRevA.17.904

17. Hugon M, Sayer B, Fournier PR, Gounand F (1982) Collisional depopulation of rubidium Rydberg levels by rare gases. 15(15):2391. http://doi.org/10.1088/0022-3700/15/15/016

18. Olson RE (1977) Theoretical excitation transfer cross sections for Rydberg Na($n^2d \rightarrow n^2f$) transitions from collision with He, Ne, and Ar. Phys Rev A 15:631. https://link.aps.org/doi/10.1103/PhysRevA.15.631

19. Gersten JI (1976) Theory of collisional angular-momentum mixing of Rydberg states. Phys Rev A 14:1354. https://link.aps.org/doi/10.1103/PhysRevA.14.1354

20. Hickman AP (1978) Theory of angular momentum mixing in Rydberg-atom-rare-gas collisions. Phys Rev A 18:1339. https://link.aps.org/doi/10.1103/PhysRevA.18.1339

21. Clark CW (1979) The calculation of non-adiabatic transition probabilities. Phys Lett A 70(4):295

22. Mihajlov AA, Srećković VA, Ignjatović LM, Klyucharev AN (2012) The chemi-ionization processes in slow collisions of Rydberg atoms with ground state atoms: mechanism and applications. J Clust Sci 23(1):47. https://doi.org/10.1007/s10876-011-0438-7

23. Miller WH (1970) Theory of penning ionization. I. Atoms. J Chem Phys 52(7):3563. https://aip.scitation.org/doi/abs/10.1063/1.1673523

24. Krupp AT, Gaj A, Balewski JB, Ilzhöfer P, Hofferberth S, Löw R, Pfau T, Kurz M, Schmelcher P (2014) Alignment of d-state Rydberg molecules. Phys Rev Lett 112(14):143008. http://dx.doi.org/10.1103/PhysRevLett.112.143008

Hybrid Atom–Ion Systems

9

Ion–neutral interactions play a major role in reactions relevant for biochemistry [1], have an impact on the chemistry of the interstellar medium [2], and contribute to different reactions in plasma physics [3]. Therefore, revealing the ultimate nature of ion–neutral interactions is crucial for chemical sciences. Luckily enough, hybrid atom–ion systems are specially designed for the task. These hybrid systems are the result of combining the best of two worlds, as shown in Fig. 9.1, ultracold atoms from ultracold physics and cold ions. Additionally, atom–ion hybrid systems, owing their versatility and degree of control, have potential applications in high-precision spectroscopy [4], quantum information [5–8], condensed matter physics [9, 10], and cold chemistry [11, 12]. In particular, cold chemistry in hybrid systems is the topic of the present chapter.

Cold chemistry is a relatively new discipline within chemical physics that focuses on the study of chemical reactions at cold temperatures, i.e., $1\,\mathrm{mK} \lesssim T \lesssim 1\,\mathrm{K}$. The cold chemistry community is divided into two different groups based on the experimental techniques:

- Experiments based on molecular beams. These experiments study low-energy collisions between molecules in crossed beam experiments.
- Atom–ion hybrid traps. These experiments rely on the possibility of having ultracold atoms and cold ions in the same spot. Therefore, the development of compatible trapping techniques is essential. These experiments are dedicated to the study of chemical reactions in which one of the reactants is an ion.

In this chapter, we focus on the physics and chemistry relevant for experiments in hybrid traps. However, it is worth noticing that these two cited techniques complement each other to reach the best possible knowledge of chemical reactions at cold temperatures.

© Springer Nature Switzerland AG 2020
J. Pérez Ríos, *An Introduction to Cold and Ultracold Chemistry*,
https://doi.org/10.1007/978-3-030-55936-6_9

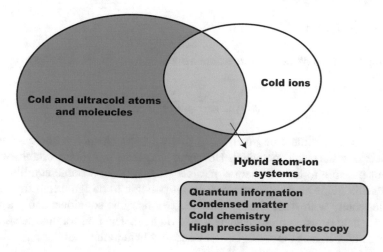

Fig. 9.1 Hybrid atom–ion systems emerge in the interplay between cold ions and ultracold atoms. Hybrid atom–ion systems may have implications in all the different research areas enumerated in the rectangle with rounded vertices

9.1   What Is a Hybrid System?

Before going into detail about the trapping of ions and how they behave in a trap, it is worth dedicating a few lines to the concept of hybrid systems, which certainly goes beyond the atom–ion systems. A hybrid system is a system composed of two or more interacting subsystems with different quantum identities (e.g., different numbers of internal degrees of freedom, or the nature of these degrees of freedom is distinct). For instance, an ultracold cloud of atoms coupled with a nanoresonator, or a gas of molecules coupled with a nanophotonic crystal, are hybrid systems. In our case, an ultracold cloud of atoms coupled with cold ions defines a hybrid system.[1] Actually, these systems are generally classified as quantum hybrid systems. The most intriguing point of hybrid quantum systems is that the interaction between the parts of the system may lead to modifications of the internal energy levels of their constituents. Therefore, a hybrid quantum system is a unique arena for studying different properties of its parts as a function of the strength of the coupling, which may be helpful for building a quantum simulator.

---

[1]This system is usually considered as a quantum hybrid system. However, this chapter mostly focuses on atom–ion systems from a classical and semi-classical framework. Thus, we have decided to use the term hybrid systems instead of the quantum hybrid system.

9.2   Ion Traps

Trapping a charged particle in three-dimensional space requires at least a quadrupole potential,

$$\phi(\mathbf{r}) = ax^2 + by^2 + cz^2, \tag{9.1}$$

where a, b and c must be positive. In this way, the charged particle should be confined in a harmonic potential. However, Earnshaw's theorem establishes that it is not possible to hold a charged particle (statically) in a stable equilibrium by electrostatic fields alone. Therefore, it is not possible to design a trap for charged particles solely with static fields. There are two possible solutions: apply a strong magnetic field or use a time-dependent electric field. The first solution leads to the Penning trap whereas the second is the basic idea behind the Paul trap.

9.2.1   Penning Trap

In the 1930s, Frans Michel Penning used an axial magnetic field in a discharge tube to increase the path length of the electrons [13]. This very same idea could be applied to trap charged particles [14, 15], as was first realized in by Dehmelt in 1959, thus establishing the foundations of the Penning trap. The main idea behind the Penning trap is to use a strong magnetic field in the axial direction to ensure the trapping of the charged particles in the radial direction. The trapping in the axial direction is achieved by using two end-cap electrodes and a ring electrode at the same potential, V_0, as shown in Fig. 9.2. This configuration leads to a quadrupole potential

$$\phi(\mathbf{r}) = A(V_0)\left(-\frac{r^2}{2} + z^2\right), \tag{9.2}$$

where the amplitude of the potential, $A(V_0)$, depends directly on the voltage applied to the electrodes.

Penning traps are extensively applied in experiments dedicated to measuring exotic nuclear decays [16–18] owing its extraordinary precession on determining the mass of trapped particles. Indeed, this makes Penning traps a great tool for studying fundamental physics [19–22]. For instance, Penning traps are an essential part of the ambitious ALPHA [23] and ATRAP [24] experiments to trap anti-hydrogen atoms.

9.2.2   Paul Trap

In the 1950s, the electric quadrupole mass filter was introduced by Wolfgang Paul. This new invention changed the paradigm of mass spectrometry into the current

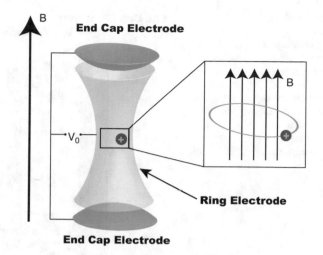

Fig. 9.2 Schematic of a Penning trap. The end-cap electrodes at the same potential V_0 confines a charged particle axially, whereas the axial magnetic field leads to the radial confinement of the charged particles. The inset shows the radial confinement due to the cyclotron motion of a charged particle in the presence of a magnetic field along the axial direction

high-precision mass spectrometry. Surprisingly, the same design is applicable to trap charged particles, as Paul recognized [25]. Since then, the use of Paul traps in atomic and molecular physics has paved the way for basal concepts of ultracold physics, e.g., laser cooling of ions[2] [26].

In a Paul trap the charged particles are trapped radially by a quadrupole RF time-dependent electric field, whereas the axial restoring force is due to the presence of end-cap electrodes with a positive DC potential.[3] The potential energy of an ion at a position $\mathbf{r} = (x, y, z)$ in a Paul trap is given by

$$\phi(\mathbf{r}, t) = \frac{U_{DC}}{2}\left(\alpha x^2 + \beta y^2 + \gamma z^2\right) + \frac{U_{RF}}{2}\cos\left(\Omega_{RF}t\right)\left(\alpha_{RF}x^2 + \beta_{RF}y^2 + \gamma_{RF}z^2\right),$$
$$(9.3)$$

where Ω_{RF} is the RF frequency, and U_{DC} and U_{RF} are factors that depend on the DC and RF potentials respectively. The first term in Eq. (9.3) corresponds with a general quadrupole static field configuration, whereas the second term is the potential due by a three-dimensional RF field. There are different Paul trap configurations, but the most common configuration employed in cold chemistry experiments is the linear Paul trap. Similarly, there are several possible configurations for a linear Paul trap, but here we work with the configuration that satisfies $\alpha = \beta = -1/2 = -\gamma/2$, $\alpha_{RF} = -\beta_{RF}$, and $\gamma_{RF} = 0$, and it is shown in Fig. 9.3.

---

[2]The first demonstration of the idea of laser cooling was done for ions in the pioneering work in Neuhauser et al. [26].

[3]This configuration is also known as a linear Paul trap.

Fig. 9.3 A linear Paul trap. A two-dimensional radio-frequency quadrupole field guarantees the radial confinement of the charged particle, whereas the axial trapping is due to the end-cap electrodes. The same direct current (DC) potential is applied to the end-cap electrodes and to the rods

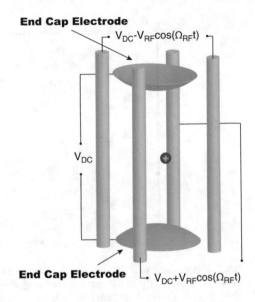

End Cap Electrode
$V_{DC} - V_{RF}\cos(\Omega_{RF}t)$

V_{DC}

End Cap Electrode
$V_{DC} + V_{RF}\cos(\Omega_{RF}t)$

Therefore, the potential energy of an ion in a linear Paul trap reads as

$$\phi(\mathbf{r}, t)$$

$$= \frac{1}{4}\left[(-U_{DC} + 2U_{RF}\cos(\Omega_{RF}t))\,x^2 + (-U_{DC} - 2U_{RF}\cos(\Omega_{RF}t))\,y^2 + 2U_{DC}z^2\right],$$

$$(9.4)$$

where the presence of an harmonic potential in z is noted that guarantees the axial confinement of the ion. However, the radial trapping is only guaranteed under particular conditions regarding the values of U_{DC}, U_{RF} and Ω_{RF}, which defines the so-called stability region [27]. For the reader interested in knowing more about the stability region for different Paul trap configurations we recommend Ref. [27].

9.3 Dynamics of Ions in a Paul Trap

Let us assume a single positively charged ion in a linear Paul trap. The ion will feel the potential given in Eq. (9.4), which defines dynamics of the ion within the trap. Thus, applying Newton's second law leads to the equations of motion of the ion as

$$\frac{dx^2}{dt^2} + (a_x + 2q_x\cos(\Omega_{RF}t))\frac{\Omega_{RF}^2}{4}x = 0;$$

$$\frac{dy^2}{dt^2} + (a_x + 2q_y\cos(\Omega_{RF}t))\frac{\Omega_{RF}^2}{4}y = 0;$$

$$\frac{dz^2}{dt^2} + \frac{a_z \Omega_{RF}^2}{4} z = 0; \tag{9.5}$$

where $a_x = a_y = -\frac{2eU_{DC}}{m\Omega_{RF}^2}$, $a_z = \frac{4eU_{DC}}{m\Omega_{RF}^2}$, $q_x = -q_y = \frac{2eU_{RF}}{m\Omega_{RF}^2}$. Here, m is the mass of the ion and e is the absolute value of the electron charge. In Eqs. (9.5) it is noted that the RF field induces a time-dependent harmonic potential in the radial direction, as was anticipated in Sect. 9.2.2. The equations of motion of the ion in a Paul trap shown in Eqs. (9.5), through the change of variables, $\xi = \Omega_{RF}t/2$, transform into the Mathieu differential equation

$$\frac{dr_i^2}{d\xi^2} + (a_i + 2q_i \cos(2\xi))r_i = 0, \tag{9.6}$$

where $i \equiv \{i, 2, 3\}$ and $r_1 = x$, $r_2 = y$, $r_3 = z$, $a_1 = a_x$, $a_2 = a_y$, $a_3 = a_z$, $q_1 = q_x$, $q_2 = q_y$ and $q_3 = q_z = 0$. Eqs. (9.6)[4] can be solved numerically, for a given configuration of a linear Paul trap, and the results for the radial and axial distance of the ion as a function of time are shown in Fig. 9.4. Panel (b) of the Figure shows the expected harmonic motion of the ion in the axial direction, whereas panel (a) shows a rather complex bounded trajectory on the radial coordinate of the ion. In particular, the radial coordinate shows a fast oscillatory motion superimposed onto a slower harmonic motion. The slow oscillatory behavior in the radial coordinate is the so-called *secular motion*, whereas the rapidly oscillatory motion is known as *micromotion*.

Let us assume an ion in a linear Paul trap, in which $q_i^2 \ll 1$ and $|a_i| \ll 1$; thus, the trajectory is stable [28]. In this scenario it is possible to solve Eqs. (9.5) in first order in q_i, which solutions read as

$$r_i(t) \approx r_i^{(0)} \cos(\omega_i t + \phi_i)\left(1 + \frac{q_i}{2} \cos(\Omega_{RF}t)\right). \tag{9.7}$$

$r_i^{(0)}$ is the initial position of the ion, ϕ_i is a phase that depends on the initial conditions of the ion, and $\omega_i \approx \frac{\Omega_{RF}}{2}\sqrt{a_i + \frac{q_i^2}{2}}$,[5] and in general $\omega_i \ll \Omega_{RF}$. The ion trajectory shown Eq. (9.7) can be seen as a harmonic trajectory at frequency ω_i with a fast oscillating amplitude at frequency Ω_{RF}. The harmonic part corresponds to the secular motion, whereas the oscillating amplitude is the micromotion, as shown in panels (a) and (b) of Fig. 9.4. The micromotion is a side effect of having an RF electric field and it is always present in a Paul trap, unless the ion is in the exact center of the trap, since the amplitude of the micromotion depends on the

---

[4]Equation (9.6) are Mathieu differential equations that can be solved analytically by means of a recurrence relation leading to a convergent continued fraction of the relevant parameters [27, 28].

[5]In general, the solution of Mathieu's equations for an ion in a Paul trap is characterized by $\beta_i = \sqrt{a_i + \frac{q_i^2}{2}}$, which is related to the stability of the solution [27, 28].

Fig. 9.4 Ion trajectory and its kinetic energy in a Paul trap. Radial distance of the ion as a function of time shown in panel (**a**). In panel (**b**), the axial distance of the ion as a function of time is presented. In panel (**c**), the kinetic energy of the ion as function of time is shown. These simulations are performed for a $^{138}\mathrm{Ba}^+$ ion in a Paul trap with $a_x = a_y = -0.0035 = -a_z/2$; $q_x = -q_y = 0.1761$ and $q_z = 0$

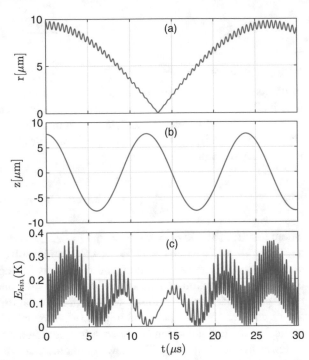

radial distance of the ion to the center of the trap, as expected from Eq. (9.7). The preceding analysis can be extended to the quantum realm to describe the quantum mechanical motion of trapped ions [28].

The presence of micromotion in the ion trap also affects the energy distribution of the ions. Panel (c) of Fig. 9.4 shows the kinetic energy of the ion as a function of time, where it is observed that the kinetic energy of the ion has a fast varying component on top of a well-defined harmonic one. The harmonic component corresponds to the frequency of motion on the axial coordinate, whereas the fast oscillatory component comes from the micromotion. Indeed, we note that the energy contribution of the micromotion depends on the radial distance of the ion, as expected from Eq. (9.7). In particular, the closer the ion from the trap center, the smaller the effect of the micromotion, although its relative effect stays the same, since the amplitude of the micromotion varies as $r_i^{(0)} \frac{q_1}{2}$. Therefore, in atom–ion hybrid trap experiments it is of capital importance to control the presence of stray electric fields to avoid the enhancement of the micromotion of the ion in a Paul trap [29, 30]. Indeed, in the presence of stray electric fields the ion trajectory $r(t)$ does not fulfil Eq. (9.7), a situation that is generally known as *excess of micromotion*.

9.3.1 Is It Possible to Define Temperature for Ions in a Trap?

Ions in a Paul trap are continuously under the action of the RF field, which leads to a time-dependent potential. Thus, the energy is not conserved and the ions are out of equilibrium. Therefore, formally, it is not possible to define the temperature of an ion, because temperature is only defined under conditions of thermal equilibrium [31]. However, it is possible to define the average kinetic energy of the ion in each coordinate as

$$\langle E_{\text{kin},i} \rangle = \frac{1}{2} m \langle \left(\frac{dr_i(t)}{dt} \right)^2 \rangle = \frac{m}{2T_i} \int_0^{T_i} \left(\frac{dr_i}{dt} \right)^2 dt, \tag{9.8}$$

where $T_i = 2\pi/\omega_i$ is the period of the secular motion, and taking into account Eq. 9.7 yields

$$\langle E_{\text{kin},i} \rangle \approx \frac{m(r_i^{(0)})^2}{4} \left(\omega_i^2 + \frac{q_i^2 \Omega_{RF}^2}{8} \right). \tag{9.9}$$

The average kinetic energy, Eq. (9.9), of the ion can be viewed as the energy available to exchange with a second body upon a collision. Therefore, the average kinetic energy of the ion can be taken as an effective "temperature" of the ion in a Paul trap. The first term of Eq. (9.9) represents the kinetic energy of the secular motion, whereas the second term stands for the kinetic energy of the micromotion associated to each coordinate (keeping in mind, $q_3 = 0$). Indeed, taking into account that $\omega_i^2 \approx \frac{\Omega_{RF}^2 q_i^2}{8}$ for $a \ll q_i^2$, the total average kinetic of an ion is given by

$$\langle E_{\text{kin}} \rangle = \frac{5}{2} k_B T. \tag{9.10}$$

A way to understand this result is to think that secular motion is defined in three dimensions, whereas micromotion is defined only in the radial direction. Actually, the equipartition theorem establishes that every degree of freedom contributes with a factor $\frac{1}{2} k_B T$ to the temperature of the system. Thus, secular motion will contribute with a total energy of $\frac{3}{2} k_B T$, whereas micromotion only contributes with two degrees of freedom as $k_B T$, which makes the total energy shown in Eq. (9.9).

One can also look at the distribution of velocities of the ion in the Paul trap and discover that the velocities of the ion do not follow the expected Maxwell–Boltzmann distribution for a system in thermal equilibrium. Therefore, this is another way of establishing that temperature is not formally well-defined for an ion in a Paul trap. However, despite these strong arguments against the use of temperature of ions or the ions, the community uses it and it has become part of the argot of the field, although it does not have the meaning of a physical temperature. Therefore, in this book, every time we refer to ions in a trap we will be talking about the average kinetic energy instead of temperature.

Fig. 9.5 False-color fluorescence images of the spatial arrangement of 25 laser-cooled Ca+ ions in a Coulomb crystal recorded under varying trapping conditions. (Figure reproduced with permission from Ref. [33] Copyright (2020). (Royal Society of Chemistry))

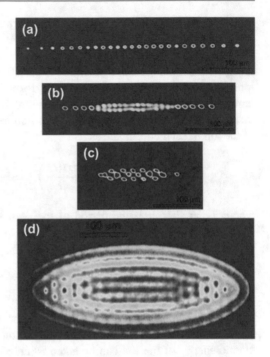

9.3.2 Coulomb Crystals

Ions in a trap behave as a plasma, but with some peculiarities [32]. First, only positively charged particles are present, which is different from plasmas where the number of positively and negatively charged particles is the same. Second, the positively charged particles are trapped and the trapping potential can be visualized as a uniform density of negatively charged particles surrounding the positively charged ions. Therefore, ions in a trap may be called single-component plasma [32].

At sufficiently low temperatures the kinetic energy of the ions becomes negligible in comparison with the Coulomb interaction between ions, and as result, the single component plasma may crystallize. This crystal phase of charged particles held by a trapping potential is known as a Coulomb crystal or Wigner crystals, which are shown in Fig. 9.5. Now, the question is, when this does this transition occur? To answer this question, it is convenient to introduce the Coulomb coupling parameter as[6]

$$\Gamma_C = \frac{V}{E_{\text{kin}}} = \frac{(Ze)^2}{4\pi\epsilon_0 r_{WS} m \langle v^2 \rangle},\tag{9.11}$$

---

[6]The Coulomb coupling parameter is generally defined in terms of T as $\Gamma_C = \frac{V}{E_{\text{kin}}} = \frac{(Ze)^2}{4\pi\epsilon_0 r_{WS} k_B T}$, which is called the temperature of the ions; however, it is clearer to use the average kinetic of the ions.

where ϵ_0 is the electric constant of vacuum, and r_{WS} is the Wigner–Seitz radius, which defines the average distance between ions and it relates to the density, ρ, as $4\pi r_{WS}^3/3\rho = 1$. Plasmas with $\Gamma_C > 2$ are considered to be in the strong-coupling regime and above 178 the crystal phase appears [34]. Interestingly enough, it is possible to laser cool ions in a trap to reach the desired Γ_C value, thus forming a Coulomb crystal. For instance, assuming that the ions are laser cooled down to an average kinetic energy of 10 mK, the crystal phase will appear when $r_{WS} \lesssim 10^{-5}$ m. Thus, the typical density of ions should be $\gtrsim 2 \times 10^{14}$ m^{-3}. These requirements are fulfilled in standard linear Paul traps, and in fact, Coulomb crystals are frequently used in cold chemistry experiments to reveal the reaction rates under different conditions [35–37].

9.4   Atom–Ion Collisions

One of the main goals of atom–ion hybrid systems is the study of charged–neutral interactions, and the different chemical processes relevant to these hybrid systems. Previously in this chapter, we have introduced ion traps, how these work, and have described the dynamics of charged particles on them. In Chap. 4, we introduced the different techniques for creating and manipulating ultracold atoms and molecules. Therefore, we are ready to study atom–ion collisions, which is a major goal in the research of atom–ion hybrid systems. Indeed, the understanding of atom–ion collisional processes is mandatory to exploit the capabilities that hybrid systems bring to the paradigm of quantum information and many-body physics [11].

9.4.1   Atom–Ion Interactions

Interactions are pivotal to understand any collisional process. In the case at hand, beyond the region where the valence electrons of the ion and atom overlap, the electric field of the ion $\mathbf{F}$ induces a dipole moment, $\mathbf{d} = \alpha\mathbf{F}$, on the atom, which is proportional to its polarizability α. From classical electromagnetism it is known that the interaction of a dipole in an electric field is given by [38]

$$V = -\mathbf{d} \cdot \mathbf{F}, \tag{9.12}$$

which implies that the atom–ion interaction energy should be $\propto F^2$ and hence $\propto R^{-4}$. This is a classical argument in understanding the R-dependence of the atom–ion interaction; however, a quantal treatment is needed to accurately describe the atom–ion interaction. In particular, quantum mechanically, the long-range interaction between an ion and an atom is due to the charge-induced dipole, and this effect can be obtained through the second-order perturbation theory leading to [39, 40]

Fig. 9.6 Potential energy curves for (LiYb)$^+$. (Figure reproduced with permission from Ref. [41] Copyright (2020) (American Physical Society))

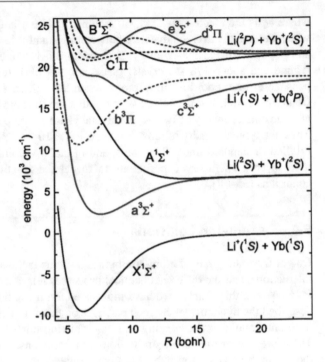

$$V(R) = -\frac{C_4}{R^4} = -\frac{e^2\alpha}{2(4\pi\epsilon_0)^2 R^4}. \tag{9.13}$$

In general we use Eq. (9.13) in atomic units, which reads as

$$V(R) = -\frac{C_4}{R^4} = -\frac{\alpha}{2R^4}. \tag{9.14}$$

At interatomic distances in which the electronic clouds from the ion and atom start to overlap, the atom–ion interaction needs to be described by quantum chemistry methods to properly account for the electronic correlation and exchange between electrons. An example for atom–ion interaction is shown in Fig. 9.6, where the different potential energy curves for (LiYb)$^+$ are displayed. It is worth noting that each of the potential energy curves correlates with a given atomic asymptote, i.e., either Li - Yb$^+$ or Li$^+$ - Yb. This rich structure of the molecular states leads to different unique reactions, as we explain in the next section.

9.4.2 Langevin Rate

The Langevin model [42] is a classical approach to ion–neutral reactions, and it is very useful for the study of cold chemistry in atom–ion hybrid systems. In particular, it assumes that only trajectories with an impact parameter smaller than a threshold value, b_L, lead, with unity probability, to the reaction under consideration. In other words, every single trajectory within a certain acceptance region, when reaching the short-range interaction, leads to a reaction. For this reason, the Langevin model is considered the first capture model for scattering problems [45].

Classically, a collision takes place if and only if the kinetic energy is higher than the centrifugal barrier. For atom–ion collisions the long-range tail of the interaction potential reads as (in atomic units)

$$V_l(R) = -\frac{\alpha}{2R^4} + \frac{l(l+1)}{2\mu R^2},$$

(9.15)

where l is the centrifugal quantum number. The potential in Eq. (9.15) shows s a maximum at

$$R^* = \sqrt{\frac{2\mu\alpha}{l(l+1)}},$$

(9.16)

and its value is

$$V_l(R^*) = \frac{l^2(l+1)^2}{8\mu^2\alpha}.$$

(9.17)

At this point, we are ready to find the maximum l value allowed for a given kinetic energy, E_k, solving $E_k = V_l(R^*)$ for l. However, the Langevin model relies on the threshold value of the impact parameter rather than the quantal angular momentum. Thus, it is necessary to find the relationship between the quantal angular momentum and the impact parameter. This relationship can be found by looking into Sect. 2.1.4, where we introduced the classical cross section, and with it, the concept of impact parameter. In particular, at large distances, the interaction energy is negligible with respect to the kinetic energy $E_k = k^2/2\mu$ (in atomic units), and since the angular momentum is conserved, one finds

$$|\mathbf{b} \times \mathbf{p}| = |\mathbf{L}|.$$

(9.18)

By definition $\mathbf{p}$ and $\mathbf{b}$ are orthogonal, thus, assuming atomic units and keeping in mind that $p = k$, Eq. (9.15) yields

$$bk = L.$$

(9.19)

Squaring both sides of this equation and employing the eigenvalues of the angular momentum operator we find

$$b^2k^2 = l(l+1).$$

(9.20)

Finally, plugging Eq. (9.20) into Eq. (9.17) and solving $E_k = V_l(R^*)$ for l yields

$$b_L^2 = 2\alpha \frac{\sqrt{\mu}}{k}, b_L^2 = \frac{\sqrt{4\alpha\mu}}{k} \tag{9.21}$$

which is the threshold value for the impact parameter and is generally called the Langevin impact parameter.

In the Langevin model, trajectories with $b \leq b_L$ lead to the reaction, whereas for $b > b_L$ the reaction does not occur. This can be formally expressed through the Langevin opacity function

$$P_L(b) = \begin{cases} 1 & b \leq b_L \\ 0 & b > b_L, \end{cases} \tag{9.22}$$

and with it, the Langevin cross section, which is calculated through Eq. (2.9) and reads as

$$\sigma_L = 2\pi \int_0^\infty P_L(b)bdb = \pi b_L^2 = \pi \sqrt{\frac{2\alpha}{E_k}}. \tag{9.23}$$

It is also possible to calculate the Langevin reaction rate as

$$k_L = \sigma_L v = 2\pi \sqrt{\frac{\alpha}{\mu}}, \tag{9.24}$$

which is energy independent. Therefore, the Langevin model predicts a rate constant independent of the collision energy. This is one of the most striking results in charge–neutral collisions, but it is also the most successful model for explaining such reactions.

Finally, we would like to point out that the Langevin model is more than a useful model. In fact, the pioneering ideas of Langevin motivated a novel approach to reaction rate theory called capture models [43–46]. Capture models apply to very different long-range interactions and even to three-body collisions (or two-step reactions).

9.4.3 Elastic Atom–Ion Collisions

In Sect. 2.2.1 we have seen that the quantum mechanical elastic cross section is defined as

$$\sigma_{el}(E) = \sum_l \sigma_l(E) = \frac{4\pi}{k^2} \sum_l (2l + 1) \sin^2(\delta_l(k)), \tag{9.25}$$

where $\delta_l(k)$ is the phase shift due to the scattering potential for a given partial wave l and momentum k. This expression (and its multi-channel analog introduced in Chap. 5) is the cornerstone of neutral–neutral collisions at cold and ultracold temperatures. However, for atom–ion interactions, at cold temperatures ($1\,\text{mK} \lesssim T \lesssim 1\,\text{K}$), a large number of partial waves contribute to the scattering observables. Therefore, it may be possible to use a semi-classical approach to describe the elastic cross section for hybrid atom–ion systems.

Within the semi-classical scattering framework the phase shift is given by [47]

$$\delta_l \simeq -\frac{\mu}{\hbar^2} \int_{R_0}^{\infty} \frac{V(R)dR}{\sqrt{k^2 - (l+1/2)^2/R^2}} = \frac{\mu}{\hbar^2} \int_{l/k}^{\infty} \frac{\alpha\, dR}{2R^4 \sqrt{k^2 - (l+1/2)^2/R^2}},$$
(9.26)

where the Langer correction[7] is assumed and $R_0 \approx l/k$ represents the outer classical turning point. Assuming an atom–ion long-range interaction as $V(R) = -\alpha/(2R^4)$, and $l \gg 1$, Eq. (9.26) can be solved yielding

$$\delta_l(k) \simeq \frac{\pi \mu \alpha k^2}{8\hbar^2 l^3}.$$
(9.27)

Finally, plugging Eq. (9.27) into Eq. (9.26) yields the semi-classical cross section. However, this step requires some thought as shown below.

Let us start by comparing the semi-classical opacity, $\sigma_l(E)$, with its quantal counterparts. The semi-classical and quantal opacities for Rb–Rb$^+$ elastic collisions are shown in Fig. 9.7; as a result, it is noted that for large values of l the semi-classical opacity agrees extremely well up to a certain value, $l = L$, where some deviations appear. Below L, the semi-classical and quantal opacities show rapid oscillatory behavior as a function of l. Therefore, for $l \leq L$, it is adequate to substitute $\sin^2(\delta_l)$ by its average, $1/2$, which is generally known as the random phase approximation (RPA) [48]. Thus, Eq. (9.26) can be approximated as

$$\sigma_{el}(E) = \frac{4\pi}{k^2} \sum_l (2l+1) \sin^2(\delta_l(k)) \simeq \frac{2\pi}{k^2} L^2 + \frac{4\pi}{k^2} \int_L^{\infty} 2l \sin^2(\delta_l(k)) dl,$$
(9.28)

and assuming that for $l > L$ the phase shifts are small, $\sin \delta_l \approx \delta_l$, is valid and Eq. (9.28) yields [49]

$$\sigma_{el}(E) \simeq \frac{2\pi}{k^2} L^2 \left(1 + \delta_L^2(k)\right).$$
(9.29)

The value of L depends on the system at hand and collision energy. However, it turns out that the solution of $\delta_L(k) = \pi/4$, for L, leads to a reasonable estimation of L [49]. Finally, the semi-classical elastic cross section reads as

---

[7]The Langer correction is the transformation $l(l+1) \to (l+1/2)^2$, which leads to more accurate semi-classical results.

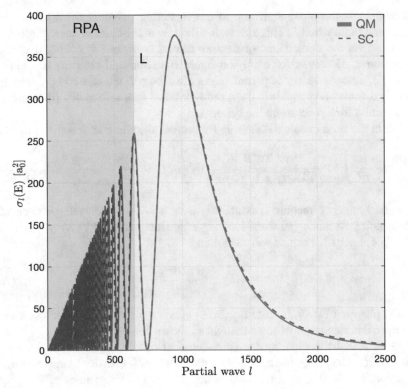

Fig. 9.7 Opacity for Rb$^+$-Rb collisions at 250 K of collision energy. The shaded gray area represents the region where the random phase approximation (RPA) for the phase shift is accurate. The dashed light green area, which starts at L, represents the region in which the semi-classical approximation for the phase shift is valid

$$\sigma_{el}(E) = \pi \left(\frac{\mu\alpha^2}{\hbar}\right)^{1/3} \left(1 + \frac{\pi^2}{16}\right) E^{-1/3}. \qquad (9.30)$$

This equation clearly reveals that the elastic cross section for atom–ion collisions shows a weaker energy dependence than the Langevin model ($\propto E^{-1/2}$, as seen in Eq. 9.23). In Fig. 9.8, the quantal elastic cross section and the results from Eq. (9.30) as a function of the collision energy for Na$^+$-Na collisions are shown. In this Figure, it is noted that the semi-classical elastic cross section describes the quantal results properly up to a collision energy of 10^{-8} a.u. or 3.15 mK, which confirms the wide range of applicability of the semi-classical approach for the elastic cross section in ion–atom systems. Finally, it is worth noting that there are two quantal results: one for the *gerade* ground electronic state and another of its analog with *ungerade* symmetry. However, the semi-classical results show a single curve. The reason for this difference is in the fact that the semi-classical cross section only depends on the

Fig. 9.8 Elastic cross section (quantal and semi-classical) as a function of the collision energy (1a.u.=3.15×10^{-5}K) for Na$^+$-Na collisions. (Figure adapted with permission from Ref. [49] Copyright (2020) (American Physical Society))

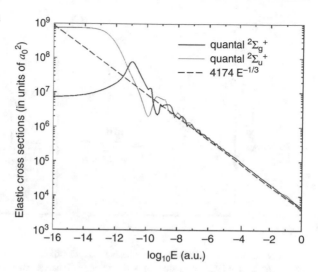

long-range tail of the atom–ion interaction, and hence leads to the same results for both symmetries for the ground state, whereas for the quantal case, for a collision such as A-A^+ the quantal elastic cross section is defined as [50]

$$\sigma_{el}^{T} = \frac{4\pi}{k^2} \sum_{l} (2l+1)\left[\sin^2\left(\delta_l^g(k)\right) + \sin^2\left(\delta_l^u(k)\right)\right];$$
(9.31)

here, T stands for the total elastic cross section and $\delta_l^{g,u}(k)$ represents the phase shift for the gerade and ungerade electronic states respectively. Therefore, both electronic states contribute to the elastic cross section.

9.4.4 Radiative Charge Transfer and Radiative Association

In general, atom–ion interactions present an intricate potential energy landscape owing to the presence of many energetically closed excited states, as shown in Fig. 9.6, and as a consequence different inelastic, or even reactive processes, may occur, depending on the initial state of the colliding partners. The possible inelastic and reactive processes are schematically represented in Fig. 9.9: nonradiative charge transfer (nRCT), radiative charge transfer (RCT), and radiative association (RA).

Nonradiative charge transfer is an inelastic process in which the charge *hops* from one atom into the other, which is represented by the transition $A + B^+ \rightarrow A^+ + B$ in Fig. 9.9. The charge hopping occurs via an avoided crossing between the relevant electronic states. However, the same transition may occur via spontaneous emission

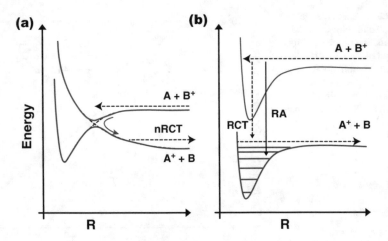

Fig. 9.9 Atom–ion collisions. Nonradiative (**a**) and radiative processes (**b**). Panel (**a**), nonradiative charge transfer (nRCT) collision, in which charge exchange reaction is produced as a consequence of the crossing between different potential energy curves. Panel (**b**) radiative charge transfer and radiative association processes. In RCT the charge exchange reaction requires the emission of a photon, whereas in RA the photon emission leads to the formation of a molecular ion

from one electronic state into another, which is known as RCT. The cross section of RCT is calculated following the logic for the derivation of the spontaneous emission rate from a molecular state and reads as [51]

$$\sigma_{RCT}(E_k) = \frac{8\pi^2\alpha^3}{3k_i^2}$$

$$\sum_J \int_0^{\epsilon_f^{max}} \omega_{i,f}^3 \left(J|\langle J-1, \epsilon_f, |D(R)|\epsilon_i, J\rangle|^2 + (J+1)|\langle J+1, \epsilon_f, |D(R)|\epsilon_i, J\rangle|^2 \right) d\epsilon_f,$$

(9.32)

where all the quantities are in atomic units and $\alpha \approx 1/137$ is the fine-structure constant. The ket $|\epsilon_i, J\rangle$ represents the initial state, which is an energy normalized state of the continua with energy $\epsilon_i = E_k + V_i(R = \infty)$, where E_k is the kinetic energy of the colliding particles in the center-of-mass frame, wave vector $k_i = \sqrt{2\mu\epsilon_i}$, and partial wave J. The final states are energy normalized continuum states with energy $\epsilon_f = E_k + V_i(R = \infty) - \omega_{i,f} - V_f(R = \infty)$ and partial wave $J+1$ or J. Finally, the probability of decaying into a given state is controlled by the transition dipole moment $D(R)$, which is calculated by means of quantum chemistry *ab initio* methods.

In the case of RA, the spontaneous emission of the initial scattering state leads to the formation of a molecular ion, i.e., $A + B^+ \rightarrow AB^+$, through decaying to one of the rovibrational states supported by $V_f(R)$. Then, following the same logic as in the case of RCT, the cross section for RA reads as [52]

$$\sigma_{RA}(E_k) = \sum_{J=0}^{J_{max}} \sum_{v=0}^{v_{max}} \sigma_{J,v}(E_k), \tag{9.33}$$

where the summation extends to all the rovibrational states within the final electronic state for the AB^+ molecule,

$$\sigma_{J,v}(E_k) = \frac{8\pi^2\alpha^3}{3k^2} \left(\omega^3_{\epsilon_i J, J-1v} J M^2_{J-1,J} + \omega^3_{\epsilon_i J, J+1v}(J+1)M^2_{J+1,J} \right), \tag{9.34}$$

$\omega_{\epsilon_i J, v J'}$, as in the RCT case, stands for the energy difference between the initial and final states. In particular, the initial state has an energy ϵ_i and angular momentum J and the final state is characterized by the vibrational quantum number v and rotational quantum number J'. The coupling terms are defined as

$$M_{J,J'} = \int_0^\infty f_J(E_k, R)D(R)\phi_{J',v}(R)dR, \tag{9.35}$$

where $f_J(E_k, R)$ stands for the initial state energy normalized continuum wave function associated with the partial wave J and $\phi_{J',v}(R)$ is the bound state wave function for a rovibrational state with vibrational quantum number v and rotational quantum number J'.

In Fig. 9.10 the cross sections for RCT and RA processes are shown as a function of the collision energy for different atom–ion systems. The cross sections for these two processes show a rich resonance structure independently of the atom–ion species at hand. Additionally, the observed structured pattern appears in both the RCT and RA cross sections. The presence of the same resonances in both cross sections implies that the resonances are related to the initial state rather than to the final state. In fact, it can be proved that the resonances are shape resonances in the entrance channel as a consequence of the existence of nearby quasi-bound states supported by the entrance centrifugal barrier (see Sect. 2.5.1). In the same Figure it is noted that the RA cross section is larger than the one for RCT cross section, and this is because the electronic states relevant for Fig. 9.10 show a more prominent Franck–Condon overlap for deeply bound states in comparison with lesser bound states or continuum states [51]. This translates into a more probable molecular ion formation to the detriment of the charge transfer process.

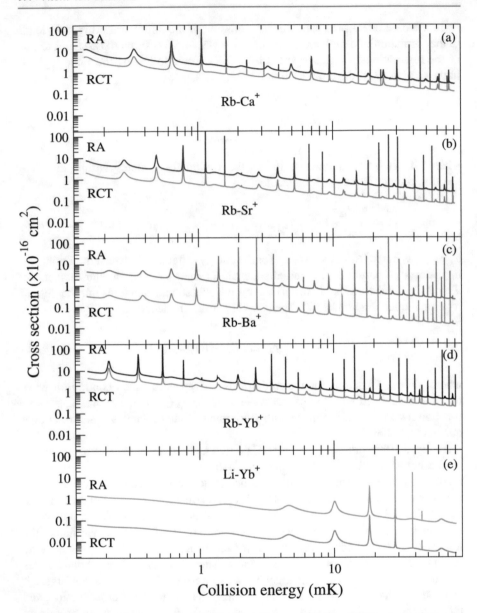

Fig. 9.10 Radiative charge transfer (RCT) and radiative association (RA) cross sections as a function of the collision energy (in mK) for different atom–ion systems. (Figure reproduced with permission from Ref. [51] Copyright (2020) (IOP Publishing, Ltd.))

References

1. Petrache HI, Zemb T, Belloni L, Parsegian VA (2006) Salt screening and specific ion adsorption determine neutral-lipid membrane interactions. Proc Natl Acad Sci 103(21):7982. https://doi.org/10.1073/pnas.0509967103. http://www.pnas.org/content/103/21/7982.abstract
2. Larsson M, Geppert WD, Nyman G (2012) Ion chemistry in space. Rep Prog Phys 75(6):066901
3. Khomenko E (2016) On the effects of ion-neutral interactions in solar plasmas. Plasma Phys Controlled Fusion 59(1):014038. https://doi.org/10.1088/0741-3335/59/1/014038
4. Brünken S, Kluge L, Stoffels A, Pérez-Ríos J, Schlemmer S (2017) Rotational state-dependent attachment of he atoms to cold molecular ions: an action spectroscopic scheme for rotational spectroscopy. J Mol Spectrosc 332:67
5. Doerk H, Idziaszek Z, Calarco T (2010) Atom-ion quantum gate. Phys Rev A 81:012708 . https://doi.org/10.1103/PhysRevA.81.012708
6. Secker T, Gerritsma R, Glaetzle AW, Negretti A (2016) Controlled long-range interactions between rydberg atoms and ions. Phys Rev A 94:013420. https://doi.org/10.1103/PhysRevA.94.013420
7. Bissbort U, Cocks D, Negretti A, Idziaszek Z, Calarco T, Schmidt-Kaler F, Hofstetter W, Gerritsma R (2013) Emulating solid-state physics with a hybrid system of ultracold ions and atoms. Phys Rev Lett 111:080501. https://doi.org/10.1103/PhysRevLett.111.080501
8. Mur-Petit J, García-Ripoll JJ, Pérez-Ríos J, Campos-Martínez J, Hernández MI, Willitsch S (2012) Temperature-independent quantum logic for molecular spectroscopy. Phys Rev A 85:022308
9. Kollath C, Köhl M, Giamarchi T (2007) Scanning tunneling microscopy for ultracold atoms. Phys Rev A 76:063602. https://doi.org/10.1103/PhysRevA.76.063602
10. Bloch I, Dalibard J, Nascimbène S (2012) Quantum simulations with ultracold quantum gases. Nat Phys 8(4):267
11. Tomza M, Jachymski K, Gerritsma R, Negretti A, Calarco T, Idziaszek Z, Julienne PS (2019) Cold hybrid ion-atom systems. Rev Mod Phys 91:035001. https://doi.org/10.1103/RevModPhys.91.035001
12. Willitsch S (2017) Chemistry with controlled ions. Adv Chem Phys 162:307
13. Penning FM (1936) Die glimmentladung bei niedrigem druck zwischen koaxialen zylindern in einem axialen magnetfeld. Physica 3(9):873. https://doi.org/10.1016/S0031-8914(36)80313-9. http://www.sciencedirect.com/science/article/pii/S0031891436803139
14. Pierce JR (1954) Theory and design of electron beams. D. Van Nostrand, Princeton
15. Gräff G, Klempt E, Werth G (1969) Method for measuring the anomalous magnetic moment of free electrons. Zeitschrift für Physik A Hadrons and Nuclei 222(3):201. https://doi.org/10.1007/BF01392119
16. Novikov YN, Vasiliev AA, Gusev YI, Nesterenko DA, Popov AV, Seliverstov DM, Seliverstov MD, Khusainov AK, Blaum K, Eliseev SA, Herfurth F, Block M, Vorobjev GK, Jokinen A, Rodriguez D, Yavor MI (2012) High-precision method of measuring short-lived nuclides by means of developed systems of ion traps for high-charge ions (mats project). Atomic Energy 112(2):139. https://doi.org/10.1007/s10512-012-9535-4
17. Chaudhuri A, Andreoiu C, Brodeur M, Brunner T, Chowdhury U, Ettenauer S, Gallant AT, Grossheim A, Gwinner G, Klawitter R, Kwiatkowski AA, Leach KG, Lennarz A, Lunney D, Macdonald TD, Ringle R, Schultz BE, Simon VV, Simon MC, Dilling J (2014) Titan: an ion trap for accurate mass measurements of ms-half-life nuclides. Appl Phys B 114(1):99. https://doi.org/10.1007/s00340-013-5618-8
18. Roux C, Blaum K, Block M, Droese C, Eliseev S, Goncharov M, Herfurth F, Ramirez EM, Nesterenko DA, Novikov YN, Schweikhard L (2013) Data analysis of q-value measurements for double-electron capture with shiptrap. Eur Phys J D 67(7):146. https://doi.org/10.1140/epjd/e2013-40110-x

19. Blaum K, Novikov YN, Werth G (2010) Penning traps as a versatile tool for precise experiments in fundamental physics. Contemp Phys 51(2):149. https://doi.org/10.1080/00107510903387652
20. Repp J, Böhm C, Crespo López-Urrutia JR, Dörr A, Eliseev S, George S, Goncharov M, Novikov YN, Roux C, Sturm S, Ulmer S, Blaum K (2012) Pentatrap: a novel cryogenic multi-penning-trap experiment for high-precision mass measurements on highly charged ions. Appl Phys B 107(4):983. https://doi.org/10.1007/s00340-011-4823-6
21. Eliseev SA, Novikov YN, Blaum K (2012) Search for resonant enhancement of neutrinoless double-electron capture by high-precision penning-trap mass spectrometry. J Phys G Nucl Part Phys 39(12):124003. https://doi.org/10.1088/0954-3899/39/12/124003
22. Sturm S, Werth G, Blaum K (2013) Electron g-factor determinations in penning traps. Annalen der Physik 525(8–9):620
23. (2019). http://alpha.web.cern.ch
24. (2019). http://gabrielse.physics.harvard.edu/gabrielse/overviews/Antihydrogen/Antihydrogen.html
25. Paul W (1990) Electromagnetic traps for charged and neutral particles. Rev Mod Phys 62:531. https://doi.org/10.1103/RevModPhys.62.531
26. Neuhauser W, Hohenstatt M, Toschek P, Dehmelt H (1978) Optical-sideband cooling of visible atom cloud confined in parabolic well. Phys Rev Lett 41:233
27. Ghosh PK (1995) Ion traps. Oxford University Press, New York
28. Leibfried D, Blatt R, Monroe C, Wineland D (2003) Quantum dynamics of single trapped ions. Rev Mod Phys 75:281. https://doi.org/10.1103/RevModPhys.75.281
29. Mohammadi A, Wolf J, Krükow A, Deiß M, Hecker Denschlag J (2019) Minimizing rf-induced excess micromotion of a trapped ion with the help of ultracold atoms. Appl Phys B 125(7):122. https://doi.org/10.1007/s00340-019-7223-y
30. Berkeland DJ, Miller JD, Bergquist JC, Itano WM, Wineland DJ (1998) Minimization of ion micromotion in a Paul trap. J Appl Phys 83(10):5025. https://doi.org/10.1063/1.367318
31. Zemansky MW, Dittman RH (1996) Heat and thermodynamics. McGraw-Hill, New York
32. Thompson RC (2015) Ion coulomb crystals. Contemp Phys 56(1):63. https://doi.org/10.1080/00107514.2014.989715
33. Willitsch S, Bell M, Gingell A, Softley TP (2008) Chemical applications of laser- and sympathetically cooled ions in traps. Phys Chem Chem Phys 10:7200
34. Jones MD, Ceperley DM (1996) Crystallization of the one-component plasma at finite temperature. Phys Rev Lett 76:4572. https://doi.org/10.1103/PhysRevLett.76.4572
35. Hall FHJ, Aymar M, Bouloufa N, Dulieu O, Willitsch S (2011) Light-assisted ion-neutral reactive processes in the cold regime: radiative molecule formation vs. charge exchange. Phys Rev Lett 107:243202
36. Hall FHJ, Willitsch S (2012) Phys Rev Lett 109:233202
37. Hall FHJ, Eberle P, Hegi G, Raoult M, Aymar M, Dulieu O, Willitsch S (2013) Ion-neutral chemistry at ultralow energies: dynamics of reactive collisions between laser-cooled Ca^+ ions and Rb atoms in an ion-atom hybrid trap. Mol Phys 111:2020
38. Jefimenko OD (1989) Electicity and magnetism. Electret Scientific Company, West Virginia
39. Stone A (2013) The theory of intermolecular forces, 2nd edn. Oxford University Press, UK.
40. Mitroy J, Safronova MS, Clark CW (2010) Theory and applications of atomic and ionic polarizabilities. J Phys B At Mol Opt Phys 43(20):202001. https://doi.org/10.1088/0953-4075/43/20/202001
41. Tomza M, Koch CP, Moszynski R (2015) Cold interactions between an Yb^+ ion and a Li atom: prospects for sympathetic cooling, radiative association, and Feshbach resonances. Phys Rev A 91:042706. https://doi.org/10.1103/PhysRevA.91.042706
42. Langevin P (1905) Une formule fodnamentale de théorie cinétique. C R Acad Sci 140:35

43. Auzinsh M, Dashevskaya EI, Nikitin EE, Troe J (2013) Quantum capture of charged particles by rapidly rotating symmetric top molecules with small dipole moments: analytical comparison of the fly-wheel and adiabatic channel limits. Mol Phys 111(14–15):2003. https://doi.org/10.1080/00268976.2013.780101
44. Nikitin EE, Troe J (2005) Dynamics of ion–molecule complex formation at very low energies and temperatures. Phys Chem Chem Phys 7(7):1540. https://doi.org/10.1039/B416401F
45. Herbst E (1979) A statistical theory of three-body ion–molecule reactions. J Chem Phys 70(5):2201. https://doi.org/10.1063/1.437775
46. Lara M, Jambrina PG, Aoiz FJ, Launay JM (2015) Cold and ultracold dynamics of the barrierless $D^+ + H_2$ reaction: quantum reactive calculations for $\sim r^{-4}$ long range interaction potentials. J Chem Phys 143(20):204305. https://doi.org/10.1063/1.4936144
47. Landau LD, Lifshitz EM (1958) Quantum mechanics. Butterworth-Heinemann
48. Levine RD, Bernstein RB (1987) Molecular reaction dynamics and chemical reactivity. Oxford University Press, New York
49. Côté R, Dalgarno A (2000) Ultracold atom-ion collisions. Phys Rev A 62:012709
50. McDaniel EW (1964) Collision phenomena in ionized gases. Wiley, New York
51. da Silva Jr H, Raoult M, Aymar M, Dulieu O (2015) Formation of molecular ions by radiative association of cold trapped atoms and ions. New J Phys 17(4):045015. https://doi.org/10.1088/1367-2630/17/4/045015
52. Zygelman G, Dalgarno A (1990) The radiative association of He^+ and H. Astrophys J 365:239

Few-Body Processes Involving Ions and Neutrals at Cold Temperatures

10

Traditionally, the term few-body physics has been mainly employed in nuclear physics, although the field recently became popular in the context of atomic physics, concretely in ultracold physics [1–8]. However, in chemical physics, and concretely in cold and ultracold chemistry, it is not frequently used, even though a chemical reaction is a few-body system per se. Motivated by this idea, we present in this chapter an introduction to the most relevant few-body processes in cold chemistry. In particular, we focus on ion–atom–atom three-body recombination, i.e., $A^+ + A + A \rightarrow A_2^+ + A$. The reader may think that we are missing a product state for the ion–atom–atom three-body recombination, i.e., $A_2 + A^+$. The reader is right, although we did this on purpose. Indeed, throughout this chapter, we invite the reader to explore the physics behind the fact that mostly molecular ions are formed, to the detriment of molecules, as a product state of ion–atom–atom three-body recombination reactions at cold temperatures.

In principle, at cold temperatures, $1\,\mathrm{mK} \lesssim T \lesssim 1\,\mathrm{K}$, only a few partial waves contribute to the scattering observables; therefore, one would be convinced that only a quantum treatment of the scattering leads to meaningful results. However, as stated in Chap. 1, the concept of cold collisions depends on the underlying potential energy between the colliding partners. Indeed, we show in this chapter that classical methods for scattering are well suited to ion–atom–atom collisions at cold temperatures.

10.1 Do We Really Need a Full Quantum Mechanical Treatment at Cold Temperatures?

As has been explained in Chap. 1, the concept of cold collisions depends on the nature of the interaction between the colliding partners. In other words, the number of partial waves contributing to the scattering observables for a given energy hinges on the strength of the interparticle interaction. For a general long-range interaction

© Springer Nature Switzerland AG 2020
J. Pérez Ríos, *An Introduction to Cold and Ultracold Chemistry*,
https://doi.org/10.1007/978-3-030-55936-6_10

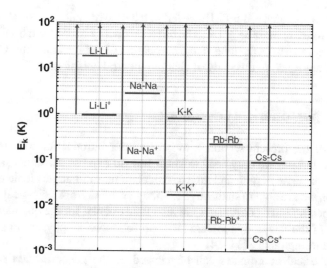

Fig. 10.1 Validity region of a classical trajectory (CT) approach for different atom–ion collisions. The red horizontal line represents the lowest collision energy where a CT method is adequate for ion–neutral collisions, whereas the horizontal dark blue line represents the same for neutral–neutral collisions. The range of applicability of CT methods does not have an upper limit, independently of the kind of interaction. (Figure adapted with permission from Ref. [9] Copyright (2020) (American Physical Society))

potential, $-C_n/r^n$, with $n \geq 2$, and a given collision energy, E_k, the partial wave that shows a barrier height equal to the collision energy is (in atomic units) [9]

$$l = \left(\frac{2}{n-2}\right)^{\frac{n-2}{2n}} \sqrt{n} E_k^{\frac{n-2}{2n}} C_n^{1/n} \mu^{1/2}, \qquad (10.1)$$

where μ stands for the reduced mass of the two-body system. This equation establishes the number of partial waves that significantly contribute to the scattering problem, within a two-body approach, at a given collision energy, E_k.

For $l = 20$ it is very probable that most systems of interest show many scattering resonances (this is even clearer in the case of heavy colliding partners). However, the large number of partial waves contributing to the scattering may wash out the resonance effects. Based on this argument, we choose $l = 20$ as the minimum number of partial waves for which a classical trajectory (CT) approach is appropriate for describing a scattering problem. From this lower bound on l it is possible, through Eq. (10.1), to infer the lowest energy at which a CT approach is reliable. In particular, for ion–neutral collisions a CT approach is reliable for collision energies $\gtrsim 1$ mK, whereas in the case of neutral–neutral collisions the collision energy must be $\gtrsim 1$ K. This discrepancy clearly reflects the role of the long-range interaction of the system.

In Fig. 10.1 the range of validity of CT for alkali–alkali and alkali–alkali ion collisions is shown. As a result, CT methods for ion–neutral collisions are reliable up to collision energies ~ 1 mK. However, in the case of neutral–neutral interactions the collision energy must be $\gtrsim 1$ K. The different energy scales for the validity of CT methods correlates with the long-range interaction of the system. Thus, CT methods are a robust approach to the study of charged–neutral collisions at $T \gtrsim 1$ mK.

10.2 Hyperspherical Coordinates

The scope of the present section is to give a brief introduction to hyperspherical coordinates. In fact, we only need to know how to express a n-dimensional vector in such a coordinate system and how to calculate its solid angle element. With this information, we are able to study any few-body problem from a classical standpoint. Indeed, hyperspherical coordinates are also fundamental for the quantum mechanical version of the few-body problem [10] and for chemical physics, in particular in reactive scattering theory [11–14].

Hyperspherical coordinates can be viewed as the generalization of spherical coordinates into a space of an arbitrary number of dimensions. A space of dimension n can be described in hyperspherical coordinates by the hyperradius ρ and by $n-1$ hyperangles $\Omega^n = (\alpha_1, \ldots, \alpha_{n-1})$. In principle, there are many ways to construct the space in hyperspherical coordinates. Here, we assume Avery's definition of hyperangles [15]; thus, a n-dimensional vector is given by

$$
\boldsymbol{x} = \begin{pmatrix} x_1 \\ x_2 \\ x_3 \\ x_n \\ \cdot \\ \cdot \\ \cdot \\ x_n \end{pmatrix} = \begin{pmatrix} \rho \prod_{i=1}^{n-1} \sin \alpha_i \\ \rho \cos \alpha_1 \prod_{i=2}^{n-1} \sin \alpha_i \\ \cdot \\ \cdot \\ \cdot \\ \rho \sin \alpha_{n-1} \sin \alpha_{n-2} \cos \alpha_{n-3} \\ \rho \sin \alpha_{n-1} \cos \alpha_{n-2} \\ \rho \cos \alpha_{n-1}, \end{pmatrix}
\tag{10.2}
$$

where x_i with $i = 1, .., n$ stands for the Cartesian coordinates of the n-dimensional space. The hyperradius is defined as

$$
\rho = \sqrt{\sum_{i=1}^{n} x_i^2},
\tag{10.3}
$$

which is similar to the moment of inertia of the system, and certainly gives the extension of the system. Finally, the element of solid angle for Avery's definition of hyperangles is given by [16]

$$
d\Omega^n = \sin^{n-2}(\alpha) d\alpha d\Omega^{n-1}.
\tag{10.4}
$$

Indeed, employing Eq. (10.4) for $n = 3$ we find $d\Omega^3 = \sin\alpha\, d\alpha\, d\Omega^2 = \sin\alpha\, d\alpha\, d\phi$, which is the well-known solid element for a sphere.

10.3   Classical Scattering Theory for Few-Body Problems

In classical mechanics, two-body collision, after neglecting the trivial center-of-mass (CM) motion, is visualized as one particle with a given velocity moving toward a scattering center. As a result of the interaction between the particles, the incoming particle experiences a deflection from its initial trajectory. As explained in Chap. 2, the cross section is the area of the plane perpendicular to the initial velocity of the incoming particle and contains the scattering center [17, 18]. Indeed, this area is a function of a single physical magnitude (apart from the initial conditions): the impact parameter b (introduced in Chap. 2), which is the component of the initial vector position lying in the perpendicular plane to the initial velocity of the incoming particle.

The intrinsic geometrical nature of the classical cross section is the key to generalizing the concept of collision into a higher dimensional space of dimension n. In particular, the cross section is given by

$$\sigma_{\text{process}}(\boldsymbol{P}_0) = \int \wp_{\text{process}}(\boldsymbol{b}, \boldsymbol{P}_0) d^{n-1}\boldsymbol{b}, \qquad (10.5)$$

which represents the effective volume of the $n - 1$ hyperplane described by the impact parameter, $\boldsymbol{b}$, perpendicular to the initial momentum $\boldsymbol{P}_0$. In Eq. 10.5, the opacity function (introduced in Chap. 2), $\wp_{\text{process}}(\boldsymbol{b}, \boldsymbol{P}_0)$, represents the probability that a given trajectory with particular initial conditions results in the process under consideration, e.g., inelastic collisions or reactive collisions. The cross section given by Eq. (10.5) depends on the orientation of the initial momentum; however, it is preferable to express the cross section as a function of the magnitude of the initial momentum by integrating out the angular degrees of freedom of the momentum as

$$\sigma_{\text{process}}(P_0) = \frac{\int \wp_{\text{process}}(\boldsymbol{b}, \boldsymbol{P}_0) d\Omega_P^n d^{n-1}\boldsymbol{b}}{\int d\Omega_P^n}, \qquad (10.6)$$

where it is assumed that the momentum is described in hyperspherical coordinates, although it is valid in any coordinate system.

The dynamics of a few-body system can be studied by mapping out the degrees of freedom of the system at hand, d, into a two-body collision in a d-dimensional space, as shown in Fig. (10.2). In this figure, the particular case when the CM motion is decoupled from the relative motion of the interacting particles is shown. Thus, any few-body system can be described as a single-particle problem in a higher dimensional space, which is one of the most notable messages of this chapter. In particular, we exploit such a powerful idea in the context of three-body collisions.

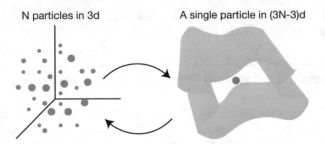

Fig. 10.2 Schematic representation of a classical approach to few-body collisions. The philosophy of the method is the mapping of the degrees of freedom of a given system, $(3N - 3)$ for N particles, into a single-particle problem in a $(3N - 3)$ dimensional space. In this way, any few-body problem can be mapped into an effective two-body problem

10.4   Classical Trajectory Calculations in Hyperspherical Coordinates for Three-Body Collisions

The classical Hamiltonian for three interacting particles under the action of the interaction potential $V(r_1, r_2, r_3)$ reads as

$$H = \frac{p_1^2}{2m_1} + \frac{p_2^2}{2m_2} + \frac{p_3^2}{2m_3} + V(r_1, r_2, r_3), \qquad (10.7)$$

where r_i and p_i stand for the vector position and momentum of the ith particle respectively. At this point, it is convenient to introduce the Jacobi coordinates, depicted in Fig. (10.3), as

$$\rho_1 = r_2 - r_1, \qquad (10.8a)$$

$$\rho_2 = r_3 - \frac{m_2 r_2 + m_1 r_1}{m_1 + m_2}, \qquad (10.8b)$$

$$\rho_{CM} = \frac{m_1 r_1 + m_2 r_2 + m_3 r_3}{M}, \qquad (10.8c)$$

where $M = m_1 + m_2 + m_3$ is the total mass of the system. It is worth noting that the relationships in Eqs. (10.8a–10.8c) represent a contact transformation [19]; therefore, they preserve the volume of the phase space. The Hamiltonian from Eq. (10.7) can be expressed in the Jacobi coordinates as [20]

$$H = \frac{P_1^2}{2m_{12}} + \frac{P_2^2}{2m_{3,12}} + \frac{P_{CM}^2}{2M} + V(\rho_1, \rho_2), \qquad (10.9)$$

Fig. 10.3 Jacobi coordinates
for the three-body problem
(ρ_1, ρ_2). (Figure reproduced
with permission from
Ref. [21] Copyright (2020)
(AIP Publishing))

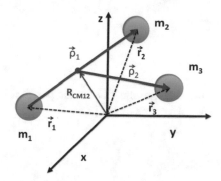

where $\frac{1}{m_{12}} = \frac{1}{m_1} + \frac{1}{m_2}$ and $\frac{1}{m_{3,12}} = \frac{1}{m_3} + \frac{1}{m_1+m_2}$. In Eq. (10.9), $V(\rho_1, \rho_2)$ is the
potential energy as a function of the Jacobi coordinates, which shows no dependence
on the CM coordinates and hence the CM momentum is a constant of motion. P_1,
P_2 and P_{CM} represent the canonical momenta that are conjugate to ρ_1, ρ_2 and ρ_{CM}
respectively. Finally, neglecting the trivial CM motion gives

$$H = \frac{P_1^2}{2m_{12}} + \frac{P_2^2}{2m_{3,12}} + V(\rho_1, \rho_2). \qquad (10.10)$$

Hamilton's equations of motion for the three-body problem as a function of the
Jacobi coordinates are

$$\frac{d\rho_{i,\alpha}}{dt} = \frac{\partial H}{\partial P_{i,\alpha}}, \qquad (10.11a)$$

$$\frac{dP_{i,\alpha}}{dt} = -\frac{\partial H}{\partial \rho_{i,\alpha}}, \qquad (10.11b)$$

where $i = 1, 2$ and $\alpha = x, y, z$ represent the Cartesian coordinates of each Jacobi
vector. Solving Hamilton's equations leads to the characterization of the dynamics
of the system at hand in the usual three-dimensional (3D) space. Alternatively,
following Sect. 10.3 the dynamics of the system can be described as a single particle
moving in a six-dimensional (6D) space. In particular, following the pioneering
method of Smith [22] a 6D position vector is constructed from the two mass-
weighted Jacobi vectors as

$$\rho_{mw} = \begin{pmatrix} \sqrt{\frac{m_{12}}{\mu}}\rho_1 \\ \sqrt{\frac{m_{3,12}}{\mu}}\rho_2 \end{pmatrix}, \qquad (10.12)$$

where

$$\mu = \sqrt{\frac{m_1 m_2 m_3}{M}}. \tag{10.13}$$

Furthermore, the 6D vector position can also be expressed in terms of the *bare* Jacobi vectors as [21]

$$\rho_{\text{bare}} = \begin{pmatrix} \rho_1 \\ \rho_2 \end{pmatrix}. \tag{10.14}$$

And associated with these representations of the vector position, the canonical momenta are given by

$$P_{\text{mw}} = \begin{pmatrix} P_1 \\ P_2 \end{pmatrix} \tag{10.15}$$

and

$$P_{\text{bare}} = \begin{pmatrix} \sqrt{\frac{\mu}{m_{12}}} P_1 \\ \sqrt{\frac{\mu}{m_{3,12}}} P_2 \end{pmatrix}, \tag{10.16}$$

respectively. The relationship between the coordinates $(\rho_{\text{mw}}, P_{\text{mw}})$ and $(\rho_{\text{bare}}, P_{\text{bare}})$ is a contact transformation [19, 23, 24], as one would expect, since the scattering observables cannot depend on the particular set of coordinates. The link between the system of three interacting particles in three dimensions and a single particle in six dimensions is the vector position ρ or ρ_{bare} and the corresponding conjugate momenta.

Here, we adopt the *bare* vector position and momentum, which are labelled as ρ and P for the sake of clarity. Thus, the Hamiltonian reads as

$$H = \frac{P^2}{2\mu} + V(\rho), \tag{10.17}$$

which represents the motion of a single particle in a 6D space. Once, the position vector and the momentum in a 6D space have been defined, the concept of impact parameter, as the component of the position vector lying in a hyperplane perpendicular to the initial momentum, is clear. Hereafter, we will employ hyperspherical coordinates for the representation of the 6D vectors, as explained in Sect. 10.2. In particular, all the vectors can be represented by means of the hyperradius, r, and five different hyperangles (α_i, $i = 1, 2, 3, 4, 5$) as

$$
r = \begin{pmatrix} r_{x_1} \\ r_{x_2} \\ r_{x_3} \\ r_{x_4} \\ r_{x_5} \\ r_{x_6} \end{pmatrix} = \begin{pmatrix} r \sin\alpha_1 \sin\alpha_2 \sin\alpha_3 \sin\alpha_4 \sin\alpha_5 \\ r \cos\alpha_1 \sin\alpha_2 \sin\alpha_3 \sin\alpha_4 \sin\alpha_5 \\ r \cos\alpha_2 \sin\alpha_3 \sin\alpha_4 \sin\alpha_5 \\ r \cos\alpha_3 \sin\alpha_4 \sin\alpha_5 \\ r \cos\alpha_4 \sin\alpha_5 \\ r \cos\alpha_5 \end{pmatrix}, \tag{10.18}
$$

where the angle ranges are $0 \leq \alpha_1 \leq 2\pi$, $0 \leq \alpha_i \leq \pi$, $i = 2, 3, 4, 5$. It is preferable to choose the 3D z axis parallel to P_2, and hence the initial momentum $\vec{P_0}$ is expressed as

$$
P_0 = \begin{pmatrix} P_0 \sin\alpha_1^P \sin\alpha_2^P \sin\alpha_5^P \\ P_0 \cos\alpha_1^P \sin\alpha_2^P \sin\alpha_5^P \\ P_0 \cos\alpha_2^P \sin\alpha_5^P \\ 0 \\ 0 \\ P_0 \cos\alpha_5^P \end{pmatrix}, \tag{10.19}
$$

where $0 \leq \alpha_1^P \leq 2\pi$, $0 \leq \alpha_2^P \leq \pi$ and $0 \leq \alpha_5^P \leq \pi$.

Let us define b as

$$
b = \begin{pmatrix} b \sin\alpha_1^b \sin\alpha_2^b \sin\alpha_3^b \sin\alpha_4^b \\ b \cos\alpha_1^b \sin\alpha_2^b \sin\alpha_3^b \sin\alpha_4^b \\ b \cos\alpha_2^b \sin\alpha_3^b \sin\alpha_4^b \\ b \cos\alpha_3^b \sin\alpha_4^b \\ b \cos\alpha_4^b \\ 0 \end{pmatrix}, \tag{10.20}
$$

where $0 \leq \alpha_1^b \leq 2\pi$, $0 \leq \alpha_i^b \leq \pi$, $i = 2, 3, 4$ and b is described as a five-dimensional (5D) vector embedded in a 6D space. To ensure that the impact parameter is perpendicular to the momentum P_0 we use the Gram–Schmidt orthogonalization procedure, and as a result the impact parameter is defined in the six-dimensional space as

$$
b = b \sin\alpha_1^b \sin\alpha_2^b \sin\alpha_3^b \sin\alpha_4^b b_1 + b \cos\alpha_1^b \sin\alpha_2^b \sin\alpha_3^b \sin\alpha_4^b b_2 +
$$
$$
b \cos\alpha_2^b \sin\alpha_3^b \sin\alpha_4^b b_3 + b \cos\alpha_3^b \sin\alpha_4^b b_4 + b \cos\alpha_4^b b_5, \tag{10.21}
$$

where

$$
\boldsymbol{b}_1 =
\begin{pmatrix}
\sqrt{\cos^2 \alpha_5^P + \left(\cos^2 \alpha_5^P + \cos^2 \alpha_1^P \sin^2 \alpha_2^P\right) \sin^2 \alpha_5^P} \\
-\dfrac{\cos \alpha_1^P \sin \alpha_1^P \sin^2 \alpha_2^P \sin^2 \alpha_5^P}{\sqrt{\cos^2 \alpha_5^P + \left(\cos^2 \alpha_2^P + \cos^2 \alpha_1^P \sin^2 \alpha_2^P\right) \sin^2 \alpha_5^P}} \\
-\dfrac{\cos \alpha_2^P \sin \alpha_1^P \sin \alpha_2^P \sin^2 \alpha_5^P}{\sqrt{\cos^2 \alpha_5^P + \left(\cos^2 \alpha_2^P + \cos^2 \alpha_1^P \sin^2 \alpha_2^P\right) \sin^2 \alpha_5^P}} \\
0 \\
0 \\
-\dfrac{\cos \alpha_5^P \sin \alpha_1^P \sin \alpha_2^P \sin \alpha_5^P}{\sqrt{\cos^2 \alpha_5^P + \left(\cos^2 \alpha_2^P + \cos^2 \alpha_1^P \sin^2 \alpha_2^P\right) \sin^2 \alpha_5^P}}
\end{pmatrix},
\tag{10.22}
$$

$$
\boldsymbol{b}_2 =
\begin{pmatrix}
0 \\
\sqrt{\dfrac{\cos^2 \alpha_5^P + \cos^2 \alpha_2^P \sin^2 \alpha_5^P}{\cos^2 \alpha_5^P + \left(\cos^2 \alpha_5^P + \cos^2 \alpha_1^P \sin^2 \alpha_2^P\right) \sin^2 \alpha_5^P}} \\
-\dfrac{\cos \alpha_1^P \cos \alpha_2^P \sin \alpha_2^P \sin^2 \alpha_5^P \sqrt{\dfrac{\cos^2 \alpha_5^P + \cos^2 \alpha_2^P \sin^2 \alpha_5^P}{\cos^2 \alpha_5^P + \left(\cos^2 \alpha_5^P + \cos^2 \alpha_1^P \sin^2 \alpha_2^P\right) \sin^2 \alpha_5^P}}}{\cos^2 \alpha_5^P + \cos^2 \alpha_2^P \sin^2 \alpha_5^P} \\
0 \\
0 \\
-\dfrac{\cos \alpha_1^P \cos \alpha_5^P \sin \alpha_2^P \sin \alpha_5^P \sqrt{\dfrac{\cos^2 \alpha_5^P + \cos^2 \alpha_2^P \sin^2 \alpha_5^P}{\cos^2 \alpha_5^P + \left(\cos^2 \alpha_5^P + \cos^2 \alpha_1^P \sin^2 \alpha_2^P\right) \sin^2 \alpha_5^P}}}{\cos^2 \alpha_5^P + \cos^2 \alpha_2^P \sin^2 \alpha_5^P}
\end{pmatrix},
\tag{10.23}
$$

$$
\boldsymbol{b}_3 =
\begin{pmatrix}
0 \\
0 \\
\sqrt{\dfrac{1}{1 + \cos^2 \alpha_2^P \tan^2 \alpha_5^P}} \\
0 \\
0 \\
-\cos \alpha_2^P \tan \alpha_5^P \sqrt{\dfrac{1}{1 + \cos^2 \alpha_2^P \tan^2 \alpha_5^P}}
\end{pmatrix},
\tag{10.24}
$$

$$
\boldsymbol{b}_4 =
\begin{pmatrix}
0 \\
0 \\
0 \\
1 \\
0 \\
0
\end{pmatrix},
\tag{10.25}
$$

and

$$
b_5 = \begin{pmatrix} 0 \\ 0 \\ 0 \\ 0 \\ 1 \\ 0 \end{pmatrix}. \tag{10.26}
$$

Thus, by means of Eq. (10.6) the classical cross section reads as

$$
\sigma_{\text{process}}(P) = \frac{\int \wp_{\text{process}}(\boldsymbol{b}, \boldsymbol{P}) d\Omega_P^6 d\Omega_b^5 b^4 db}{\int d\Omega_P^6}. \tag{10.27}
$$

The magnitude of the initial vector position $\rho_0 = R$ is such that the interaction potential is negligible in comparison with the collision energy. In this way, the trajectory shows a rectilinear motion, as expected before interacting with any force or field. Then, Eq. (10.17) yields the magnitude of the initial momentum $E_k = P_0^2/2\mu$, where E_k is the collision energy, which is equal to the initial kinetic energy. The hyperangles α_i^P with $i = 1, 2, 5$, and α_j^b with $j = 1, 2, 3, 4$, are randomly generated for a given magnitude of the impact parameter b. In particular, the angles are generated by means of the probability density function associated with each of them [21]. As the initial momentum $\boldsymbol{P}_0$ and the impact parameter $\boldsymbol{b}$ are orthogonal, the initial vector position can be written as

$$
\rho_0 = \boldsymbol{b} - \frac{\sqrt{R^2 - b^2}}{P_0} \boldsymbol{P}_0. \tag{10.28}
$$

Equation (10.28) generates ρ_0 from R, $\boldsymbol{b}$ and $\boldsymbol{P}_0$.

A given set of initial conditions ρ_0, R, $\boldsymbol{P}_0$ and b, are transformed into the usual 3D space by means of Eqs. (10.14) and (10.16), where Hamilton's equations of motion are numerically solved [21, 25]. The momenta and positions resulting from these equations, at a given final time of propagation, are mapped back into the six-dimensional space where the classical three-body cross section is defined by Eq. (10.27), as shown in Fig. (10.4).

10.5 Classical Three-Body Recombination for Ion–Neutral–Neutral Systems at Cold Temperatures

The hyperspherical classical trajectory method, explained in the previous section, is an appropriate approach to describing ion–neutral–neutral three-body recombination at cold temperatures. This is a consequence of the large number of partial waves contributing to the cross section, as explained in Sect. 10.1. In particular, the ion–atom–atom three-body recombination cross section, following Eq. (10.27), is defined as

Fig. 10.4 Method for the classical trajectory calculations for three-body collisions. The three-body problem is mapped out into a 6D space where the initial conditions of the trajectories are defined. In step I, these initial conditions are translated back into the 3D space where the Hamilton's equations are solved (step II). In step III, the final conditions of the trajectories are transformed back into the 6D space where the cross section is properly defined. (Figure adapted with permission from Ref. [10] Copyright (2020) (American Physical Society))

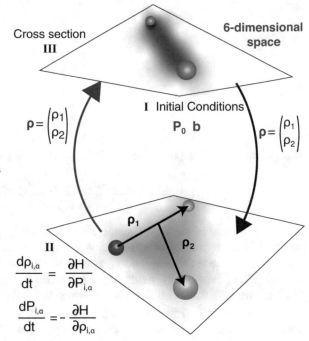

$$\sigma_{\text{rec}}(P_0) = \frac{\int \wp_{\text{rec}}(\boldsymbol{b}, \boldsymbol{P}_0) d\Omega^6_{P_0} d\Omega^5_b b^4 db}{\int d\Omega^6_{P_0}}, \tag{10.29}$$

where $d\Omega^6_{P_0}$ and $d\Omega^5_b$ represent the differential elements for the hyperangular degrees of freedom associated with the initial momentum P_0 and impact parameter $\boldsymbol{b}$ respectively. The opacity function for three-body recombination, $\wp_{\text{rec}}(\boldsymbol{b}, \boldsymbol{P})$, is the probability that an ion–atom–atom collision results in a molecule plus an ion or a molecular ion plus an atom. As an example, the opacity function for $Ba^+ + Rb + Rb$ three-body recombination is shown in Fig. 10.5, where the different panels stand for different collision energies. In this figure it is noted that the higher the collision energy the smaller the relevant impact parameter is. This behavior is related to the fact that the deflection of the trajectories with higher collision energies occur at shorter distances where the interparticle interaction is strong enough.

In general, the opacity function shows a stereochemical dependence, i.e., the orientation of the reactants affect the reaction rate,[1] although the hyperangular degrees of freedom can be averaged out, yielding

$$\wp_{\text{rec}}(b, P_0) = \frac{\int \wp_{\text{rec}}(\boldsymbol{b}, \boldsymbol{P}_0) d\Omega^6_{P_0} d\Omega^5_b}{\int d\Omega^5_b d\Omega^6_{P_0}}, \tag{10.30}$$

---

[1] Stereochemistry is the study of chemical reactions that depend on the spatial arrangement of the atoms within the reactants, i.e., the orientation of the reactants.

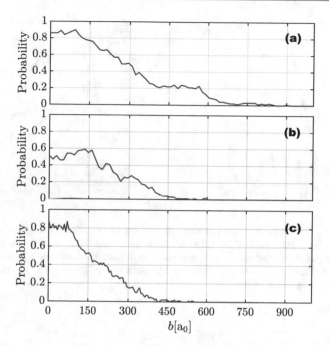

Fig. 10.5 Opacity function for Ba$^+$ + Rb + Rb three-body recombination as a function of the collision energy. Panel (**a**) $E_k = 10$ mK; panel (**b**) $E_k = 50$ mK; and $E_k = 100$ mK in panel (**c**). The calculations have been performed assuming a pair-wise additive potential for the ion–atom–atom interaction. In particular, for the Ba$^+$-Rb interaction the potential reads as $V(r) = -\frac{\alpha}{2r^4}\left[1 - \frac{1}{2}\left(\frac{r_m}{r}\right)^4\right]$, where $r_m = 9.27$ a$_0$ is the equilibrium distance for the atom–ion system and $\alpha = 320$ a.u. is the polarizability of the Rb. For the Rb–Rb interaction it is assumed that the atoms are spin polarized and hence they interact through the triplet potential energy curve

which only depends on the magnitude of the initial momentum and impact parameter. The integral is evaluated by means of the Monte Carlo technique by sampling different initial conditions and impact parameters. The sampling is performed following the probability distribution function in each degree of freedom. Within this approach, for a given momentum P_0, it is possible to find the maximum impact parameter at which the three-body recombination occurs, $b_{max}(P_0)$. In other words, $\wp_{rec}(b, P_0) = 0$ for $b > b_{max}(P_0)$. Therefore, the three-body recombination cross section can be expressed as

$$\sigma_{rec}(P_0) = \Omega_b^5 \int_0^{b_{max}(P_0)} \wp_{rec}(b, P_0)b^4 db, \qquad (10.31)$$

where $\Omega_b^5 = 8\pi^2/3$ is the solid hyperangle associated with b, and the integral is preferably solved using the Monte Carlo approach [26]. Finally, the energy-dependent three-body rate constant is introduced as

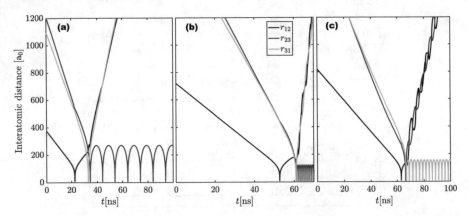

Fig. 10.6 Classical trajectories for ion–atom–atom three-body collisions at 10 mK and different impact parameters. $Ba^+ + Rb + Rb$ for $b = 0a_0$ in panel (**a**); for $b = 222a_0$ in panel (**b**) and $b = 444a_0$ in panel (**c**). The calculations are based on the same method and potential interaction as in Fig. 10.5. Here, the ion–neutral distances are characterized by r_{23} and r_{31}. Here, the ion is identified with the particle labelled as 3

$$k_3(P_0) = \frac{P_0}{\mu} \sigma_{\text{rec}}(P_0). \tag{10.32}$$

Three-body recombination in ion–neutral–neutral collisions leads to two different product states: a molecular ion or a neutral molecule. However, as shown in Fig. (10.6), displaying some characteristic trajectories for $Ba^+ + Rb + Rb$ three-body collisions, in the three examples, the product state is a molecular ion. Therefore, it seems that ion–atom–atom collisions mainly lead to the formation of molecular ions; indeed, this hypothesis is the main topic of the next section. Finally, we would like to point out that the study of ion–neutral–neutral three-body recombinations at cold temperatures plays a major role in ion–neutral hybrid trap experiments. Indeed, the three-body recombination reaction is the principal loss mechanism for certain ionic species when they are brought into contact with an ultracold cloud of atoms [27–30].

10.6 Universality in Ion–Atom–Atom Three-Body Recombination Reactions

The existence of threshold laws for elastic and inelastic collisions is very familiar in quantum mechanics: the so-called Wigner threshold laws, introduced in Chap. 2. These threshold laws explain how the elastic and inelastic cross sections behave at low collision energies. In the same vein, there are also classical threshold laws, such as the Langevin cross section introduced in Chap. 9, which establishes the behavior of the ion–neutral collisions at low collision energies [31]. Several years after that, Wannier founded the classical threshold law for the interaction between three charged particles [32], although following a different approach than Langevin proposed. Interestingly enough, the Wannier threshold law predicts that the rate constant for three charged particles shows a power-law dependence on the collision energy in which the exponent is an irrational number. Indeed, this unexpected dependency has been experimentally corroborated by looking at the double photoionization spectra of helium [33, 34] and to the scape properties of two electrons from a single charged ion [35, 36]. Recently, following a classical capture model [18], the study of classical threshold laws have been generalized to the three-body problem involving neutral atoms as well as two neutral atoms and one single charged ion [21, 37]. These threshold laws are studied in detail below.

At low collision energies the scattering properties are highly influenced by the long-range two-body interaction, $V(R) \rightarrow C_s/R^s$, with $s > 2$. In this scenario, the maximum impact parameter $\tilde{b}_{max}$ can be estimated as the distance at which the interaction potential is of the same order of magnitude as the collision energy, i.e.,

$$E = \frac{C_s}{b_{max}^s}. \tag{10.33}$$

In other words, the maximum impact parameter is the distance at which the motion of the particle starts to deviate from a rectilinear uniform trajectory (see Chap. 2). This distance has a similar physical meaning to the classical capture radius employed for the derivation of the Langevin cross section [17, 18, 31], but at zero angular momentum. As in the classical capture model, here, it is assumed that all the trajectories with $b \leq b_{max}$ will result in a three-body recombination event. Thus, the three-body recombination cross section is then expressed as (taking into account Eq. (10.29))

$$\sigma_{rec}(E_k) = \frac{8\pi^2}{3} \int_0^{b_{max}(E_k)} b^4 db \propto b_{max}^5(E_k). \tag{10.34}$$

By means of Eqs. (10.33) and (10.34), and by including the relationship between momentum and energy ($P \propto E_k^{1/2}$), we obtain

$$k_3(E_k) \propto E_k^{1/2} \frac{1}{E_k^{5/s}} = E_k^{\frac{s-10}{2s}}. \tag{10.35}$$

The same classical threshold law can be derived from a different standpoint. Let us assume that the maximum impact parameter b_{max} defines a capture volume around each particle $V_b = 4/3\pi b_{max}^3$. The probability p of having a three-body event is given by the probability of finding two particles within the capture volume of the third one, i.e., $p = n^2 V_{b_{max}}^2$, where n represents the number density of the system. Thus, the mean free path is given by [18]

$$\lambda = \frac{b_{max}}{p} = \frac{9}{16\pi^2 n^2 b_{max}^5}, \tag{10.36}$$

and taking into account that $\lambda = 1/(n^2 \sigma_{rec}(E))$,[2] it is finally found $\sigma_{rec}(E) \propto b_{max}^5(E)$, and hence $k_3(E) \propto E^{\frac{s-10}{2s}}$, as it has been obtained by means of the three-body cross section in a 6D space.

10.6.1 Threshold Law for Cold Ion–Atom–Atom Collisions

In the previous derivations it has been assumed that all the two-body interactions show the same long-range behavior; thus, a reasonable question arises: what about different two-body interactions? This question has been partially answered by studying the threshold law of ion–neutral–neutral three-body recombination [37]. In such a scenario, the two neutral atoms interact through the expected van der Waals potential $V(R) = -C_6/R^6$, whereas the two ion–neutral interactions will be dominated by the charge-induced dipole moment interaction $V(R) = -C_4/R^4$. Then, a classical capture model is employed, in analogy to the derivation for the threshold law for neutral three-body recombination, but in this case the capture radius is given by

$$E = \frac{C_4}{b_{max}^4}, \tag{10.37}$$

where it is assumed, as one would expect, that the ion–neural interaction will dominate over the neutral–neutral interaction. Plugging Eq. (10.37) into Eq. (10.34) we find that the three-body recombination cross section reads as

$$\sigma(E_k) \propto E_k^{-5/4}, \tag{10.38}$$

and the energy-dependent rate constant reads as

$$k_3(E_k) \propto E_k^{-3/4}. \tag{10.39}$$

The results from Eqs. (10.38) and (10.39) are the threshold behaviors of the ion–neutral–neutral three-body recombination cross section and rate constant respectively.

The numerical results for ion–neutral–neutral three-body recombination at low collision energies have been fitted to $k_3(E_k) = \gamma E_k^\beta$, and the fitting parameters are shown in Table 10.1. The results of the Table clearly support the classical threshold law for ion–atom–atom collision. Therefore, the ion–atom interaction dominates the ion–atom–atom dynamics at cold temperatures. This threshold law implies that in processes like $A^+ + A + A$, the dominant product state would be $A_2^+ + A$ instead

---

[2]This is the definition of the mean free path for three-body collisions, where the density appears squared since the cross section has units of length to the fifth power.

Table 10.1 Classical threshold law for the three-body recombination cross section. The three-body recombination cross section as a function of the collision energy, based on classical trajectory calculations. The fitting assumes a power law functional form, as explained in the text. The error on the fitting parameters is associated with a confidence interval of 95%. (Table adapted with permission from Ref. [37] Copyright (2020) (AIP Publishing))

System	γ (a_0^5)	β (dimensionless)
^{87}Rb$^+$ - ^{87}Rb - ^{87}Rb	$(7.94 \pm 2.72)\,10^{11}$	-1.178 ± 0.068
^{138}Ba$^+$ - ^{87}Rb - ^{87}Rb	$(3.57 \pm 0.07)\,10^{11}$	-1.269 ± 0.132
Classical threshold law		-1.25

of $A_2 + A^+$. Similarly, $A_2^+ + A$ collisions would lead to $A_2^+ + A$ as the main product state. Therefore, the threshold law has important implications in cold chemistry.

10.6.2 Threshold for Cold Ion–Atom–Atom Collisions: From the Cold to the Thermal Regime

The previous results based on the CT method indicate that for ion–atom–atom three-body recombination processes the ion–atom interaction mainly controls the nature of the product states, since the ion–atom interaction is more attractive than the usual van der Waals forces between atoms. In the Langevin capture model, an inelastic or reactive collision occurs for every trajectory that shows a barrier height lower than its kinetic energy, E_k. Indeed, for a given kinetic energy there is a maximum impact parameter, the so-called Langevin impact parameter, $b_L = (2\alpha/E_k)^{1/4}$, (introduced in Chap. 9) where α is the atom polarizability, which determines the maximum impact parameter for a given inelastic or reactive ion–atom collision, whereby Eq. (10.34) reads as

$$\sigma_3(E_k) = \frac{8\pi^2}{3} \int_0^{b_L} b^4 db, \tag{10.40}$$

and performing the integration yields

$$\sigma_3(E_k) = \frac{8\pi^2}{15} \left(\frac{2\alpha}{E_k}\right)^{5/4}. \tag{10.41}$$

From the cross section it is possible to define the energy-dependent three-body recombination rate as

$$k_3(E_k) = \sigma_3(E_k) \left(\frac{2E_k}{\mu}\right)^{1/2}, \tag{10.42}$$

and performing its thermal average through the Maxwell–Boltzmann distribution yields

$$k_3(T) = \frac{1}{2(k_B T)^3} \int_0^\infty k_3(E_k) E_k^2 e^{-\frac{E_k}{k_B T}} dE_k$$

$$= \frac{8\pi^2}{15} \frac{\Gamma(9/4)3^{1/4}}{\sqrt{2}(k_B T)^{3/4}} \frac{(2\alpha)^{5/4}}{\sqrt{m}}. \tag{10.43}$$

Therefore, we find that the three-body recombination rate depends on the temperature as $T^{-3/4}$ [10, 37], as expected. The same dependence on the temperature was obtained by Smirnov [41] back in the 1960s, but assuming that the three-body recombination can be described as a two-step chemical process, i.e., as two different two-body collisions. The first leads to the formation of a complex or quasi-bound state and the second may stabilize it, thus leading to the formation of a molecular ion.

The predictions for ion–neutral–neutral three-body recombination based on Eq. (10.43) for noble gas atoms are confronted with experiment data in Fig. 10.7. In particular, panel (a) presents results for data at 78 K, whereas panel (b) shows data

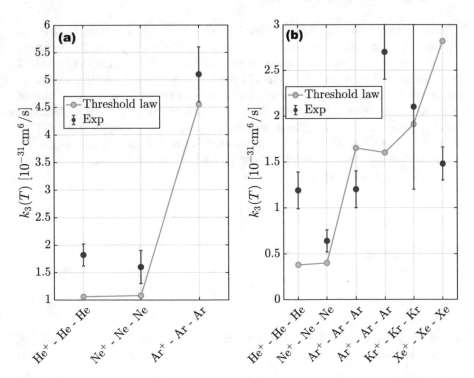

Fig. 10.7 Ion–neutral–neutral three-body recombination rate for different rare gas atoms as a function of the temperature. In panel (**a**) the temperature is 78 K. In panel (**b**) the temperature for all the atoms is 300 K except the second Ar$^+$-Ar-Ar data point, which is for 320 K. The experimental data points of panel (**a**) have been obtained from Ref. [38]; for panel (**b**), the data points are taken from Ref. [38] for He, Ne, and Ar, whereas Kr and Xe data are taken from Ref. [39]. In panel (**b**) there are two data points for Ar, the first one (from left to right) corresponds to 300 K [39] and the second to 320 K [38]. (Figure adapted with permission from Ref. [40] Copyright (2020) (American Physical Society))

for 300 K. The presented theory agrees well with the room temperature data, and it does even better at lower temperatures. Furthermore, the threshold law describes qualitatively how the rate depends on the atomic properties of the colliding partners, independently of the temperature. Therefore, the threshold law is confirmed beyond the cold regime, and at the same time, reinforces the fact that the ion–neutral interaction is the dominant interaction in ion–atom–atom three-body recombination.

10.7   Shadow Scattering: A Universal Feature of Few-Body Collisions

It is well-known that quantal elastic two-body cross section at high energies is twice the classical one, which is the geometric cross section [42–44]. This universal behavior of the high-energy elastic cross section is known as shadow scattering. At high collision energies most of the incoming particles are reflected to the backward direction; however, some of them close to the edge of the potential will diffract with a very small angle inducing a shadow in the forward direction as sketched in Fig. (10.8). This shadow can be observed under the right conditions [45, 46], although it will disappear at some point in the forward direction owing to the diffraction with the edge of the potential. In such a scenario, the quantum mechanical cross section will be given by the flux of particles reflected in the backward direction plus the ones deflected in the forward direction, both being equal to the geometric cross section [42], and in this way, explaining the nature of the shadow scattering from a phenomenological approach.

A more fundamental explanation of the shadow scattering relies on the inherent quantum mechanical nature of scattering. Motivated by this statement we go back to

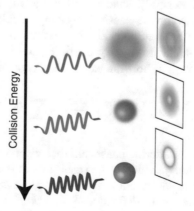

Fig. 10.8 Shadow scattering illustration. At high collision energies, the quantum mechanical cross section is twice the expected classical cross section, i.e., the geometric cross section. This phenomenon can be explained assuming that half of the flux of the incoming particles are scattered in the backward direction, whereas the remainder are scattered in the forward direction. Forward scattering leads to the effect of shadow scattering

the definition of the elastic cross section introduced in Sect. 2.2.1, but in the present case we choose its general definition as

$$\sigma_e = \frac{\pi}{k^2} \sum_{l=0}^{\infty} (2l+1)|1 - S_l|^2, \tag{10.44}$$

where $S_l = e^{i2\delta_l}$ denotes the S-matrix for the partial wave l, with δ_l representing the two-body scattering phase shift, and k stands for the wave vector. Apart from elastic processes, two-body collision may show absorption channels, fragmentation channels, and inelastic channels; all these processes are included in the so-called reaction cross section [42, 47]

$$\sigma_r = \frac{\pi}{k^2} \sum_{l=0}^{\infty} (2l+1) \left(1 - |S_l|^2\right). \tag{10.45}$$

Equations (10.44) and (10.45) clearly show that the existence of elastic processes is accompanied by the presence of inelastic processes, and vice versa. Also, from such equations it is noted that only σ_e can vanish if $S_l = 1$, implying the absence of inelastic scattering too through Eq. (10.45). This relation can be viewed as the origin of shadow scattering, since it reflects the fact that elastic and inelastic are always together. Indeed, it can be shown that in the case of high-energy collisions, where $S_l = 0$ below certain maximum l, $\sigma_e = \sigma_r$ and they are the same as the classically predicted [42].

Shadow scattering has been studied and characterized in two-body collisions. However, the few-body analog has not been explored, even though Eqs. (10.44) and (10.45) can be generalized for any few-body collisions, and hence shadow scattering eventually will emerge. Recently, shadow scattering in three-body recombination has been studied [21] based on the classical trajectory calculations in hyperspherical coordinates and the quantal three-body elastic cross section.

The quantum mechanical three-body elastic cross section for distinguishable particles interacting through a short-ranged two-body potential is given by

$$\sigma^{3B} = \frac{32\pi^2}{k^5} \sum_{\lambda} N_\lambda |e^{2i\delta_\lambda} - 1|^2 = \frac{32\pi^2}{k^5} 4 \sum_{\lambda} N_\lambda \sin^2 \delta_\lambda. \tag{10.46}$$

Here, δ_λ is the three-body scattering phase shift for a given value of λ (it is equivalent to l for two-body collisions), which is the eigenvalue of the squared grand angular momentum operator (it is equivalent to the total angular momentum for two-body collisions), k is the wave vector defined as $E = \hbar^2 k^2/2\mu$, and N_λ is a numerical factor that takes into account the degeneracy associated with the grand angular momentum

$$N_\lambda = \frac{4 + 2\lambda}{4} \frac{(\lambda + 3)!}{\lambda! 3!}. \tag{10.47}$$

The summation in Eq. (10.46) is extended over all the values of λ (partial waves). But the finite range nature of the potential translates to $\delta_\lambda \to 0$ for high values of λ, and therefore the summation in practice has a finite number of terms.

For a high-energy collision a large number of partial waves must be included for the computation of the cross section. In general, at high energies the phase shift, δ_λ, is a rapidly oscillating function of λ; hence, the approximation $\sin^2 \delta_\lambda \approx 1/2$ is adequate (random phase approximation, see Chap. 9), leading to

$$\sigma^{3B} \approx \frac{32\pi^2}{k^5} 2 \sum_{\lambda=0}^{\lambda_{max}} N_\lambda, \tag{10.48}$$

where λ_{max}; it is the highest value of λ satisfying the random phase criterion. In general, $\lambda_{max} \gg 1$ then

$$\sum_{\lambda=0}^{\lambda_{max}} N_\lambda \approx \sum_{\lambda=0}^{\lambda_{max}} \frac{\lambda^4}{12} \approx \frac{\lambda_{max}^5}{5 \times 12}, \tag{10.49}$$

and hence the quantum mechanical three-body elastic cross section is expressed as

$$\sigma^{3B} \approx \frac{16\pi^2}{3k^5} \frac{\lambda_{max}^5}{5}. \tag{10.50}$$

The classical three-body elastic cross section at high energies is given by Eq. (10.6)

$$\sigma(P_0) = \frac{8\pi^2}{3} \int_0^R b^4 db = \frac{8\pi^2 R^5}{3 \times 5}, \tag{10.51}$$

where a hard-sphere collision model has been assumed for the opacity function associated with the elastic collisions.

The initial grand angular momentum is defined as

$$\Lambda = \rho_0 \times P_0, \tag{10.52}$$

where the cross product is defined in a 6D space. By definition, P_0 and b are orthogonal; thus,

$$|\vec{\Lambda}|^2 = b^2 P_0^2. \tag{10.53}$$

The eigenvalues of the grand angular momentum are $\hbar^2 \lambda(\lambda + 4)$ [15], and for large values of λ they can be approximated by $\hbar^2 \lambda^2$; thus, Eq. (10.53) determines the relation between the grand angular momentum and the impact parameter yielding

$$\lambda_{\max} = kR, \tag{10.54}$$

and plugging Eq. (10.54) in Eq. (10.51) one finds

$$\sigma^{3B} = 2 \times \sigma(P_0). \tag{10.55}$$

In Eq. (10.55) it is noticed that the high-energy collision limit of the quantum three-body elastic cross section is twice the classical one, as would be expected in the case of two-body collisions [42, 43]. In other words, Eq. (10.55) established the existence of shadow scattering in three-body collisions. Furthermore, this result establishes the universality of the shadow scattering beyond the conventional two-body collisions.

References

1. Hess HF, Bell DA, Kochanski GP, Cline RW, Kleppner D, Greytak TJ (1983) Phys Rev Lett 51:483
2. Hess HF, Bell DA, Kochanski GP, Kleppner D, Greytak TJ (1984) Phys Rev Lett 52:1520
3. de Goey LPH, vd Berg THM, Mulders N, Stoof HTC, Verhaar BJ, Glöckle W (1986) Phys Rev B 34:6183
4. Esry BD, Greene CH, Burke JP (1999) Phys Rev Lett 83:483
5. Burt EA, Ghrist RW, Myatt CJ, Holland MJ, Cornell EA, Wieman CE (1997) Coherence, correlations, and collisions: what one learns about bose-einstein condensates from their decay. Phys Rev Lett 79:337. https://doi.org/10.1103/PhysRevLett.79.337
6. Weber T, Herbig J, Nägerl HC, Grimm R (2003) Phys Rev Lett 91:123201
7. Fedichev PO, Reynolds MW, Shlyapnikov GV (1996) Phys Rev Lett 77:2921
8. Suno H, Esry BD, Greene CH (2003) Phys Rev Lett 90:053202
9. Pérez-Ríos J (2019) Vibrational quenching and reactive processes of weakly bound molecular ions colliding with atoms at cold temperatures. Phys Rev A 99:022707. https://doi.org/10.1103/PhysRevA.99.022707
10. Greene CH, Giannakeas P, Pérez-Ríos J (2017) Universal few-body physics and cluster formation. Rev Mod Phys 89:035006. https://doi.org/10.1103/RevModPhys.89.035006
11. Linderberg J, Vessal B (1987) Reactive scattering in hyperspherical coordinates. Int J Quantum Chem 31(1):65. https://doi.org/10.1002/qua.560310108. https://onlinelibrary.wiley.com/doi/pdf/10.1002/qua.560310108, https://onlinelibrary.wiley.com/doi/abs/10.1002/qua.560310108
12. Pack RT, Parker GA (1987) Quantum reactive scattering in three dimensions using hyperspherical (aph) coordinates. theory. J Chem Phys 87(7):3888. https://doi.org/10.1063/1.452944
13. Kuppermann A, Kaye JA, Dwyer JP (1980) Hyperspherical coordinates in quantum mechanical collinear reactive scattering. Chem Phys Lett 74(2):257. https://doi.org/10.1016/0009-2614(80)85153-0. http://www.sciencedirect.com/science/article/pii/0009261480851530
14. Schatz GC (1988) Quantum reactive scattering using hyperspherical coordinates: results for h+h2 and cl+hcl. Chem Phys Lett 150(1):92. https://doi.org/10.1016/0009-2614(88)80402-0. http://www.sciencedirect.com/science/article/pii/0009261488804020
15. Avery J (1986) Hyperspherical harmonics: applications in quantum theory. Kluwer, Dordrecht, The Netherlands.
16. Avery JE, Avery JS (2018) Hyperspherical harmonics and their physical applications. World Scientific, Singapore
17. Levine RD (2005) Molecular reaction dynamics. Cambridge University Press, Cambridge
18. Levine RD, Bernstein RB (1987) Molecular reaction dynamics and chemical reactivity. Oxford University Press, New York

19. Whittaker ET (1937) A trataise on the analytical dynamics of particles and rigid bodies. Cambridge University Press, Cambridge
20. Karplus M, Porter RN, Sharma RD (1965) J Chem Phys 43:3259
21. Pérez-Ríos J, Ragole S, Wang J, Greene CH (2014) Comparison of classical and quantal calculations of helium three-body recombination. J Chem Phys 140:044307
22. Smith FT (1962) Discuss Faraday Soc 33:183
23. Maslov VP, Fedoriuk MV (1981) Semi-classical approximation in quantum mechanics. D. Reidel Publishing Company, London
24. Landau LD, Lifshitz EM (1976) Mechanics. Elsevier Butterworth-Heinemann, Burlington
25. Press WH, Teukolsky SA, Vetterling WT, Falnnery BP (1986) Numerical recipes in Fotran 77. Cambridge Univeristy Press, Cambridge
26. Shui VH (1972) Thermal dissociation and recombination of hydrogen according to the reactions $H_2 + H \rightarrow H + H + H$. Doctoral Tesis, Deparment of mechanical ingeneering MIT
27. Härter A, Krükow A, Brunner A, Schnitzer W, Schmid S, Denschlag JH (2012) Phys Rev Lett 109:123201
28. Härter A, Krükow A, DeißM, Drews B, Tiemann E, Denschlag JH (2013) Population distribuion of product staets foloowing three-body recombination in an ultracold atomic gas. Nat Phys 9:512
29. Härter A, Denschlag JH (2014) Cold atom-ion experiemtns in hybrid traps. Contemp Phys 55:33
30. Krükow A, Mohamadi A, Härter A, Denschlag JH, Pérez-Ríos J, Greene CH (2016) Energy scaling of cold atom-atom-ion three-body recombination. Phys Rev Lett 116:193201
31. Langevin P (1905) Une formule fodnamentale de théorie cinétique. C R Acad Sci 140:35
32. Wannier GH (1953) The threhold law for single ionization of atoms of ions by electrons. Phys Rev 90:817
33. van der Wiel MJ (1972) Threshold behavior of double photoionisation in He. Phys Lett A 41:389
34. Kossmann H, Schmidt V, Andersen MA (1988) Test of Wannier threshold laws: double-photoionization cross section in helium. Phys Rev Lett 60:1266
35. Cvejanovic S, Read FH (1974) Studies of the threshold electron impact ionization of helium. J Phys B 7:1841
36. Donahue JB, Gram PAM, Hynes MV, Hamm RW, Frost CA, Bryant HC, Butterfield KB, Clark DA, Smith WW (1982) Observation of two-electron photoionization of the H^- ion near threshold. Phys Rev Lett 48:1538
37. Pérez-Ríos J, Greene CH (2015) Communication: classical threhold law for ion-neutral-neutral three-body recombination. J Chem Phys 143:041105
38. Johnsen R, Chen A, Biondi MA (1980) Three-body association reactions of He^+, Ne^+, and Ar^+ ions in their parents gases from 78 to 300 K. J Chem Phys 73:1717
39. Neves PNB, Conde CAN, Tavora LMN (2010) The $x^+ + 2x \rightarrow x_2^+ + x$ reaction rate constant for Ar, Kr and Xe, at 300 k. Nucl Inst Meth Phys Res A 619:75
40. Pérez-Ríos J, Greene CH (2018) Universal temperature dependence of the ion-neutral-neutral three-body recombination rate. Phys Rev A 98:062707
41. Smirnov BM (1967) Transitions between atomic and molecular ions. JETP 24:1180
42. Bethe HA, Morrison P (2006) Elementary nuclear theory. Dover, New York
43. Child MS (1974) Molecular collision theory. Academic, New York
44. Serber R (1965) Shadow scattering at large angles. Proc Natl Acad Sci USA 54:692
45. Geiger J, Morón-León D (1979) Electrom-atom shadow scattering. Phys Rev Lett 42:1336
46. Goggi G, Sfora MC, Conta C, Fraternali M, Livan M, Montovani GC, Pastore F, Rossini B, Alberi G (1978) Observation of a narrow minimum and of inelastic shadow effects in deuteron-deuteron elastic scattering at $\sqrt{s}$= 53 GeV. Phys Lett B 77:433
47. Landau LD, Lifshitz EM (1958) Quantum mechanics. Butterworht Heinemann, Oxford

Cold Chemical Reactions Between Molecular Ions and Neutral Atoms

11

11.1 Introduction

In previous chapters we have studied physical and chemical processes in atom–ion hybrid systems. However, the field of hybrid atom–ion systems is evolving toward the study of molecular ion–atom interactions. This emerging field brings a new degree of control into play, i.e., the internal degrees of freedom in one of the colliding partners. The presence of internal degrees of freedom is crucial to elucidating the ultimate nature of ion–neutral collisions, including stereochemical effects, and the possibility of sympathetic cooling of molecular ions. Furthermore, molecular ion–atom hybrid systems are interesting for the development of novel high-precision spectroscopy techniques for rotational states of molecular ions [1–3].

The control over the internal degrees of freedom of a molecular ion is complicated by multiple decoherence processes that depend on the collision energy, and among them, the two most relevant are chemical reactions and relaxation [4–7]. Relaxation processes are all sorts of energy transfer mechanisms between different degrees of freedom in a molecular system. For instance, vibrational relaxation is the transfer of vibrational energy into translational degrees of freedom and vice versa. Relaxation processes have largely been studied in chemical physics [4–7] and, in a few cases, for molecular ion–atom interactions [8–14].

In this chapter, we introduce the relaxation and reactive processes attached to a cold molecular ion when it is brought in contact with an ultracold gas of atoms. In particular, we introduce a semi-classical framework, applicable at cold temperatures, as we explain throughout the chapter. In addition, some quantal results are presented to understand the different relaxation mechanisms of molecular ions in ultracold gases.

© Springer Nature Switzerland AG 2020
J. Pérez Ríos, *An Introduction to Cold and Ultracold Chemistry*,
https://doi.org/10.1007/978-3-030-55936-6_11

11.2   Quasi-Classical Trajectory Method

The quasi-classical trajectory (QCT) method has become a standard theoretical approach for the study of molecular reaction dynamics since the pioneering work of Karplus et al. for the study of vibrational relaxation of the hydrogen molecule induced by collisions with hydrogen atoms [15, 16]. In the QCT methodology, the dynamics of the nuclei is described by means of Newton's laws of classical mechanics. The initial conditions of every trajectory are selected from the initial quantum state of the system through the Wentzel, Kramers, and Brillouin (WKB) or semi-classical approximation. The same logic is ulteriorly used at the end of the propagation of the Newton's or Hamilton's equations to map the momenta and coordinates of a trajectory into the quantal state of the system, as shown in Fig. 11.1.

The QCT method is an approximation of the inherent quantal nature of every scattering process. Therefore, it may only be applicable under particular conditions where its validity is guaranteed. In particular, the QCT method is suitable for scenarios where quantum mechanical effects are washed out or hard to recognize. For instance, the QCT method describes better the reaction dynamics for heavier nuclei and high collision energies. The first situation refers to collisions with negligible zero point energy where quantum effects are highly suppressed. The second scenario is related to high-energy collision where a large number of partial waves contributes to the collision. Moreover, QCT is especially suited to deal with situations in which complex time-dependent potentials difficult the application of pure quantum mechanical approaches. The range of applicability of the QCT method is the same as the classical trajectory (CT) method presented in the previous

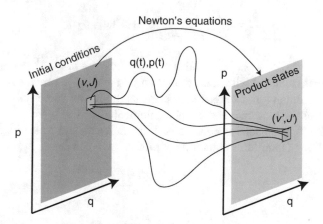

Fig. 11.1 Schematic representation for the quasi-classical trajectory method. The semi-classical quantization rule is applied to transform the initial collisional state (v, J) into a set of different trajectories that correlates with the continuum phase space in classical mechanics. The evolution of different trajectories correlating with the initial states is performed using Newton's equations until their final state, in which the semi-classical quantization rule is applied (v', J')

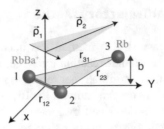

Fig. 11.2 Atom–molecule collision from a classical perspective. The cross section is character-ized by the impact parameter, b, whereas the dynamics is described with the Jacobi vectors (ρ_1, ρ_2), which are related to vectors joining the different atoms, r_{ij}, with (i, j)=1,2,3 and $i \neq j$

chapter (see Sect. 10.1). Finally, it is important to note that the QCT method is unable to predict any resonance or pure quantal effects in scattering observables.

The Hamiltonian (this Hamiltonian has been previously introduced in Chap. 10) of three particles with masses m_i, with $i = 1, 2, 3$ subjected to the potential energy surface $V(r_1, r_2, r_3)$ is given by

$$H = \frac{p_1^2}{2m_1} + \frac{p_2^2}{2m_2} + \frac{p_3^2}{2m_3} + V(r_1, r_2, r_3), \qquad (11.1)$$

where r_i and p_i are the vector position and momentum of the i-th atom respectively. This Hamiltonian is further simplified in Jacobi coordinates [16–18], shown in Fig. 11.2, and neglecting the trivial center of mass (CM) motion yields

$$H = \frac{P_1^2}{2m_{12}} + \frac{P_2^2}{2m_{3,12}} + V(\rho_1, \rho_2), \qquad (11.2)$$

where $m_{12} = (m_1^{-1} + m_2^{-1})^{-1}$ and $m_{3,12} = (m_3^{-1} + m_{12}^{-1})^{-1}$. Here, ρ_1 is the Jacobi vector associated with the orientation and size of the molecule and P_1 is its conjugate momentum, whereas ρ_2 represents the Jacobi vector joining the atom with the CM of the molecule and P_2 is its momentum.

The motion of the nuclei is considered fully classical, and hence its evolution is found by solving Hamilton's equations of motion that read as

$$\frac{d\rho_{i,\alpha}}{dt} = \frac{\partial H}{\partial P_{i,\alpha}} \qquad (11.3)$$

$$\frac{dP_{i,\alpha}}{dt} = -\frac{\partial H}{\partial \rho_{i,\alpha}}, \qquad (11.4)$$

where $i = 1, 2$, and α represent the different Cartesian components of the Jacobi vectors.

11.2.1 Initial Conditions

Let us assume an atom–molecule collision at a particular collision energy E_k in the CM frame in which the atom is placed along the z-axis at a distance R from the CM of the molecule. The CM of the molecule corresponds to the origin of the coordinates. In this scenario, after the impact parameter b has been specified, the vectors $\boldsymbol{\rho}_2$ and $\boldsymbol{P}_2$ are fully characterized [16].

However, $\boldsymbol{\rho}_1$, and its conjugate momentum, $\boldsymbol{P}_1$, are selected according to the rovibrational state, (v, j), of the molecule through the WKB approximation. In particular, the size of the molecule, $|\rho_1|$, is taken as the outer classical turning point r_+.[1] The orientation of the molecule is described in spherical coordinates after specifying its azimuthal angle, θ, with respect to the z-axis and its polar angle in the $x - y$ plane, ϕ. The momentum of the molecule, $\boldsymbol{P}_1$, is determined through the rovibrational energy of the molecule by imposing the conservation of energy and angular momentum. In particular, at r_+ the available momentum is in the form of angular momentum since the radial component of the momentum is zero. As a consequence, the radial component of the angular momentum, $\boldsymbol{J} = \boldsymbol{r} \times \boldsymbol{p} \equiv \boldsymbol{\rho}_1 \times \boldsymbol{P}_1$, is null too, which implies that the momentum and the radial position need to be orthogonal and one finds $P_1 = \hbar\sqrt{j(j + 1)}/r_+$. Finally, the $\boldsymbol{P}_1$ is determined by the angle η between the angular momentum, $\boldsymbol{J}$, and a normal vector to the molecular axis, $\boldsymbol{K} = \boldsymbol{J} \times \boldsymbol{e}_z$, as [16]

$$\boldsymbol{P}_1 = \begin{pmatrix} P_1 \left(\sin \phi \cos \eta - \cos \theta \cos \phi \sin \eta\right) \\ -P_1 \left(\cos \phi \cos \eta + \cos \theta \sin \phi \sin \eta\right) \\ P_1 \sin \theta \sin \eta \end{pmatrix} \qquad (11.5)$$

To be sure that rotation and vibration of the molecule are described properly during the course of a collision, it is better to take the initial distance between the atom and the molecule as

$$R = R_0 + \frac{\chi P_2 \tau_{v,j}}{\mu_{3,12}}, \qquad (11.6)$$

where R_0 is some fixed atom-molecule distance in which the potential energy is negligible in comparison with the collision energy, $\chi \in [0, 1]$ is a uniform random variable and

---

[1] In general, the same condition applies to r_-, which can be used to change from trajectory to trajectory where the vibrational motion starts [16].

$$\tau_{v,j} = \sqrt{2m_{12}} \int_{r_-}^{r_+} \frac{dr}{\sqrt{2\mu \left[E_{\text{int}} - V(r) - \frac{\hbar^2 j(j+1)}{2m_{12}r^2} \right]}}, \tag{11.7}$$

is the vibrational period associated with a rovibrational state (v, j). In Eq. (11.7), E_{int} is the rovibrational energy, $V(r)$ stands for the molecular potential energy[2] and r_- represents the inner classical turning point for a given rovibrational energy of the molecule.

11.2.2 Reaction Products

The different types of reaction products depend on the characteristic reaction under consideration, whereas the branching ratio of the product states depends on the nature of the reactants. For molecular ion–atom reactions, $AB^+ + C$, assuming that the collision occurs in a single potential energy curve,[3] there are only two possible reactions. The first is an inelastic process, a change of the rovibrational state of the molecular ion, the so-called vibrational relaxation or quenching. Vibrational quenching is fully characterized by the change in the vibrational quantum number of the molecular ion, which is determined through the WKB quantization rule as

$$v' = -\frac{1}{2} + \frac{1}{\pi \hbar} \int_{r_-}^{r_+} \sqrt{2m_{12} \left[E'_{\text{int}} - V(r) - \frac{\hbar^2 j'(j'+1)}{2m_{12}r^2} \right]} dr, \tag{11.8}$$

where E'_{int} represents the internal energy of the molecule, j' is the rotational quantum number given by

$$j' = -\frac{1}{2} + \frac{1}{2}\sqrt{1 + 4\frac{\vec{J}' \cdot \vec{J}'}{\hbar^2}}, \tag{11.9}$$

where $\boldsymbol{J}' = \boldsymbol{\rho}_1 \times \boldsymbol{P}_1$ at the final propagation time.[4] The solution Eq. (11.8) for v' leads to a real number; however, quantum mechanics dictates that v' must be an integer number. However, this apparent inconsistency is overcome through the binning methods, which are discussed below.

---

[2]Here, $r = r_{12}$ based on Fig. 11.2.

[3]This statement is equivalent to saying that charge-transfer processes are not considered.

[4]When the Langer approximation is used for the rotational quantum number, i.e., $j(j + 1) = (j + 1/2)^2$, the final rotational quantum number is given as $j' = -1/2 + \sqrt{\frac{\vec{J}' \cdot \vec{J}'}{\hbar^2}}$.

The second possible reaction for $AB^+ + C$ is a reactive process in which old bonds are substituted by a new one. In particular, two reactive pathways are plausible:

- Dissociation, $AB^+ + C \rightarrow A + B^+ + C$, the final state are three atoms moving apart from each other. This process, classically, becomes operative when the collision energy is larger than the binding energy of the molecular ion.
- Molecular formation or reactive process, $AB^+ + C \rightarrow AC + B^+$, or, $AB^+ + C \rightarrow CB^+ + A$, that is, the formation of a new molecule not present in the reactants.

11.2.3 Binning Methods

QCT methods rely on the semi-classical quantization rule to transform the quantum numbers of the reactants into valid initial conditions to propagate the classical trajectories. The same quantization rule is applied at the final time of propagation to translate from the momentum and position of a trajectory into the proper quantum numbers for the products. In fact, the assignment of the quantum numbers for the trajectories at the final propagation time can be viewed as a discretization of the continuous classical phase space. And the first approach to this discretization would be to histogram the phase space for different quantum numbers by counting the number of trajectories that end up in each bin. Of course, the binning method is not unique and one may use different criteria.

The most straightforward and frequently used binning method is the so-called standard or uniform binning, in which each trajectory has the same weight

$$W(v', v_t) = \begin{cases} 1 & |v' - v_t| \leq 1/2 \\ 0 & \text{Otherwise,} \end{cases} \tag{11.10}$$

where v_t is the target final vibrational state and v' is the result of Eq. (11.8). In other words, the uniform binning method rounds v' to its closest integer [19]. A more advanced binning method is the so-called Gaussian binning [20]. In this method, it is assumed that the trajectories that begin from the initial quantum number and end in the final quantum number are the best classical representation of a quantum transition. In particular, the trajectories are weighted following a Gaussian distribution as

$$W(v', v_t) = \frac{1}{\sigma\sqrt{2\pi}} e^{-\frac{|v' - v_t|^2}{2\sigma^2}}. \tag{11.11}$$

The performance of these binning methods depends on the problem at hand, but in general Gaussian seems to work better than uniform binning for predicting product distributions of vibrational states, although a large number of trajectories are required owing to the narrow weight function needed to reach accurate results

[21–23]. However, there is no strong physical reason why Gaussian binning should be better than uniform binning for all sorts of situations.

11.2.4 Cross Section

Classically, the cross section of a given process:

- Quenching (q); $AB^+(v) + C \rightarrow AB^+(v') + C$.
- Dissociation (d); $AB^+(v) + C \rightarrow A + B^+ + C$.
- Reaction (r); $AB^+(v) + C \rightarrow CB^+(v'') + A$ or
  $AB^+(v) + C \rightarrow AC(v''') + B^+$

is given by

$$\sigma_{q,r,d}(E_k) = 2\pi \int_0^{b_{max}^{q,r,d}(E_k)} P_{q,r,d}(b, E_k) b \, db, \qquad (11.12)$$

where $b_{max}^{q,r,d}(E_k)$ represents the maximum impact parameter of trajectories resulting in a particular reaction channel, and $P_{q,r,d}(b, E_k)$ is the opacity function, introduced in Chap. 2. The opacity function, as stated before in this book, is the probability that a reaction channel (q, r or d) occurs as a function of the impact parameter, b, and collision energy, E_k, and in the present case reads as

$$P_{q,r,d}(b, E_k) = \int P_{q,r,d}(b, E_k, \theta, \phi, \eta, \xi) d\Omega$$

$$d\Omega = \sin \theta d\theta d\phi d\eta d\xi. \qquad (11.13)$$

This integral is evaluated by means of a Monte Carlo method yielding

$$P_{q,r,d}(b, E_k) = \frac{N_{q,r,d}(b, E_k)}{N} \pm \delta_{q,r}(b, E_k). \qquad (11.14)$$

The evaluation of the integral in Eq. (11.13) proceeds as follows:

- We launch N trajectories, sampling the different variables of the integrand.
- Each trajectory evolves in time following Hamilton's equation.
- Some of the trajectories correlate with each of the possible product states $N_{q,r,d}(b, E_k)$ for a given b and E_k.

The error bar due to the inherent statistical nature of the Monte Carlo method can be estimated as the square root of the variance of $N_{q,r,d}(b, E_k)/N$, which is given by

$$\delta_{q,r,d}(b, E_k) = \frac{\sqrt{N_{q,r,d}(b, E_k)}}{N} \sqrt{\frac{N - N_{q,r,d(b,E_k)}}{N}} \tag{11.15}$$

where a one-standard deviation rule (68% confidence level) has been applied.

11.3   Vibrational Quenching of Molecular Ions by Ultracold Atoms

Understanding the quenching mechanisms of a molecular ion in the presence of a gas of ultracold atoms is crucial for exploring the possibility of sympathetic cooling of molecular ions and the physics of charged impurities in ultracold gases. In general, vibrational quenching shows the slower rate among all the different degrees of freedom in atom–molecule collisions. However, for some systems the situation may be a more involved and hence it becomes mandatory to reveal the underlying physical mechanism of vibrational quenching.

The physics behind the vibrational quenching of molecular ions in ultracold gases is generally studied in a full quantum mechanical framework [8–10, 24]. In particular, by solving the atom–molecule scattering problem by means of the coupled channel approach discussed in Chap. 5 and in Sect. 5.4.3. Namely, after a truncation of the Hilbert space (basis), the coupled differential equations for the scattering channels are numerically solved up to the long-range region, where its solutions are compared with the asymptotic ones, yielding the S-matrix and with it, the cross section, as shown in Chap. 2.

The results for vibrational quenching rate coefficients for $BaCl^+(v, j) + Ca$ are shown in Fig. 11.3, where it is noted that the theoretical quenching rates are in good agreement with the experimental ones (shaded square on the Figure). At low temperatures, the quenching rate coefficient tends toward a constant following the Wigner threshold law for inelastic collisions, given by Eq. (2.70).[5] The theoretical quenching rate coefficient shows that the efficiency of the vibrational quenching enhances as the temperature increases and reaches the Langevin rate, defined by Eq. (9.24) with α taken as the polarizability of Ca and $\mu = (1/m_{Ca} + 1/m_{BaCl^+})^{-1}$. Such behavior can be explained classically too.

Let us introduce the adiabaticity parameter as [25]

$$\xi = \frac{\tau_c}{\tau_{v,j}}, \tag{11.16}$$

where $\tau_{v,j}$ is the vibrational period introduced in Eq. (11.7). However, for the sake of simplicity we assume $\tau_{v,j} = \frac{1}{\omega_e}$, where ω_e is the harmonic frequency of the molecular ion. The collision time is estimated as

---

[5]It is worth recalling that the rate coefficient is defined as $\langle \sigma v \rangle$, where $\rangle\langle$ stands for the thermal averaged.

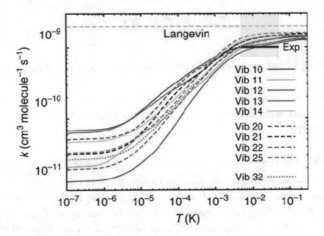

Fig. 11.3 Comparison between the vibrational quenching rate coefficients of several excited rovibrational levels (v, j) of $BaCl^+$ resulting from collisions with Ca with the experimental results and with the Langevin law. The first and second numbers designate, respectively, the initial vibrational and rotational quantum number of $BaCl^+$. (Figure adapted with permission from Ref. [8] Copyright (2020) (SpringerNature))

$$\tau_c = \frac{b_L}{v_k}, \qquad (11.17)$$

where b_L is the Langevin impact parameter given by Eq. (9.21), and v_k represents the collision velocity. As shown in panel a of Fig. 11.4, large values of ξ imply that the molecule vibrates many times before the collision happens; therefore, the atom feels an average rather than the complete effect of the vibration of the molecule–atom interaction, which corresponds to an inefficient vibrational–translational energy transfer. For $\xi \sim 1$ the vibrational–translational energy transfer becomes efficient because now the atom feels the vibration of the molecule during the course of a collision. Finally, at $\xi \ll 1$ the molecule shows a negligible vibrational motion during the collision; thus, in every collision a particular orientation and size of the molecule is being probed, which translates into a highly efficient vibrational–translational energy transfer [25].

In panel b of Fig. 11.4 the adiabaticity parameter for the vibrational–translational energy transfer for $BaCl^+$ + Ca collisions as a function of the collision energy is shown. For this calculation, for the sake of simplicity, we have assumed $\tau_{v,j} = \frac{1}{\omega_e}$, where ω_e is the harmonic frequency of $BaCl^+$ and its value is of $334 \, cm^{-1}$ [8]. The results in panel b of Fig. 11.4 indicate that for lower collision energies the vibrational quenching is less efficient, as is observed from quantum mechanical simulations in Fig. 11.3. However, is worth noticing that this comparison is appropriate for collision energies $\gtrsim 1$ mK.

Quantum mechanical scattering calculations for different molecular ion–atom systems have been paving the way for a general understanding of the vibrational

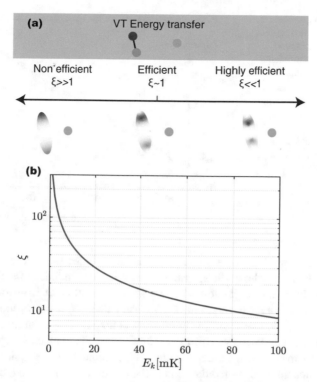

Fig. 11.4 Vibrational–translation energy transfer based on the adiabaticity parameter. Panel (**a**), an atom–molecule collision (top of the figure) has three distinct regimes based on the adiabaticity parameter, ξ, and regarding the efficiency of vibrational–translational energy transfer. The bottom part of the figure represents the probability of finding the atoms in the molecule within a collision. Panel (**b**), adiabaticity parameter for vibrational–translational energy transfer for $BaCl^+$ + Ca collisions as a function of the collision energy

quenching mechanism, which can be conveyed by a statistical capture model [26]. Within the statistical capture model vibrational quenching can be understood as a two-step mechanism. First, the long-range tail of the charge–neutral interaction pushes the atom toward the molecular ion, which captures the atom with a rate given by the Langevin rate k_L. Second, at short distances, the molecular ion–atom forms a three-body complex, in which the initial internal energy of the molecular ion plus the collision energy is shared among the different vibrational states of the complex. The energy is equally shared among every degree of freedom, which is the core assumption of any statistical model. The number of accessible vibrational states of the complex will be proportional to $D_e^{complex}/\omega_e$; therefore, the statistical capture model yields the following rate

$$k^{SC} \sim k_L \frac{D_e^{complex}}{\omega_e},$$

(11.18)

Fig. 11.5 Comparison between the full quantum mechanical vibrational quenching rate coefficients $k_{10}^{CC}(T)$ with the statistical capture rate $k_{10}^{SC}(T)$ for different colliding systems with molecular ions in the initial state ($v = 1, j = 0$). (Figure reproduced with permission from Ref. [8] Copyright (2020) (SpingerNature))

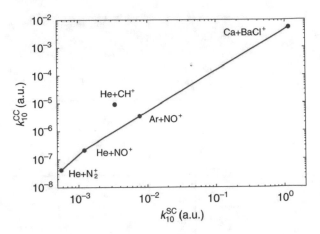

which can be viewed as a Langevin rate, but it includes the role of the internal degrees of freedom of the colliding partners, and is independent of the temperature. Figure 11.5 shows a comparison between the rate coefficient from the statistical capture model and the full quantum mechanical calculations for various molecular ion–atom systems. In that figure, is interesting to note that the quantum mechanical results correlate qualitatively with the predictions of the statistical capture model, which implies that such a model captures approximately the physics behind the vibrational quenching of molecular ions in the presence of ultracold atoms.

It is worth mentioning that the statistical approach is only suitable for molecular ions in deeply vibrational states, because as soon as $\omega_e \lesssim D_e^{complex}$, the rate overpasses k_L, and k_L is the largest rate possible for a charged–neutral collision. In particular, this condition is fulfilled when the molecular ion is in a highly vibrational state,[6] since in that case the energy spacing between energy levels is very small owing to the inherent anharmonicity of the molecular ion interaction. Indeed, the study of the vibrational relaxation of molecular ions in highly vibrational states is the topic of the next section, where QCT appears as a suitable method.

11.4 A Cold Highly Vibrational Excited Molecular Ion Immersed in an Ultracold Gas: QCT Method in Action

In the previous section, we have seen that quantum mechanical simulations accurately revealed the behavior of the vibrational quenching rate coefficient as a function of the temperature. However, the situation changes drastically if the molecular ion is in a highly excited vibrational state, i.e., $v \gg 1$. In such a scenario, a large number of rovibrational states need to be included in the Hilbert space

---

[6]The statistical capture model also presents some difficulties when the three-body system shows a very shallow well depth.

for the quantum mechanical simulations, thus leading to an enormous number of channels that make the quantum mechanical simulations computationally expensive and impractical. Moreover, as long as we study collisional properties at collision energies in which several partial waves contribute, a classical approach to the reaction dynamics should describe the major features of the scattering observables.

As we have seen in the previous chapter, the main product state of ion–atom–atom three-body recombination is the creation of a molecular ion. In particular, the molecular ion appears in highly vibrational states. However, the molecular ion is immersed in an ultracold *sea* of atoms, and hence further reaction may occur such as vibrational quenching, dissociation, or reactive processes. In this section, we focus on the scattering BaRb$^+$ + Rb system, as an example, to show the intriguing reaction network of a single ion in high-density media.

The starting point of the reaction of interest are the product states after ion–atom–atom three-body recombination, i.e., BaRb$^+(v_i)$ + Rb, where $v_i \gg 1$. For BaRb$^+$-Rb at 1 mK at least 18 partial waves contribute to the scattering; therefore, the QCT method is a reasonable choice for BaRb$^+(v_i)$ + Rb collisions at cold temperatures (for a more detailed discussion see Sect. 10.1). Therefore, BaRb$^+(v_i)$ + Rb collisions are studied by means of the QCT method, assuming that collision occurs in a unique potential energy surface, and hence charge transfer reactions are not included. Here, the PES is assumed to be described as pair-wise additive potentials as

$$V(r_1, r_2, r_3) = V(r_{12}) + V(r_{13}) + V(r_{23}), \tag{11.19}$$

where the charge is assumed to be localized on the Ba atom and the vectors involved have been described in Fig. 11.2. The localization of the charge is a consequence of considering that the reaction happens in a unique PES, because charge transfer can only occur in the presence of several potential energy surfaces, which correlate with different ion–atom–atom asymptotic states. The ion–atom interaction is described by means of the generalized Lennard-Jones potential

$$V(r) = -\frac{\alpha}{2r^4} \left[1 - \frac{1}{2} \left(\frac{r_m}{r} \right)^4 \right], \tag{11.20}$$

where $r_m = 9.27$ a$_0$ is the equilibrium distance for the atom–ion system and $\alpha = 320$ a.u. is the polarizability of the Rb(5S) atom. The model potential in Eq. (11.20) cannot describe the realistic position of the bound states of BaRb$^+$; however, it properly accounts for the density of states, since it accounts for the realistic long range atom–ion interaction. In other words, the model potential accurately reproduces the energy difference between consecutive states. For the Rb(5S)–Rb(5S) interaction it is assumed that the atoms are spin polarized, and hence they interact through the triplet potential energy curve.

The most relevant vibrational states for BaRb$^+$ from the potential of Eq. (11.20) are shown in Table 11.1. The first column in Table 11.1 stands for the vibrational ordering from the bottom of the potential minimum, whereas the second stands for

Table 11.1 Vibrational bound states (quantum mechanical calculation) for BaRb$^+$ (in mK) assuming the generalized Lennard-Jones potential described in Eq. (11.20)

v	v_i	Binding energy (E_v)
198	−1	0.01
197	−2	0.11
196	−3	0.43
195	−4	1.13
194	−5	2.48
193	−6	4.77
192	−7	8.37
191	−8	13.70
190	−9	21.24
189	−10	31.52
188	−11	45.15
187	−12	62.84
186	−13	85.12
185	−14	113.03

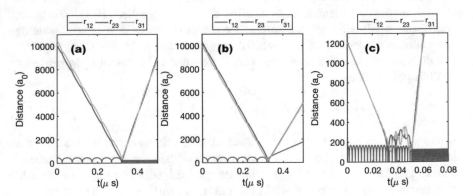

Fig. 11.6 Trajectories for RbBa$^+$(v_i) + Rb collisions. Panel (**a**): trajectory for $b = 0a_0$, $v = -4$, and E$_k$=10 mK, leading to a quenching process where $v_f = -8$. Panel (**b**) a trajectory for $b = 90\,a_0$, $v_i = -4$, and E$_k$=10 mK, leading to a dissociation process. Panel (**c**): a collision for $b = 0a_0$, $v_i = -12$ and E$_k$=10 mK leading to a quenching collision with $v_f = -15$. In this collision the quenching is produced through an exchange of the Rb atom forming the initial molecular ion and the colliding Rb atom. For this simulation the triplet potential for Rb$_2$ is taken from Ref. [27]. (Figure adapted with permission from Ref. [28] Copyright (2020) (American Physical Society))

the ordering from the dissociation limit, i.e., $v_i = -1$ is the the closest vibrational state to the dissociation asymptote.

Figure 11.6 shows a few characteristic trajectories associated with BaRb$^+$(v_i) + Rb collisions. The trajectory in panel (b) is related to a dissociation process, whereas the trajectories of panels (a) and (c) describe quenching events. Nevertheless, the trajectory of panel (c) shows a more interesting scenario: after the collision an ion–atom–atom three-body complex appears and eventually decays into BaRb$^+$(v_f) + Rb in a time $\sim$10 ns. This complex formation resembles the so-called roaming resonances in chemical physics [29]. However, the calculated lifetime is three orders of magnitude larger than the typical lifetime for He–C complexes (where C is a large organic molecule) [30]. It is possible that the observed lifetime of these complexes is a consequence of the low temperature of the collisions, as Li and Heller pointed out [30]. Furthermore, the presence of highly vibrational states of the molecular ion leads to longer vibrational periods that may definitively increase the lifetime of the three-body complex. Finally, we would like to emphasize that classically, the process in panel (c) of Fig. 11.6 can be viewed as an exchange reaction, since the incoming atom and the atom within the molecule are exchanged after the collision. However, quantum mechanically the atoms are identical particles and hence the exchange between them does not affect the quantum nature of the system at hand. Therefore, we consider the reaction in panel (c) to be a vibrational quenching process.

The QCT method has been applied to the study of the reaction $BaRb^+(v_i)$ + Rb at cold temperatures, and the results are presented in Fig. 11.7. In panel a of the figure, the quenching cross section is shown as a function of the initial vibrational state of the molecular ion ($v_i = -12, \ldots, -4$) at different collision energies. As a result, the vibrational quenching cross section is almost independent of the initial vibrational state of the molecular ion, which is an indication that the reaction mechanism depends on the long-range tail of the molecular ion–atom interaction, as is suggested by the Langevin capture model. Indeed, we see that the QCT results agree with the Langevin cross section for the atom–ion interaction, given by Eq. (9.23). These results indicate that the reaction mechanism behind vibrational quenching is controlled by the ion–atom interaction in accordance with the ion–atom–atom three-body recombination processes presented in the previous chapter.

The cross section shown in panel (a) of Fig. 11.7 appears as a function of the collision energy in panel (b) of the same figure. This Figure shows that at larger collision energies the agreement between the QCT and Langevin predictions for the quenching cross section depend on the initial vibrational state of the molecular ion. Moreover, the deviations appear at different collision energies for different vibrational states, which means that it is a vibrational-dependent mechanism. Indeed, the novel mechanism is a new product state: dissociation of the molecular ion. The most important hint comes from the fact that only when the collision energy is larger than the binding energy of the molecular ion (represented by the vertical dashed lines) does the quenching cross section deviate from the Langevin model.

The dissociation channel, $BaRb^+(v_i)$ + Rb $\rightarrow$ Ba^+ + Rb + Rb, opens for $E_k > E_{v_i}$; otherwise, the collision energy is not sufficient to break the bond within the molecular ion. The dissociation cross section as a function of the collision energy is shown in Fig. 11.8, where it is demonstrated that only for $E_k > E_{v_i}$ (see Table 11.1) does this channel become relevant to the dynamics, as expected. Furthermore, we observe that the higher the collision energy, the larger the cross section, which reaches the Langevin prediction.

It is interesting that independently of the reaction under study both tend toward the Langevin prediction. In other words, for a highly vibrational molecular ion–atom collision the vibrational–translational energy transfer is very efficient. Indeed, this can be checked by calculating the adiabaticity parameter for different vibrational

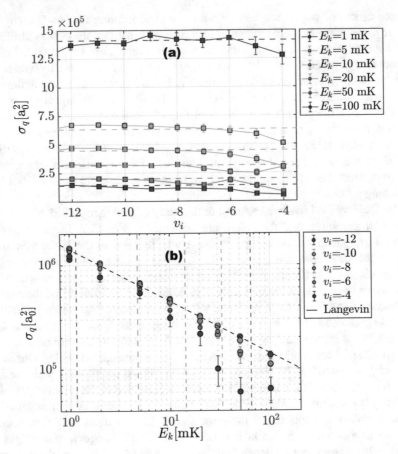

Fig. 11.7 Quenching cross section for a molecular ion colliding with an atom at cold temperatures. In particular, we show the results for $BaRb^+(v_i)$ + Rb collisions as a function of the vibrational state of the molecular ion, v_i, panel (**a**), and as a function of the collision energy E_k, panel (**b**). The Langevin cross section is represented by the black dashed lines. For this calculation $j = 0$ was taken. The error bars are related to one standard deviation as introduced in Eq. (11.15). For these calculations we run batches of 10^4 trajectories per collision energy covering 100 values uniformly distributed of the impact parameter

states, and the results are shown in Fig. 11.9. In this Figure, we find that $\xi \ll 1$ for all the relevant vibrational and collision energies studied. Finally, taking into account the sketch of panel a of Fig. 11.4, one can conclude that the vibrational–translational energy transfer is very efficient for the system at hand (Fig. 11.9).

The collision $BaRb^+(v_i)$ + Rb has another possible product state, the formation of a neutral molecule, which was called reaction process in Sect. 11.2.4. The results based on the QCT method for the cross section regarding the reactive process $BaRb^+(v)$ + Rb $\rightarrow$ $Rb_2(v'')$ + Ba^+ are shown in Fig. 11.10. In this Figure, in comparison with Figs. 11.7 and 11.8, it is noted that quenching and dissociation

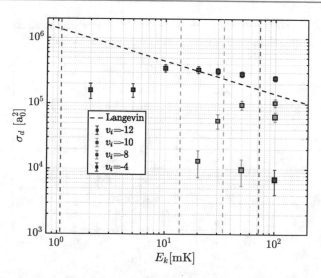

Fig. 11.8 Dissociation cross section for the collision $BaRb^+(v) + Rb \rightarrow Ba^+ + Rb + Rb$ as a function of the collision energy E_k. The different initial vibrational states v are denoted by the different colors. The black dashed line represents the Langevin cross section. For this calculation, we took $j = 0$. The vertical dashed colored lines stand for the binding energy of the vibrational states of the molecular ion. The different colors are related to different initial vibrational states, as indicated in the legend. (Figure adapted with permission from Ref. [28] Copyright (2020) (American Physical Society))

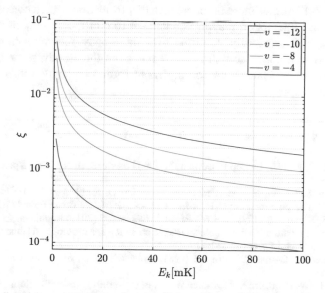

Fig. 11.9 Adiabaticity parameter as a function of the collision energy associated with $BaRb^+ + Rb$ collisions. Different colors correspond with different initial vibrational states of the molecular ion, as indicated in the legend of the figure

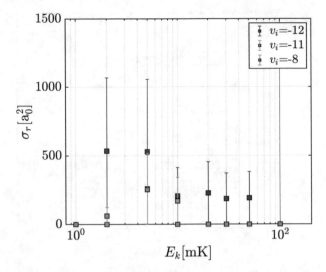

Fig. 11.10 Reaction cross section for the collision $BaRb^+(v) + Rb \rightarrow Rb_2(v'') + Ba^+$ as a function of the collision energy E_k. The initial vibrational states v are denoted by colors related to different initial vibrational states, as indicated in the legend. For these calculations we took $j = 0$ and ran batches of 10^4 trajectories per collision energy covering 100 values of the impact parameter. (Figure reproduced with permission from Ref. [28] Copyright (2020) (American Physical Society))

of $BaRb^+$ is $\sim 10^3$ times more probable than Rb_2 formation, which corroborates the charged–neutral interaction playing a dominant role in the system, as anticipated in Chap. 10, and previously confirmed for three-body recombination [31, 32]. Therefore, we can conclude that in ion–atom–atom three-body recombination or molecular ion–atom systems the ion prefers to form a molecule rather than to be free after a collision.

References

1. Schlemmer S, Kuhn T, Lescop E, Gerlich D (1999) Laser excited N_2^+ in a 22-pole ion trap. Int J Mass Spectrom 185:589
2. Schlemmer S, Lescop E, von Richthofen J, Gerlich D, Smith MA (2002) Laser induced reactions in a 22-pole ion trap: $C_2H_2^* + h\nu_3 + H_2 \rightarrow C_2H_3^+ + H$. J Chem Phys 117:2068
3. Brünken S, Kluge L, Stoffels A, Pérez-Ríos J, Schlemmer S (2017) Rotational state-dependent attachment of he atoms to cold molecular ions: an action spectroscopic scheme for rotational spectroscopy. J Mol Spectrosc 332:67
4. Hirschfelder JO, Curtiss CF, Bird RB (1954) Molecualr theory of transport in gases. Wiley, New York
5. McCourt FRW, Beenaker JJM, Köhler WE, Kuscer I (1991) Nonequilibrium phenomena in polyatomic gases. Oxford University Press/Clarendon, Oxford
6. Zhdanov VM (2002) Transport processes in multicomponent plasma. Taylor and Francis, London

7. Montero S, Pérez-Ríos J (2014) Rotational relaxation in molecular hydrogen and deuterium: theory versus acoustic experiments. J Chem Phys 141:2014
8. Sotecklin T, Halvick P, Gannouni MA, Holchaf M, Kotochigova S, Hudson ER (2016) Explanation of efficient quenching of molecular ion vibrational motion by ultracold atoms. Nat Commun 7:11234
9. Tacconi M, Gianturco FA, Yurtsever E, Caruso D (2011) Cooling and quenching of $^{24}MgH^4(x^1\sigma^+)$ by $^4He(^1s)$ in a coulomb trap: a quantum study of the dynamics. Phys Rev A 84:013412
10. Hauser D, Lee S, Carelli F, Spieler S, Lakhmanskaya O, Endres ES, Kumar SS, Gianturco F, Wester R (2015) Rotational state-changing cold collisions of hydroxyl ions with helium. Nat Phys 11:467
11. Stoecklin T, Voronin A (2005) Phys Rev A 72:042714
12. Stoecklin T, Voronin A (2008) Eur Phys J D 46:259
13. Stoecklin T, Voronin A (2011) Vibrational and rotational cooling of NO^+ in collisions with He. J Chem Phys 134:204312
14. P'erez-Ríos J, Robicheaux F (2016) Rotational relaxation of molecular ions in a buffer gas. Phys Rev A 94:032709
15. Karplus M, Porter RN, Sharma RD (1965) Exchange reactions with activation energy. I. Simple barrier potential for (H,H$_2$. J Chem Phys 43:3259
16. Truhlar DG, Muckerman JT (1979) Atom-molecule collision theory: a guide for the Experimentalist, chapter Reactive scattering Cross sections III: quasiclassical and semiclassical methdos. Plenum Press, New York, pp 505–561
17. Pérez-Ríos J, Ragole S, Wang J, Greene CH (2014) Comparison of classical and quantal calculations of helium three-body recombination. J Chem Phys 140:044307
18. Greene CH, Giannakeas P, Pérez-Ríos J (2017) Universal few-body physics and cluster formation. Rev Mod Phys 89:035006. https://doi.org/10.1103/RevModPhys.89.035006
19. Pattengill MD (1979) Rotational excitation III: classical trajectory methods. Springer US, Boston, pp 359–375. https://doi.org/10.1007/978-1-4613-2913-8_10
20. Bonnet L, Rayez JC (1997) Quasiclassical trajectory method for molecular scattering processes: necessity of a weighted binning approach. Chem Phys Lett 277(1):183. https://doi.org/10.1016/S0009-2614(97)00881-6. http://www.sciencedirect.com/science/article/pii/S0009261497008816
21. Balucani N, Casavecchia P, Bañares L, Aoiz FJ, Gonzalez-Lezana T, Honvault P, Launay J-M (2006) Experimental and theoretical differential cross sections for the n(2d) + h2 reaction. J Phys Chem A 110(2):817. https://doi.org/10.1021/jp054928v
22. Bañares L, Aoiz FJ, Honvault P, Bussery-Honvault B, Launay JM (2002) Quantum mechanical and quasi-classical trajectory study of the c(1d)+h2 reaction dynamics. J Chem Phys 118(2):565. https://doi.org/10.1063/1.1527014
23. Czakó G, Bowman JM (2009) Quasiclassical trajectory calculations of correlated product distributions for the f+chd3(v1=0,1) reactions using an ab initio potential energy surface. J Chem Phys 131(24):244302. https://doi.org/10.1063/1.3276633
24. Achymski K, Meintert F (2020) Vibrational quenching of weakly bound cold molecular ions immersed in their parent gas. Appl Sci 10:2371. https://www.mdpi.com/2076-3417/10/7/2371
25. Levine RD, Bernstein RB (1987) Molecular reaction dynamics and chemical reactivity. Oxford University Press, New York
26. Lara M, Jambrina PG, Aoiz FJ, Launay JM (2015) Cold and ultracold dynamics of the barrierless D^+ + H$_2$ reaction: quantum reactive calculations for $\propto r^{-4}$ long range interaction potentials. J Chem Phys 143(20):204305. https://doi.org/10.1063/1.4936144
27. Strauss C, Takekoshi T, Winker K, Grimm R, Denschlag JH (2010) Phys Rev A 82:052514
28. Pérez-Ríos J (2019) Vibrational quenching and reactive processes of weakly bound molecular ions colliding with atoms at cold temperatures. Phys Rev A 99:022707. https://doi.org/10.1103/PhysRevA.99.022707
29. Bowman JM (2014) Roaming. Mol Phys 112:2516

30. Li Z, Heller EJ (2012) Cold collision of complex polyatomic molecules. J Chem Phys 136:054306
31. Pérez-Ríos J, Greene CH (2015) Communication: classical threhold law for ion-neutral-neutral three-body recombination. J Chem Phys 143:041105
32. Krükow A, Mohamadi A, Härter A, Denschlag JH, Pérez-Ríos J, Greene CH (2016) Energy scaling of cold atom-atom-ion three-body recombination. Phys Rev Lett 116:193201
33. Härter A, Denschlag JH (2014) Cold atom-ion experiemtns in hybrid traps. Contemp Phys 55:33

Ultracold Physics and the Quest of New Physics

12

At ultracold temperatures (T$\lesssim$ 1 mK), atoms and molecules move very slowly, which translates into negligible Doppler broadening of spectra. In addition, owing to the diluteness of typical ultracold systems, the pressure broadening of the spectroscopy lines is negligible. In this scenario, atoms and molecules are in a well-defined rovibrational state and in a specific translational state that can be further tuned. Therefore, ultracold systems are a perfect arena for developing high-precision spectroscopic tools. Indeed, ultracold physics has played a major role in the evolution of high-precision spectroscopy to levels hardly imaginable at the beginning of the twenty-first century.

The spectroscopy of a quantum system reveals the particular energy levels' structure as a consequence of the light-matter interaction. Thus, high-precision spectroscopy can be used to test the validity of quantum electrodynamics, i.e., the fundamental theory behind the electromagnetic interaction. Indeed, the weak and strong nuclear forces, and the gravitational interaction, may lead to extra energy shifts between levels that may be observed in high-precision spectroscopy of atoms and molecules as well. In the same vein, having extremely precise spectroscopic measurements helps to elucidate differences between the predictions of the Standard Model of particle physics and the measurements, which can be used to constraint the parameter space for different models regarding the existence of physics beyond the Standard Model. Therefore, precision in spectroscopy measurements may lead to a better understanding of the fundamental nature of the universe, which is the topic of the present chapter.

12.1 The Fundamental Laws of Physics

The fundamental theories of nature can be described in terms of the values of three fundamental constants, as depicted in Fig. 12.1

© Springer Nature Switzerland AG 2020
J. Pérez Ríos, *An Introduction to Cold and Ultracold Chemistry*,
https://doi.org/10.1007/978-3-030-55936-6_12

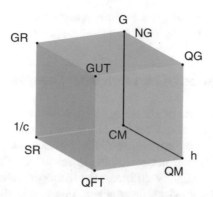

Fig. 12.1 The physical theories in terms of the fundamental constants. At the origin stands the part of Newtonian mechanics (NM) that does not take gravity into account. NG, QM, and SR stand for Newtonian gravity, quantum mechanics, and special relativity respectively. Special relativity is applied in Newtonian gravity to give rise to general relativity (GR), whereas when applied to quantum mechanics it leads to quantum field theory (QFT). Bringing quantum mechanics and Newtonian gravity together leads to nonrelativistic quantum gravity, and all theories together give a grand unified theory (GUT), a theory that explains the four fundamental forces of nature emerging form a unique underlying field theory. (Figure adapted with permission from Ref. [6] Copyright (2020) (American Physical Society))

- The speed of light in vacuum c, which defines the highest speed at which information may travel.
- Planck constant h, which defines the realm of quantum mechanics.
- Newton's constant of gravity G, which defines the realm of gravity, one of the four fundamental forces in nature.

The most basic physical theories are classical mechanics and electromagnetism, which is neither quantal nor relativistic. From this point, it is possible to include quantum mechanical effects through h, and hence enter into the realm of quantum mechanics (QM). The inclusion of relativistic effects in classical mechanics leads to the special theory of relativity (SR), and by the addition of gravitation effects we recover the Newton's gravity (NG) law. At the next level, merging SR and QM leads to quantum field theory (QFT), which applied to the electromagnetic, weak and strong forces, lead to the Standard Model of particle physics. The Standard Model is the most complete and successful theory in particle physics and explains most of the observations to date. The addition of relativistic effects on NG leads to the general theory of relativity (GR), the most complete and accurate description of the gravitational interaction. Finally, the most refined physical theories are those that merge gravity and quantum mechanics, leading to the quantum theory of gravity (QG), and the addition of gravity to the Standard Model, which evolves into a grand unified theory (GUT), or theory of everything. It is worth mentioning that the well-known string theory [1, 2] is a candidate for quantum gravity and for grand unified theories. However, string theories require many different conditions that hardly seem to be realistic based on current observations [1, 2].

12.2   Time Variation of Fundamental Constants

Since the very first moment of attending physics lectures we are told that in nature some magnitudes have a fixed value and are called constants, and the value of these constants is fixed by nature itself. For instance, why is the electron mass $9.10938356 \times 10^{-31}$ Kg and not something different? One answer may be that if the electron mass were slightly different, the hydrogen atom would not be stable and chemistry would not exist. Nevertheless, this is a justification rather than an explanation. In particular, it assumes that our perception of nature is the reason for the universe instead of assuming that our perception is a cause of the universe. To find a *proper* answer to the question one may invoke the anthropic principle [3] or even multi-verse theory [4, 5], which is beyond the scope of this book.

A question more in line with the scope of the present chapter would be: are the constants of nature really constants? In principle, one may think that this question is ill-formulated, since by definition a constant is something that does not change in time. However, the values of the constants of physics are based on measurement during a very short period of time compared with the typical time scale of the universe; hence, we must question whether the constants of nature remain the same throughout the evolution of the universe.

The constants of nature have a value based on a given system of units. As a consequence, the result of the time variation of the constants is linked to the same system of units. Therefore, it only makes sense to look for the time variation of dimensionless ratios of fundamental constants [6], that is, functions of different constants of nature that do not have units; among them, the electron-to-proton mass ratio $\mu \equiv \frac{m_e}{m_p}$ should show the largest variation [6]. The typical way of determining the time variation of μ is to compare the molecular spectra of distant cosmological sources ($\sim 10^{10}$ years ago) with the same spectra in the laboratory [7–10], which leads to a variation of $\sim 10^{-15}$/year. Nevertheless, the systematic errors attached to cosmological observations, such as determination of the distance of a far object, compromises the accuracy of the time variation measurements.

Instead of using cosmological observables that are hard to characterize, it is possible to determine the time variation of μ directly in the laboratory by measuring the energy shift between different vibrational states of a diatomic molecule as a function of time. The basic idea of the method is to prepare an ensemble of ultracold molecules in two given well-defined vibrational states that show a different energy sensitivity to the change of μ and interrogate the energy shift between the states as a function of time [11]. This method is only possible thanks to the high degree of control of the internal states of ultracold molecules and to the development of high-precision spectroscopy.

The vibrational states of a diatomic molecule have a characteristic energy given by

$$\int_{R_i}^{R_f} \sqrt{2m(E_v - V(R))} dR = (v + 1/2) \pi, \tag{12.1}$$

where the WKB quantization condition has been assumed. In Eq. (12.1), $V(R)$ is the potential energy curve of a given electronic state of the molecule and m is the reduced mass of the molecule, which is assumed to be proportional to the proton mass, m_p, and m_e is presumed constant, i.e., $m \propto \mu$. The energy sensitivity to a change in μ is defined as

$$|\partial_\mu E_v| \equiv \left| \frac{\partial E_v}{\partial (\ln \mu)} \right| = \left| \mu \frac{\partial E_v}{\partial \mu} \right|, \tag{12.2}$$

which in the present case is calculated by differentiating Eq. (12.1) yielding

$$\int_{R_i}^{R_f} \left((E_v - V(R)) + \mu \frac{\partial E_v}{\partial \mu} \right) \frac{dR}{\sqrt{2\mu(E_v - V(R))}} = 0 \rightarrow$$

$$-\int_{R_i}^{R_f} \frac{1}{2\mu} \sqrt{2\mu(E_v - V(R))} dR = \mu \frac{\partial E_v}{\partial \mu} \int_{R_i}^{R_f} \frac{dR}{\sqrt{2\mu(E_v - V(R))}} \rightarrow$$

$$-\frac{v + 1/2}{2\mu} = \mu \frac{\partial E_v}{\partial \mu} \int_{R_i}^{R_f} \frac{dR}{\sqrt{2\mu(E_v - V(R))}}. \tag{12.3}$$

Equation (12.3) is further simplified by taking into account that

$$\int_{R_i}^{R_f} \mu \frac{\partial E_v}{\partial v} \frac{dR}{\sqrt{2\mu(E_v - V(R))}} = \pi \rightarrow$$

$$\left(\frac{\partial E_v}{\partial v} \right)^{-1} = \frac{\mu}{\pi} \int_{R_i}^{R_f} \frac{dR}{\sqrt{2\mu(E_v - V(R))}}, \tag{12.4}$$

arriving at

$$\mu \frac{\partial E_v}{\partial \mu} = -\frac{(v + 1/2)}{2 \left(\frac{\partial E_v}{\partial v} \right)^{-1}}. \tag{12.5}$$

Therefore, the energy sensitivity is given by

$$\partial_\mu E_v = \frac{(v + 1/2)}{2\rho(E_v)}, \tag{12.6}$$

where $\rho(E_v) = \left(\frac{\partial E_v}{\partial v} \right)^{-1}$, represents the density of states around E_v.

The energy sensitivity for different vibrational states in Cs_2 is shown in Fig. 12.2, and as a result the largest energy sensitivity appears in intermediate vibrational states

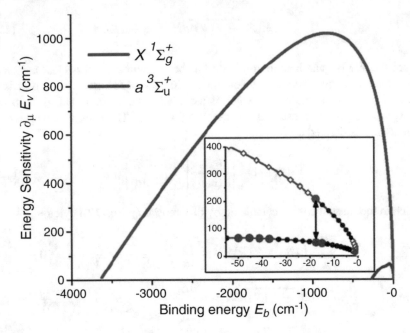

Fig. 12.2 Energy sensitivity as a function of the binding energy of the vibrational states of the $X^1\Sigma_g$ and $a^3\Sigma_u$ electronic states of Cs_2. The inset is a zoom into the region close to the dissociation threshold. (Figure reproduced with permission from Ref. [11] Copyright (2020) (American Physical Society))

as a consequence of the best ratio between the vibrational quantum number, v, and the density of states, as Eq. (12.6) establishes. In the same Figure it is noted that vibrational states with the same binding energy but in different electronic states show a distinct energy sensitivity, which is exploited to observe a change in the energy separation as a function of time between the involved states. Indeed, using Cs_2 molecules it is possible to put constraints on the time variance of $\Delta\mu/\mu \sim 10^{-17}$ [11].

12.3 Ultracold Molecules and Their Role in the Existence of Extra Dimensions

It is well known that gravity is the weakest of the four fundamental forces of nature; however, it is not known why this is the case. In fact, this is one of the main open questions in particle physics. In particular, the mass of the electroweak gauge bosons (M_W and M_Z) is of the order of 100 GeV, whereas the Planck mass, which represents the energy scale for the gravitational force, is approximately 10^{19} GeV. This large difference between the mass and energy scales is generally known as the hierarchy problem in particle physics. The Planck mass is defined as

$$M_{\text{Pl}} = \sqrt{\frac{\hbar c}{G}}, \tag{12.7}$$

where c is the speed of light in a vacuum, $\hbar$ is the reduced Planck constant, and G is Newton's constant. The Planck mass represents the upper bound for the validity, energy-wise, of the Standard Model of particle physics or it is the threshold energy where a quantum theory of gravity is required. It also represents the smallest mass that a black hole could have.

Different theories have been proposed to circumvent the energy difference between the weak force and gravity, and one of the most interesting ones, from a theoretical point of view, is the proposal of Arkani-Hamed, Dimopoulos, and Dvali [12, 13], in which extra dimensions are invoked to explain the weakness of gravity.

12.3.1 The Arkani-Hamed, Dimopoulos, and Dvali Model

One of the solutions to the hierarchy problem is the Arkani-Hamed, Dimopoulos and Dvali (ADD) model [12, 13]. In particular, the ADD model solves the hierarchy problem assuming that gravity is in reality as strong as the weak force, although it is diluted owing to the existence of extra dimensions, as shown in Fig. 12.3. The electromagnetic, weak, and strong nuclear forces live in a four-dimensional brane (three spatial dimensions and one temporal one, our regular space-time), whereas the gravitons (the quanta carriers of the gravitational force) live in a bulk, including n extra compactified dimensions with range R_n.

The Newtonian gravitational potential between two bodies of masses m_1 and m_2 is given by

$$V_N(r) = -G\frac{m_1 m_2}{r}, \tag{12.8}$$

which can be expressed in terms of the Planck mass as

$$V_N(r) = -\frac{m_1 m_2}{M_{\text{Pl}}^2}\frac{\hbar c}{r}, \tag{12.9}$$

and using natural units ($\hbar = c = 1$) yields

$$V_N(r) = -\frac{m_1 m_2}{M_{\text{Pl}}^2}\frac{1}{r}. \tag{12.10}$$

In the ADD model the presence of n extra dimensions modifies the gravitational potential, which needs to be consistent with the Gauss law and reads as

$$V_{\text{ADD}}(r) \sim -\frac{m_1 m_2}{M_{(4+n)}^{n+2}}\frac{1}{r^{n+1}}, \tag{12.11}$$

Fig. 12.3 Arkani-Hamed, Dimopoulos, and Dvali (ADD) theory. The strong, weak, and electromagnetic forces lie in a brane with (3+1) dimensions, i.e., three spatial dimensions and one temporal, which represent the regular space-time. However, the force of gravity leaks into n extra dimensions, explaining why it is much weaker than the rest of the forces in our brane

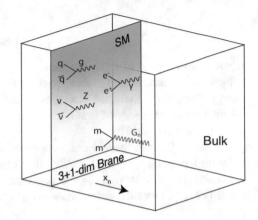

where $M_{(4+n)}$ is the fundamental mass in 3+n spatial dimensions and one temporal one. However, when two interacting bodies are at distances larger than the range of the extra dimensions (for $r > R_N$) the ADD potential must show the expected $1/r$ law characteristic of the Newtonian potential and hence

$$V_{\text{ADD}}(r) \sim -\frac{m_1 m_2}{M_{(4+n)}^{n+2} R_n^n} \frac{1}{r}. \tag{12.12}$$

where the factor R_n^n is proportional to the volume of the compactified extra dimensions and the particular factor (of the order unity) that depends on the geometry of the extra dimensions.

The Planck mass, within the ADD model, is related to the higher-dimensional fundamental mass $M_{(4+n)}$ by

$$M_{\text{Pl}}^2 = M_{(4+n)}^{n+2} R_n^n. \tag{12.13}$$

Thus, choosing a particular value for the fundamental mass $M_{(4+n)}$ and n it is possible to determine the size of the extra dimensions, in particular, making $M_{(4+n)} \sim M_{\text{EW}}$, where $M_{\text{EW}} = 159\,\text{GeV}$ is the electro-weak energy scale, for $n = 1$, $R_1 = 10^{10}\,\text{km}$, which is ruled out by observations, but for $n = 2$, $R_2 \lesssim 10^{-4}\,\text{m}$, which is a range unexplored by conventional techniques [14]. Therefore, the extra dimensions are small and that is the reason why we do not feel its effects in our world (Fig. 12.3).

12.3.2 High-Precision Spectroscopy in Cold Molecules

Newton's law of gravity has been shown to be correct at large distances, but at short distances it has hardly been studied and robust results only exist at the 1 cm scale [14] (owing to the presence of van der Waals forces). Therefore, very little is known about gravity at short distances. However, molecules are bound atoms at distances of the order of the Angstrom, and they may be a good platform for constraining the size of compactified extra dimensions. In particular, if they exist and the size is comparable with the molecular size, then we should be able to see a shift in the energy levels of the molecule that is inexplicable according to the Standard Model of particle physics.

To work with diatomic molecules it is preferable to introduce the coupling constant

$$\alpha_G = \frac{Gm_p^2}{\hbar c}, \tag{12.14}$$

since it is assumed that the two interacting bodies are both of the nuclei of the molecule, and m_p is the proton mass. Assuming that the mass of the proton and neutron are equal $m_p \approx m_n$, Newton's potential reads as

$$V_N(R) = -\alpha_G N_1 N_2 \frac{1}{R}, \tag{12.15}$$

where N_1 and N_2 represent the number of nucleons of the nucleus 1 and 2 of the molecule respectively. In the same vein, the ADD potential reads as

$$V_{ADD}(R) \sim -\alpha_G N_1 N_2 R_n^n \frac{1}{R^{n+1}}. \tag{12.16}$$

The presence of extra dimensions will lead to an energy shift between the two vibrational states, which is given by

$$\Delta E_{ij}(n, R_n) = \langle \psi_j | V_{ADD}(R) \psi_j \rangle - \langle \psi_i | V_{ADD}(R) \psi_i \rangle$$
$$= \int_0^\infty \left(|\psi_j(R)|^2 - |\psi_i(R)|^2 \right) V_{ADD}(R) R^2 dR, \tag{12.17}$$

which can be recast as

$$\Delta E_{ij}(n, R_N) \sim -\alpha_G N_1 N_2 \int_0^{R_n} \frac{R_n^n}{R^{n+1}} \left(|\psi_j(R)|^2 - |\psi_i(R)|^2 \right) R^2 dR, \tag{12.18}$$

where it has been assumed that the gravitational force between the nuclei is negligible for $R > R_n$.

The energy shifts between different vibrational states can be estimated accurately by quantum electrodynamics ($\mathcal{O}(\alpha^4)$) with uncertainty δE_{Th} for a few systems [15]. The same energy shifts are experimentally determined with uncertainty δE_{Exp}. Therefore, the total uncertainty is characterized by $\delta E = \sqrt{\delta E_{Exp}^2 + \delta E_{Th}^2}$, which is used to constrain the ADD model through

$$\Delta E_{ij}(n, R_N) < \delta E. \tag{12.19}$$

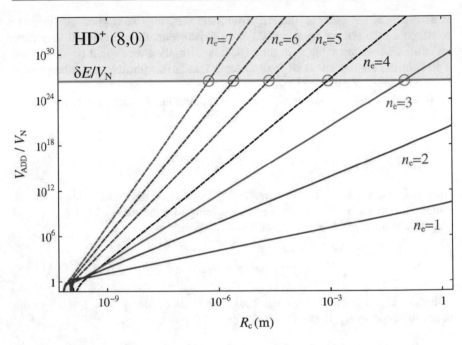

Fig. 12.4 Upper bounds on the size of the extra dimensions based on high-precision spectroscopy of HD^+ (8,0) band, i.e., the energy shift between the ground vibrational state, the $v = 8$ state. The strength of the ADD potential in comparison with the Newtonian one as a function of the size of the extra dimensions R_c and the number of extra dimensions $n_e \equiv n$. The shaded yellow is excluded following Eq. (12.20). (Figure reproduced with permission from Ref. [15] Copyright (2020) (Elsevier))

Plugging Eq. (12.18) into Eq. (12.19) yields

$$R_n^n < \frac{\delta E}{\alpha_G N_1 N_2 \Delta_n}, \tag{12.20}$$

with

$$\Delta_n = \int_0^{R_n} \frac{1}{R^{n+1}} \left(|\psi_j(R)|^2 - |\psi_i(R)|^2 \right) R^2 dR. \tag{12.21}$$

From Eq. (12.20) it is possible to constrain the existence of extra dimensions within the ADD model as long as the vibrational wave functions are known. Figure 12.4 shows the constraints on the ADD model based on high-precision spectroscopy on cold HD^+ for the energy shifts between the ground vibrational state and the $v = 8$ vibrational state. In the figure we consider up to seven extra dimensions, which is the number of extra dimensions required by M-theory within the string theory framework.

12.4 Ultracold Molecules and the Quest of Physics Beyond the Standard Model

The Standard Model of particle physics is the most precise theory of nature to date, as stated in Sect. 12.1. However, the Standard Model is far from being a complete theory, since it does not account for the existence of very solid cosmological observations such as dark matter, dark energy, and the fact that neutrinos have mass, and each of these phenomena are classified within the label of physics beyond the Standard Model.

The spectra of molecules are a direct consequence of the electromagnetic inter-action (mainly), and ultimately of the Standard Model. Thus, observing deviations from the measured energy shift between rovibrational states and the theoretically predicted values based upon the Standard Model can be used as a tool to constraint the existence of physics beyond the Standard Model. The procedure is quite general

- Choose a model for the interaction of the new particle with the Standard Model particles.
- Calculate the effective potential, $V^{\mathrm{BSM}}(\boldsymbol{\beta}, R)$, that the molecule feels as a consequence of the interaction between the new particle and the Standard Model particles, where $\boldsymbol{\beta}$ represents the different parameters that play a role in the interaction.
- Calculation of the energy shift as a consequence of the effective potential $V(\boldsymbol{\beta}, R)$ as

$$
\Delta E_{ij}^{\mathrm{BSM}} = \langle \psi_j | V^{\mathrm{BSM}}(\boldsymbol{\beta}, R) | \psi_j \rangle - \langle \psi_i | V^{\mathrm{BSM}}(\boldsymbol{\beta}, R) | \psi_i \rangle =
$$
$$
\int_0^\infty \left(|\psi_j(R)|^2 - |\psi_i(R)|^2 \right) V^{\mathrm{BSM}}(\boldsymbol{\beta}, R) \quad (12.22)
$$

- Calculation of the constraints on the model by comparing the results of the previous equation with the total uncertainty of the energy shift δ_E

$$
\Delta E_{ij}^{\mathrm{BSM}} < \delta E, \quad\quad\quad\quad\quad (12.23)
$$

where δE is the same as in Sect. 12.3.2.

12.4.1 The Fifth Force

The fifth force occurs through a Yukawa potential; therefore, the effective potential reads as [16]

$$
V^{\mathrm{BSM}}(\boldsymbol{\beta}, R) = \alpha_5 N_1 N_2 \frac{e^{-R/\lambda}}{R}, \quad\quad\quad\quad (12.24)
$$

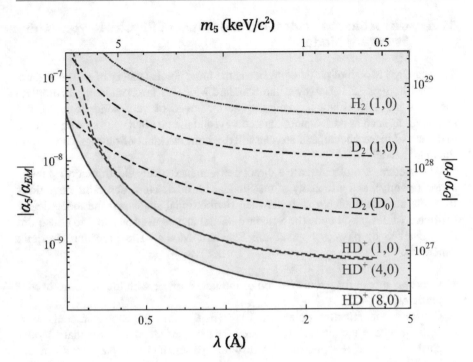

Fig. 12.5 Limits on the strength of a fifth force relative to the fine structure constant, α_{EM} (left scale), or the gravitational coupling, α_G (right scale) as a function of the range of the effective potential λ (lower axis). The shaded (yellow) area is the region that is excluded by the molecular precision experiments, with the result obtained in the (8,0) band of $HD^+ +$. The top axis denotes the mass of the boson carrier of the fifth force. (Figure reproduced with permission from Ref. [15] Copyright (2020) (Elsevier))

where $\beta = (\alpha_5, \lambda)$, λ is the range of the force, α_5 is the strength of the force, and N_1 and N_2 stand for the number of nucleons in each of the nuclei of the molecule. The Yukawa potential is the natural modification of the typical $1/R$ for the gravitational interaction, and it emerges as a consequence of the exchange of massive bosons (with null spin), $m_5 = \hbar/(c\lambda)$, between hadrons, within the present model. Figure 12.5 displays the constraints on the fifth force given by the Yukawa potential of Eq. (12.24) for different vibrational states and molecules. In particular, the spectroscopy of HD^+ is performed at cold temperatures. As a result, the constraints are tighter for $\lambda \sim 2\text{Å}$, which is the size of the vibrational state, and the constraints are weaker for small values of the range of the Yukawa potential, since at such short distances the vibrational wave functions involved do not change much.

This is just an example of how high-precision spectroscopy of cold and ultracold molecules may help to elucidate some of the physics beyond the Standard Model. With this last section and with the previous section of the chapter, we hope to convince the reader that ultracold molecules are essential for facing some of the most profound and open questions of modern physics.

References

1. Polchinski J (1998) String theory: volume 1: an introduction to the bosonic string, vol 1. Cambridge University Press, Cambridge. https://doi.org/10.1017/CBO9780511816079. https://www.cambridge.org/core/books/string-theory/30409AF2BDE27D53E275FDA395AB667A

2. Green MB, Schwarz JH, Witten E (2012) Superstring theory: 25th anniversary edition: volume 1: introduction, vol 1. Cambridge University Press, Cambridge. https://doi.org/10.1017/CBO9781139248563. https://www.cambridge.org/core/books/superstring-theory/35152940E2D9C3B0BD327EB7FF74DBBE

3. Barrow JD, Tipler FJ (1986) The anthropic cosmological principle. Oxford University Press, New York

4. Steinhardt P (2014) Big bang blunder bursts the multiverse bubble. Nature 510(7503):9. https://doi.org/10.1038/510009a

5. Carr B (2007) Universe or multiverse? Cambridge University Press, Cambridge. https://doi.org/10.1017/CBO9781107050990. https://www.cambridge.org/core/books/universe-or-multiverse/38972284ED1AECC0A692CF429DF57D53

6. Uzan J-P (2003) The fundamental constants and their variation: observational and theoretical status. Rev Mod Phys 75:403. https://doi.org/10.1103/RevModPhys.75.403

7. Hudson ER, Lewandowski HJ, Sawyer BC, Ye J (2006) Cold molecule spectroscopy for constraining the evolution of the fine structure constant. Phys Rev Lett 96:143004. https://doi.org/10.1103/PhysRevLett.96.143004

8. Kanekar N, Carilli CL, Langston GI, Rocha G, Combes F, Subrahmanyan R, Stocke JT, Menten KM, Briggs FH, Wiklind T (2005) Constraints on changes in fundamental constants from a cosmologically distant oh absorber or emitter. Phys Rev Lett 95:261301. https://doi.org/10.1103/PhysRevLett.95.261301

9. Reinhold E, Buning R, Hollenstein U, Ivanchik A, Petitjean P, Ubachs W (2006) Indication of a cosmological variation of the proton-electron mass ratio based on laboratory measurement and reanalysis of h_2 spectra. Phys Rev Lett 96:151101. https://doi.org/10.1103/PhysRevLett.96.151101

10. Flambaum VV, Kozlov MG (2007) Limit on the cosmological variation of m_p/m_e from the inversion spectrum of ammonia. Phys Rev Lett 98:240801. https://doi.org/10.1103/PhysRevLett.98.240801

11. DeMille D, Sainis S, Sage J, Bergeman T, Kotochigova S, Tiesinga E (2008) Enhanced sensitivity to variation of m_e/m_p in molecular spectra. Phys Rev Lett 100:043202. https://doi.org/10.1103/PhysRevLett.100.043202

12. Arkani-Hamed N, Dimopoulos S, Dvali G (1998) The hierarchy problem and new dimensions at a millimeter. Phys Lett B 429(3):263. https://doi.org/10.1016/S0370-2693(98)00466-3. http://www.sciencedirect.com/science/article/pii/S0370269398004663

13. Antoniadis I, Arkani-Hamed N, Dimopoulos S, Dvali G (1998) New dimensions at a millimeter to a fermi and superstrings at a tev. Phys Lett B 436(3):257. https://doi.org/10.1016/S0370-2693(98)00860-0. http://www.sciencedirect.com/science/article/pii/S0370269398008600

14. Adelberger EG, Heckel BR, Nelson AE (2003) Tests of the gravitational inverse-square law. Annu Rev Nucl Part Sci 53(1):77. https://doi.org/10.1146/annurev.nucl.53.041002.110503

15. Ubachs W, Koelemeij J, Eikema K, Salumbides E (2016) Physics beyond the standard model from hydrogen spectroscopy. J Mol Spectrosc 320:1

16. Salumbides EJ, Koelemeij JCJ, Komasa J, Pachucki K, Eikema KSE, Ubachs W (2013) Bounds on fifth forces from precision measurements on molecules. Phys Rev D 87:112008. https://doi.org/10.1103/PhysRevD.87.112008

Quasistatic Theory for Line Broadening

A

The core idea of the quasistatic theory is well defined by the words of Kuhn: "It should therefore be possible to investigate the potential function of the interaction of two atoms (especially of the polarization forces) experimentally by measuring the intensity curve towards the wings of the broadened lines" [1].

In the quasistatic theory for line broadening, the absorber (which in our case is a Rydberg atom) is assumed to be at rest and its absorption properties are modified by the surrounding perturber atoms that are also considered at rest. In particular, the presence of atoms around the absorber leads to the absorption of photons with frequency $\omega_0 + \Delta\omega$, where ω_0 is the excitation frequency, whereas $\Delta\omega = \Delta V/\hbar$ is a shift due to the perturber–atom interaction V. Additionally, only atoms located within a given region contribute to the line shape of the target excitation.

In the particular case of a Rydberg excitation in a ultracold dense gas, ω_0 represents the excitation frequency to a Rydberg state and atoms located within the Rydberg volume, V_{Ryd}, are the perturbers, inducing a shift of the Rydberg excitation proportional to the Rydberg electron–perturber interaction. The line shape of the Rydberg excitation in such a scenario is simulated by randomly placing atoms at positions r_i, according to the measured distribution. The Bose–Einstein condensation is treated within the Thomas–Fermi approximation, having a parabolic shape, whereas the thermal cloud has a Gaussian profile, according to the trap frequencies and temperature. All atoms are treated as spatially uncorrelated, assuming $T = 0$. For every atom in the sample, the shift, δ_i, from the atomic Rydberg resonance is calculated by

$$\delta_i = \sum_{j \neq i} V(|r_j - r_i|). \tag{A.1}$$

The atoms are located at r_j surrounding the Rydberg atom at r_i. These atoms are treated as point-like particles in the Rydberg–ground-state atom interaction potential. The spectrum $S(\delta)$ is finally calculated by summing over all Lorentzian

© Springer Nature Switzerland AG 2020
J. Pérez Ríos, *An Introduction to Cold and Ultracold Chemistry*,
https://doi.org/10.1007/978-3-030-55936-6

contributions of each atom, taking into account its shift δ_i, the excitation bandwidth Γ, and the spatially varying excitation laser intensity $I_i = I(r_i)$, relating to the incoherent excitation rate in our system.

From a purely experimental perspective, Rydberg blockade effects (see Sect. 7.2) are neglected, since the mean ion count rate is well below one in the area of interest. Each simulated spectrum is averaged over several atom configurations and normalized with a factor α to have the same $\int S(\delta)d\delta$ compared with the experimental data [2]

$$S(\delta) = \alpha \sum_i \frac{1}{\pi} \frac{\Gamma^2/4}{\delta_i^2 + \Gamma^2/4} \times I_i. \tag{A.2}$$

The quasistatic picture works as long as collisions between the perturber and the excited atom are negligible during the excitation process, i.e., when the collision time is larger than the inverse detuning, $\tau_c \sim (\Delta\omega)^{-1}$ the collision time can be estimated as $\tau_c = b/\langle v \rangle$, where the impact parameter b is assumed to be of the same order of magnitude as the Rydberg orbit size, i.e., b $2n^2$. As an example, consider a thermal cloud of hydrogen at $T \sim 50\,\mu K$, $\tau_c \sim 20$–$200\,ns$. Thus, for a thermal cloud of hydrogen the line shapes in the range $\Delta\Omega \gtrsim 25\,MHz$ can be described using the quasistatic approach [3].

The quasistatic theory describes several details of the observed line shapes fairly well by sampling the energy shifts in the frozen nuclear Rydberg levels associated with different configurations of the perturbers [2]. This approach readily shows the effect of the Rydberg–neutral potential energy in the observed line shapes, and hence it can be a robust technique for probing and testing different Rydberg–neutral potential energy curves within the range of validity of the quasistatic approximation, which is on the line of Kuhn's quotation at the very beginning of the Appendix.

References

1. Kuhn H (1934) XCIII. Pressure shift and broadening of spectral lines. London Edinburgh Dublin Philos. Mag J Sci 18(122):987. https://doi.org/10.1080/14786443409462572.
2. Schlagmüller M, Leibisch TC, Nguyen H, Lochead G, Engel F, Böttcher F, Westphal KM, Kleinbach KS, Löw R, Hofferberth S, Pfau T, Pérez-Ríos J, Greene CH (2016) Probing and electron scattering resonance using Rydberg molecules within a dense and ultracold gas. Phys Rev Lett 116:053001
3. Pérez-Ríos J, Eiles MT, Greene CH (2016) Mapping trilobite state signatures in atomic hydrogen. J Phys B Atom Mol Opt Phys 49(14):14LT01. https://doi.org/10.1088/0953-4075/49/14/14LT01.

Fluid Dynamics of Supersonic Expansions

B

In this appendix we will briefly outline the most important equations of fluid dynamics and apply them to the variation of thermodynamic quantities in a compressible fluid (gas) [1–4]. With this, we intend to give the basic theoretical background to understanding the physics of supersonic expansions, which is the key ingredient of molecular beams and the starting point for many experiments in cold and ultracold chemistry.

B.1 Brief Introduction to Fluid Dynamics

In fluid dynamics, fluids are considered to be a continuous medium, i.e., any volume element, no matter how small it is, contains a sufficiently large number of particles. Fluids can be described from two different approaches:

- The Eulerian description, in which the fields (vector or scalar) are calculated based on the flux. This is the main descriptive method in fluid mechanics.
- The Lagrangian description, in which the fields (vector or scalar) are calculated based on the changes experienced by each of the particles of the fluid.

From the two approaches, the Eulerian description seems more tangible for fluid dynamics. As an example, when a pressure probe is introduced into an experimental flow, the measurement occurs at a particular point of the fluid (x, y, z). Therefore, the value will be given by the pressure field at that point, which will be a consequence of the continuous particles of the fluid that pass through that point.

The mathematical description of a moving fluid is characterized by the velocity field $\mathbf{v} = \mathbf{v}(x, y, z, t)$ and by any two thermodynamic quantities, for example, the

© Springer Nature Switzerland AG 2020

J. Pérez Ríos, *An Introduction to Cold and Ultracold Chemistry*,

https://doi.org/10.1007/978-3-030-55936-6

pressure, $p(x, y, z, t)$, and density, $\rho(x, y, z, t)$.[1] Therefore, we calculate the scalar and vector fields of the flux instead of the changes that a particle experiences during its motion.

B.1.1   Continuity Equation

The conservation of mass is a general principle of nature; therefore, it must also be fulfilled in fluids, although in this case, it is better to work with density and volume rather than mass. To understand the equation that describes such a conservation law, let us consider a certain volume, V_0, in the fluid. From the density of the fluid, $\rho = dm/dV$, it is possible to determine the mass contained in that volume as $\int \rho \, dV$, where the integral extends to the volume V_0. The mass per unit of time flowing through an element $d\mathbf{S}$ (the surface limiting the volume V_0) is $\rho \mathbf{v} \cdot d\mathbf{S}$. The magnitude of the vector $d\mathbf{S}$ is equal to dS and its direction parallel to the normal of the surface. We assume that the direction of the normal points outward from the surface, and hence $\rho \mathbf{v} \cdot d\mathbf{S}$ is positive if the flow is going out and negative if it is coming in. The total mass of fluid leaving the volume V_0 per unit of time is equal to the variation of the mass of fluid inside the volume per unit of time, which is expressed as

$$\frac{\partial}{\partial t} \int \rho \, dV = - \oint \rho \mathbf{v} \cdot d\mathbf{S}. \qquad (\text{B.1})$$

And, applying the Gauss theorem, Eq. (B.1) can be recast as

$$\frac{\partial \rho}{\partial t} + \vec{\nabla} \cdot (\rho \mathbf{v}) = 0. \qquad (\text{B.2})$$

This is the so-called continuity equation and represents the law of conservation of mass for fluids. Let us imagine a fluid flowing in a pipe, then the continuity equation implies that the amount of fluid that crosses every section of the pipe per unit of time must be constant. The equation Eq. (B.2) can be written in terms of the current, $\mathbf{j} = \rho \mathbf{v}$ as

$$\frac{\partial \rho}{\partial t} + \vec{\nabla} \cdot \mathbf{j} = 0. \qquad (\text{B.3})$$

The structure of this equation should sound familiar to the reader, since it appears in other fields of physics, such as quantum mechanics, where $\mathbf{j}$ is the probability current density [5, 6], or in electromagnetism, where $\mathbf{j}$ represents the current flow in a medium [7, 8].

---

[1]From the equation of state it is well known that the value of any thermodynamic magnitude is related to the values of either of the other two magnitudes. For instance, the ideal gas satisfies the equation of state, $p = \rho k_B T$, where k_B is the Boltzmann constant.

B.1.2   Conservation of Linear Momentum or Newton's Second Law for Fluids

Let us consider a certain volume element of the fluid. As is well known, the pressure is the ratio of the applied force to the surface to which the force is applied. The force acting on a fluid element can be expressed as

$$\mathbf{F} = -\oint p d\mathbf{S} = -\int \vec{\nabla} p dV, \tag{B.4}$$

where we have employed the Gauss theorem (the negative sign is because the force is toward the center of the fluid element). On the other hand, taking into account Newton's second law

$$\mathbf{F} = m\mathbf{a} = \int \rho \frac{d\mathbf{v}}{dt} dV, \tag{B.5}$$

we find

$$\rho \frac{d\mathbf{v}}{dt} = -\vec{\nabla} p. \tag{B.6}$$

Since fluids are a continuous medium, the derivative with respect to time refers to the variation in speed of a given infinitesimal element of volume that moves with the fluid. That is, the velocity field of the flow is given by $\mathbf{v} = \mathbf{v}(x(t), y(t), z(t), t)$, where it has been emphasized that the coordinates depend on time. Therefore, to evaluate the derivative of the velocity field we have to apply the chain rule (in its vector form), which reads as

$$\frac{d\mathbf{v}}{dt} = \frac{\partial \mathbf{v}}{\partial t} + \left(\mathbf{v} \cdot \vec{\nabla} \right) \mathbf{v}, \tag{B.7}$$

where the first term on the right is the so-called local term (it only accounts for variations at a given point) and the second is the so-called convective term (takes into account the variation of a given magnitude within the volume element). Thus, plugging Eq. (B.6) into Eq. (B.7) leads to

$$\frac{\partial \mathbf{v}}{\partial t} + \left(\mathbf{v} \cdot \vec{\nabla} \right) \mathbf{v} = -\frac{\vec{\nabla} p}{\rho}. \tag{B.8}$$

This equation was first obtained by L. Euler in 1755, and is called the Euler equation, also known as the equation of the conservation of linear momentum.

B.1.3   Energy Conservation

Energy conservation for fluids is expressed as

$$h + \frac{1}{2}\mathbf{v}^2 = \kappa \tag{B.9}$$

where h is the enthalpy per unit of mass, $\mathbf{v}$ is the flow velocity and κ is a constant. The enthalpy per unit of mass is defined as [9, 10]

$$h = u + p\mathsf{V} \tag{B.10}$$

where p is the pressure, V is the specific volume, and u is the internal energy. Equation (B.9) has two contributions: one of them is internal (h) and the other refers to the kinetic energy content of the fluid ($\frac{v^2}{2}$). It is worth pointing out that the term $p\mathsf{V}$ that contributes to the enthalpy represents the work that the fluid performs on the walls of the container through the pressure.

B.2   Gas Dynamics

When a fluid moves at speeds comparable with or greater than the speed of sound within the fluid, density variations become important. In such a scenario the flow is referred to as compressible flow [11]. The compressible flow is the cornerstone of molecular fluid dynamics. It is so important that the study of compressible flow in gases is a whole discipline in fluid dynamics called gas dynamics.

B.2.1   The Ideal Gas

Ideal gases are defined as those whose equation of state is

$$pV = nRT, \tag{B.11}$$

where p is the pressure, V is the volume, n is the number of moles, $R = 8.3145 \, \text{JK}^{-1}\text{mol}^{-1}$ is the so-called constant of gases, and T is the temperature. An ideal gas can also be defined through the kinetic theory of gases (a microscopic description of the gas), as a gas in which its particles do not interact with each other. In other words, the ideal gas is a compound of non-interacting particles.

The constant of gases can be expressed in terms of the heat capacity per unit mass at constant volume, c_V, and the heat capacity per unit mass at constant pressure and at constant pressure c_p as

$$c_p - c_\mathsf{V} = \frac{R}{W}, \tag{B.12}$$

where W denotes the molecular weight of the gas being studied. The equation of state of Eq. (B.11) implies that the internal energy and enthalpy of the gas are unique functions of temperature[2] [9]. This implies that both c_V and c_p are unique functions of temperature

$$c_V = \left(\frac{\partial u}{\partial T}\right)_V = c_V(T),\tag{B.13}$$

where it has been taken into account that $dh = du + pdV + dpV = du + \frac{R}{W}dT = c_p(T)dT$. In a diatomic gas, it can be shown from statistical mechanics that $c_V = \frac{5}{2W}R$ and $c_p = \frac{7}{2W}R$ [10,12], provided that $T > \Theta_R$, where Θ_R is the characteristic rotational temperature [10, 13]. Thus, from the definition of the so-called adiabatic index, γ, as the quotient between the specific heats, $\gamma = c_p c_V$, we obtain $\gamma = 7/5$ for diatomic molecules. This index is fundamental in the study of compressible fluids, as we shall see. In what follows, we will assume that both c_V and c_p remain constant as a function of temperature.

B.2.2   Isentropic Processes

Let us assume that we have an ideal gas and the flow is isentropic (entropy is constant, $ds = 0$), then a differential change in the enthalpy may be expressed as

$$dh = Tds + \frac{dp}{\rho} = \frac{dp}{\rho}.\tag{B.14}$$

Taking into account that enthalpy is a function of the temperature $dh = c_p dT$, and comparing this expression with Eq. (B.14), we arrive at the following relation for an isentropic process in an ideal gas

$$\frac{p}{p_0} = \left(\frac{T}{T_0}\right)^{\frac{\gamma}{\gamma-1}} = \left(\frac{\rho}{\rho_0}\right)^{\gamma}.\tag{B.15}$$

The isentropic approach leads to acceptable results in supersonic flows when there is no large deviation from an equilibrium situation. When the energy transfer is much more favorable in some molecular degrees of freedom than in others this leads to an abrupt rupture of equilibrium where the isentropic approximation does not work properly.

---

[2]The thermal motion of the particles is the only source of energy for the ideal gas, which explains why the internal energy only depends on the temperature.

B.2.3   Speed of Sound and Mach Number

The speed of sound, a, is the velocity of propagation of an infinitesimal pressure
pulse through a fluid at rest, and it is a thermodynamic property of the fluid [1, 4].
The speed of sound in a fluid is defined as [1–4]

$$a = \sqrt{\left(\frac{\partial p}{\partial \rho}\right)_s},$$ (B.16)

which for an ideal gas and isentropic flow reads as

$$a = \sqrt{\gamma R T},$$ (B.17)

where we have made use of the Eq. (B.15). From Eq. (B.17), it is clear that the speed
of sound of an ideal gas only depends on the temperature. Thus, the speed of sound
is a local variable and varies from point to point in the fluid.

In a fluid, when we compare the speed of sound with the speed of the flow itself,
we obtain a significant parameter for characterizing the degree of compressibility of
a fluid and it is known as the Mach number M_a.

$$M_a = \frac{v}{a}.$$ (B.18)

The Mach number is the most important parameter in compressible flow theory. It
could be said that the Mach number is to the dynamics of compressible fluids what
the refractive index is to electromagnetic optics.

B.2.4   Single-Dimensional Flow Equations for Compressible Fluids

A one-dimensional flow is considered a flow that is directed toward a direction
that shows a larger size than the rest of the dimensions of a given vessel. In a
one-dimensional flow the magnitudes characterizing the flow depend only on the
direction of the flow.

To study the evolution of the thermodynamic magnitudes of an expanding gas, it
is convenient to choose an origin, or reference point, where we know precisely the
value of these magnitudes. This point of reference is called the stagnation point [1–
4]. By definition, the speed of the fluid at the stagnation point is zero, and hence all
the fluid energy is in the form of enthalpy h_0 (the thermodynamic magnitudes with
the subscript 0 refer to their values at the stagnation point).

Taking into account the equation for the conservation of energy for a fluid given by Eq. (B.9), it is possible to express the energy at any point in the fluid as a function of the energy at the stagnation point as

$$h_0 = \frac{1}{2}v^2 + h \rightarrow c_p T_0 = \frac{1}{2}v^2 + c_p T, \rightarrow \frac{T_0}{T} = 1 + \frac{v^2}{2c_p T}. \tag{B.19}$$

By substituting the value of c_p in Eq. (B.12), taking into account the ratio of specific heats and making use of the expression of the speed of sound for an ideal gas Eq. (B.17) yields

$$\frac{T_0}{T} = 1 + \frac{v^2}{2c_p T} = 1 + \frac{(\gamma - 1)v^2}{2a^2} = 1 + \frac{(\gamma - 1)M_a^2}{2}. \tag{B.20}$$

Plugging Eq. (B.15) into Eq. (B.20), it is possible to find similar expressions for the rest of the thermodynamic properties as

$$\left(\frac{p_0}{p}\right) = \left(\frac{T_0}{T}\right)^{\gamma/(\gamma-1)} = \left(1 + \frac{(\gamma - 1)M_a^2}{2}\right)^{\gamma/(\gamma-1)} \tag{B.21}$$

$$\left(\frac{\rho_0}{\rho}\right) = \left(\frac{T_0}{T}\right)^{1/(\gamma-1)} = \left(1 + \frac{(\gamma - 1)M_a^2}{2}\right)^{1/(\gamma-1)}. \tag{B.22}$$

From Eqs. (B.20), (B.21) and (B.22), it is possible to determine the thermodynamic magnitudes of a fluid at any point from their values at the stagnation point and the Mach number. Such a relationship can be determined by a mathematical relation obtained from an empirical fitting [14] or through tabulated values for certain substances [1].

The behavior of temperature, density, and pressure as a function of the Mach number is shown in Fig. B.1, where one notices that the higher the adiabatic index, the faster the values of the thermodynamic magnitudes of the fluid decrease. This explains why mixtures of noble gases and molecules are used to increase the speed of flow in a supersonic expansion, constituting what are called "seeded molecular beams". These have been and are widely used for the study of collisions for rotational, vibrational, and electronic energy transfer, and for determination of total cross sections.

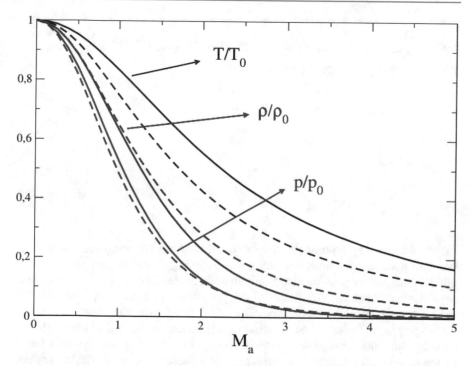

Fig. B.1 Temperature T, density ρ, and pressure p in a supersonic expansion. The magnitudes are presented as a function of their values in the stagnation point and Mach number for a monoatomic gas, $\gamma - 5/3$, represented by the dashed line and a diatomic gas, $\gamma = 7/5$, depicted by the solid black line

References

1. White FM (1979) Fluid mechanics. McGraw-Hill, New York
2. Zucrow MJ, Hoffman JD (1976) Gas dynamics. Wiley, New York
3. Batchelor GK (1967) An introduction to fluid dynamics. Cambridge University Press, Cambridge
4. Landau LD, Lifshitz EM (1959) Fluid mechanics. Butterworth-Heinemann, Oxford
5. Landau LD, Lifshitz EM (1958) Quantum mechanics. Butterworth-Heinemann, Oxford
6. Schwabl F (2007) Quantum mechanics. Springer, Berlin
7. Panofsky WKH, Phillips M (1983) Classical electricity and magnetism. Dover, New York
8. Jefimenko OD (1989) Electricity and magnetism. Electret Scientific Company, Star City, WV
9. Zemansky MW, Dittman RH (1996) Heat and thermodynamics. McGraw-Hill, New York
10. Landau LD, Lifshitz EM (1980) Statistical mechanics. Butterworth-Heinemann, Oxford
11. Feynman RP (1971) Lectures on physics, vol 2. Addison Wesley, New York
12. Maté B, Tejeda G, Montero S (1998) J Chem Phys 108:2676
13. Pathria RK (1996) Statistical mechanics. Butterworth-Heinemann, Oxford
14. Tejeda G, Maté B, Fernández-Sánchez JM, Montero S (1996) Temperature and density mapping of supersonic jet expansions using linear Raman spectroscopy. Phys Rev Lett 76:34

Computational Techniques for Quantum Mechanical Scattering

C

In principle, it may be thought that the usual methods for integration of second-order differential equations are valid, but in trying them we find that they are quite tedious and very ineffective, computationally speaking. To solve this problem, sophisticated computational methods have been developed, the most intuitive being the Numerov's method, introduced in Chap. 5 [1]. In addition, there are methods that do not *propagate* the wave function, but its logarithmic derivative, which are the so-called log-derivative propagator methods. Here, propagation means that the solution of the differential equation in a given point is a function of the results in previous points. Thus, a correlation between actual value and past values is employed to solve complex differential equations. Within the log-derivative methods one finds Johnson's method [2] and the improvement of this method proposed by Manolopoulos [3]. Finally, we would like to mention the hybrid methods [4], which consist in applying two propagators, one more efficient in the short-range region, and the other more convenient in the long range of the interaction.

In this appendix we focus on the log-derivative–Airy hybrid propagator, since it is one of the most efficient for molecular and atomic collisions. We begin with a more detailed study of the coupled channel equations [Eq. (5.4)], introducing the log-derivative propagator, and studying how the transition matrix can be obtained from this propagator. First, we introduce the log-derivative propagator of Manolopoulos, followed by the Airy propagator, to end with the hybrid log-derivative–Airy propagator.

C.1 Coupled-Channel Equations: Propagators and Asymptotic Conditions

Let us start by writing the coupled-channel equations in matrix form through Eq. (5.4)

© Springer Nature Switzerland AG 2020

J. Pérez Ríos, *An Introduction to Cold and Ultracold Chemistry*,

https://doi.org/10.1007/978-3-030-55936-6

$$\left(\bar{I}\frac{d^2}{dR^2} + \bar{W}(R)\right)\bar{g}(R) = 0, \tag{C.1}$$

where all matrices have dimension $N \times N$, where N is the total number of channels γ (open and closed) included in the calculation. $\bar{I}$ is the identity matrix, $\bar{g}(R) \equiv g(R)_{\gamma'\gamma}$ is the matrix of the radial coefficients of the wave function, where each column represents the wave function associated with a given entrance channel γ.[1] The potential matrix $\bar{W}(R)$ is expressed as

$$\bar{W}(R) = \bar{k}^2 - \bar{l}^2(R) - 2\mu\bar{V}(R), \tag{C.2}$$

with $\bar{k}^2$ and $\bar{l}^2(R)$ diagonal arrays.

The solutions of (C.1) have to be regular at the origin and in addition their asymptotic form must fulfilled [5]

$$\bar{g}(R \to \infty) = \bar{J}(R) + \bar{N}(R)\bar{K}, \tag{C.3}$$

This asymptotic form is more convenient in practice as it only involves real matrices. The matrices $\bar{J}$ and $\bar{R}$ are diagonal, and are given by the Riccati–Bessel and Riccati–Neumann functions respectively, for open channels ($k_\gamma^2 > 0$),

$$[\bar{J}(R)]_{\gamma\gamma} = \frac{j_l(k_\gamma R)}{\sqrt{k_\gamma}} \tag{C.4}$$

$$[\bar{N}(R)]_{\gamma\gamma} = \frac{n_l(k_\gamma R)}{\sqrt{k_\gamma}}. $$

For the case of closed channels ($k_\gamma^2 < 0$, in other words, the internal energy of the γ fragment is greater than the total energy), these matrix elements behave like the Riccati–Bessel functions of the first and third type respectively

$$[\bar{J}(R)]_{\gamma\gamma} = \left(k_\gamma R\right)^{1/2} I_{l+1/2}(k_\gamma R) \tag{C.5}$$

$$[\bar{N}(R)]_{\gamma\gamma} = \left(k_\gamma R\right)^{1/2} K_{l+1/2}(k_\gamma R). \tag{C.6}$$

There are different ways of comparing the asymptotic solution of Eq. (C.3) with the numerical solution of Eq. (C.1), to determine the $\bar{K}$ matrix, which depends on the propagator used in solving the coupled equations. In the Numerov [1] algorithm (see Chap. 5), the matrix linking the wave function between consecutive points of the collision coordinate is propagated, and the $\bar{K}$ matrix is extracted by considering Eq. (C.3) at two consecutive points in the asymptotic region.

---

[1] We have not made explicit the dependence of $\bar{R}$ on the total angular momentum J and other symmetry indexes for the sake of clarity

A widely used propagator is that of the logarithmic derivative

$$\bar{Y}(R) = \bar{g}'(R)\bar{g}^{-1}(R), \tag{C.7}$$

where $\bar{Y}$ is the so-called matrix of the logarithmic derivative (log-derivative matrix for short), [2, 3]. The equation Eq. (C.7) allows us to obtain $\bar{K}$ at a point in the asymptotic zone R_N by comparison with $\bar{g}'(R_N)^{-1}(R_N)$, using Eq. (C.3)

$$\bar{K} = \left(\bar{Y}(R_N)\bar{N}(R_N) - \bar{N}'(R_N)\right)^{-1}\left(\bar{Y}(R_N)\bar{J}(R_N) - \bar{J}'(R_N)\right). \tag{C.8}$$

The $\bar{K}$ matrix obtained in this way has N dimension (including elements connecting to closed channels). The $\bar{K}_{OO}$ submatrix of $\bar{K}$, of dimension $N_{op} \times N_{op}$ (where N_{op} is the number of open channels) is needed to evaluate the $\bar{S}$ matrix or the $\bar{T}$ [5]:

$$\bar{T} = \frac{1}{2\iota}\left(\bar{I} - (\bar{I} + \iota\bar{K}_{OO})(\bar{I} - \iota\bar{K}_{OO})^{-1}\right), \tag{C.9}$$

C.1.1   Log-Derivative Propagator of Manolopoulos

The log-derivative propagator was first developed by Johnson [2] and later improved by Manolopoulos [3]. It is based on the propagation of the logarithmic derivative Eq. (C.7), which eliminates stability problems that arise when propagating the wave function in the classically forbidden region (linear dependence issues due to closed channels). The coupled-channel equations, Eq. (C.1), are transformed under the new matrix $\bar{Y}(R)$, yielding a Riccati-type equation

$$\bar{Y}'(R) = \bar{W}(R) - \bar{Y}^2(R). \tag{C.10}$$

From Eq. (C.7), one notes that the log-derivative matrix $\bar{Y}(R)$ has singularities in each point of the mesh in R in which $\bar{g}(R)$ is null or its determinant is null. These singularities require the use of numerical techniques beyond conventional numerical methods for the solution of Eq. (C.10). Even so, the invariant imbedding technique can be used to propagate the log-derivative matrix. An imbedding propagator can be defined in a segment of the collision coordinate $[a, b]$ as [3]

$$\begin{pmatrix} \bar{g}'(a) \\ \bar{g}'(b) \end{pmatrix} = \begin{pmatrix} \bar{\mathcal{Y}}_1(a, b) & \bar{\mathcal{Y}}_2(a, b) \\ \bar{\mathcal{Y}}_3(a, b) & \bar{\mathcal{Y}}_4(a, b) \end{pmatrix} \begin{pmatrix} -\bar{g}(a) \\ \bar{g}(b) \end{pmatrix}, \tag{C.11}$$

with $\bar{\mathcal{Y}}_1, \bar{\mathcal{Y}}_2, \bar{\mathcal{Y}}_3$, and $\bar{\mathcal{Y}}_4$ the blocks of the imbedding propagator. After some algebra, a recurrence relation is obtained for the log-derivative ($\bar{Y}(b)$ is obtained from $\bar{Y}(a)$),

$$\bar{Y}(b) = \bar{\mathcal{Y}}_4(a, b) - \bar{\mathcal{Y}}_3(a, b)\left(\bar{Y}(a) + \bar{\mathcal{Y}}_1(a, b)\right)^{-1}\bar{\mathcal{Y}}_2(a, b). \tag{C.12}$$

This relationship is the most important equation for the log-derivative propagator. First, a mesh will be chosen for the R coordinate (collision coordinate), dividing this coordinate into sectors, in which a reference potential will be properly defined. The blocks of the imbedding propagator matrix are calculated in each of the sectors, and from them and by means of the recurrence relation in Eq. (C.12) the log-derivative is obtained in each of the sectors.

Let us assume that we want to propagate the log-derivative matrix $(\bar{Y})$ in a sector $[a, b]$. To do this, we divide the sector into two subsectors $[a, c]$ and $[c, b]$ of integration step $h = (b - a)/2$ and where $c = (b + a)/2$ is the midpoint of the sector. Manolopoulos' idea is to include a reference potential $\bar{W}_{ref}(R)$, which is taken as the diagonal values of the potential matrix, $\bar{W}(c)$, at the midpoint of the c interval

$$[\bar{W}_{ref}(R)]_{ij} = \delta_{ij}[\bar{W}(c)]_{jj} = \delta_{ij}p_j^2. \quad R \in [a, b]. \tag{C.13}$$

With this potential we can obtain the analytical solution of Eq. (C.1) in the intervals $[a, c]$ and $[c, b]$. Then, comparing the solutions with the definition of the imbedding propagator, Eq. (C.11), we obtain each of the blocks of that propagator in the same interval

$$\bar{y}_1(a, c) = \bar{y}_4(a, c) = \delta_{ij} \begin{Bmatrix} |p_j| \coth(|p_j|h), & p_j^2 \leq 0, \\ |p_j| \cot(|p_j|h), & p_j^2 \geq 0, \end{Bmatrix}$$

$$\bar{y}_2(a, c) = \bar{y}_3(a, c) = \delta_{ij} \begin{Bmatrix} |p_j| \operatorname{csch}(|p_j|h), & p_j^2 \leq 0, \\ |p_j| \csc(|p_j|h), & p_j^2 \geq 0. \end{Bmatrix} \tag{C.14}$$

We thus obtain the value of the log-derivative matrix in c, if the potential of our problem were $\bar{W}_{ref}(R)$. Even so, this is very important, since the Schrödinger equation, Eq. (C.1), can be redefined as an integral equation in the sector $[a, b]$, so that the homogeneous part of the equation is given by the solution Eq. (C.14), and the residual potential of that equation is

$$\bar{U}(R) = \bar{W}(R) - \bar{W}_{ref}(R). \tag{C.15}$$

The effect of the potential $\bar{U}(R)$ on the blocks of the imbedding propagator in the interval $[r_1, r_2]$ (where $[r_1, r_2]=[a, c]$ and $[c, b]$, in turn) is evaluated by means of a three-point quadrature[2] yielding

$$\bar{\mathcal{Y}}_1(r_1, r_2) = \bar{y}_1(r_1, r_2) + Q(r_1) \tag{C.16}$$

$$\bar{\mathcal{Y}}_2(r_1, r_2) = \bar{y}_2(r_1, r_2)$$

---

[2]The quadrature method $Q(R)$ is applicable when the integration step is very small, thus avoiding the appearance of possible singularities for the locally open channels [3]

$$\bar{\mathcal{Y}}_3(r_1, r_2) = \bar{y}_3(r_1, r_2)$$
$$\bar{\mathcal{Y}}_4(r_1, r_2) = \bar{y}_4(r_1, r_2) + Q(r_2),$$

where $Q(r)$ at the three points in the sector is given by[3]

$$\bar{Q}(a) = \frac{h}{3}\bar{U}(a) \qquad (\text{C}.17)$$

$$\bar{Q}(c) = \frac{1}{2}\left(\bar{I} - \frac{h^2}{6}\bar{U}(c)\right)^{-1}\frac{4h}{3}\bar{U}(c)$$

$$\bar{Q}(b) = \frac{h}{3}\bar{U}(b).$$

At this point, it is convenient to describe the *recipe* of the algorithm for its implementation:

1. A grid is defined for the collision coordinate R. Each sector $[a, b]$ shall contain three points of the mesh $R_i = a$, $R_{i+1} = c$, $R_{i+2} = b$.
2. A value of the log-derivative matrix is chosen at the starting point of the grid (where all channels are closed locally). Following the work of Manolopoulos it is convenient to choose the value of the log-derivative matrix at the first point of the mesh as [3]

$$[\bar{Y}(R_0)]_{ij} = \delta_{ij}\sqrt{|\bar{W}(R_0)|}. \qquad (\text{C}.18)$$

3. The reference potential is calculated at the mid-point in each sector $[a, b]$, by (ref. d-4). This reference value changes from one sector to another, as well as the residual potential $\bar{U}(R)$, which is used for the calculation of the quadratures.
4. The blocks of the imbedding propagator matrix are calculated using Eq. (C.14).
5. $\bar{\mathcal{Y}}_i(a, c)$ are calculated[3] ($i = 1, \ldots, 4$) using Eq. (C.16) and taking into account Eq. (C.17), for the value of the quadratures.
6. The log-derivative matrix, $\bar{Y}(a)$, is propagated to c, yielding $\bar{Y}(c)$.
7. $\bar{\mathcal{Y}}_i(c, b)$ ($i = 1, \ldots, 4$) are calculated by means of Eq. (C.16) and taking into account Eq. (C.17), for the value of the quadratures.
8. The log-derivative matrix, $\bar{Y}(c)$, is propagated to b by means of Eq. (C.12) yielding $\bar{Y}(b)$.
9. All the previous points are made in an iterative way, until arriving at a zone where the potential of interaction $\bar{V}(R)$ is negligible. In this zone, the asymptotic conditions can be applied by extracting the reaction matrix $\bar{K}$.

---

[3]Note that $\bar{\mathcal{Y}}_2(a, c)$ and $\bar{\mathcal{Y}}_3(a, c)$ are constant in a given sector.

C.1.2   Airy Propagator

The basic idea of the Airy propagator is extremely simple, it consists of approximating the $\bar{W}(R)$ potential of Eq. (C.1) by a diagonal matrix with linear terms. The coupled-channel equations then support analytical solutions: Airy functions [6, 7]. This method is very suitable for the long-range tail of most of the intermolecular potentials, in particular the van der Waals. Therefore, it is usually combined with another propagation method for the short-range region of the potential.

In this method the log-derivative matrix is also propagated, as in the previously discussed algorithm (see Sect. C.1.1). However, in the case at hand in each sector $[R_n, R_{n+1}]$, it is better to work on a local basis obtained upon diagonalization of the potential matrix and evaluated at the midpoint of the sector, $R_{n+1/2}$, as

$$\bar{\mathcal{T}}_n \bar{W}(R_{n+1/2}) \bar{\mathcal{T}}_n^T = \bar{k}_n^2, \tag{C.19}$$

where the matrices $\bar{\mathcal{T}}_n$ and $\bar{\mathcal{T}}_n^T$ allow the transformation of the matrix of the wave function, as well as the log-derivative of the global base (the one given by Eq. (5.2) of the main text) into the local one, which is labeled following the order of the sector

$$\bar{g}_n(R) = \bar{\mathcal{T}}_n \bar{g}(R), \tag{C.20}$$
$$\bar{Y}_n(R) = \bar{\mathcal{T}}_n \bar{Y}(R) \bar{\mathcal{T}}_n^T.$$

The algorithm of this propagator uses the same expression to propagate the log-derivative between the sector boundaries, although in this case expressed in the local base as

$$\bar{Y}_n(R_{n+1}) = \bar{\mathcal{Y}}_4^{(n)} - \bar{\mathcal{Y}}_3^{(n)} \left(\bar{Y}_n(R_n) + \bar{\mathcal{Y}}_1^{(n)} \right)^{-1} \bar{\mathcal{Y}}_2^{(n)}. \tag{C.21}$$

The main differences lie in the expressions for $\bar{\mathcal{Y}}_i^{(n)}$ $(i = 1, \ldots, 4)$ since these are derived from the solutions of the system of decoupled equations in the sector

$$\left(\bar{I} \frac{d^2}{dR^2} + \bar{k}_n^2 + (R - R_{n+1/2}) \, \bar{W}_n' \right) \bar{g}_n(R) = 0, \tag{C.22}$$

whose linearly independent solutions are the Airy $Ai(R)$ and $Bi(R)$ functions [8] (whose Wronskian equals π^{-1}) and are known through the theory of differential equations [9, 10]. In the work of Alexander and Manolopoulos, the reader may find the explicit form taken by the propagators of $\bar{\mathcal{Y}}_i$, as well as many other technical details. We will only add here the optimal form for the linear reference potential, proposed by Alexander and Gordon [7, 11]. Based on Gauss–Legendre's two points in a given sector, with $\Delta_n = R_{n+1} - R_n$,

$$R_{n\pm} = R_{n+1/2} \pm \frac{\Delta_n}{2\sqrt{3}}, \tag{C.23}$$

the matrix of the potential to be diagonalized is given by

$$\bar{W}(R_{n+1/2}) \equiv \frac{1}{2}\left(\bar{W}(R_{n+}) + \bar{W}(R_{n-})\right), \tag{C.24}$$

and for the evaluation of the log-derivative matrix, the following matrix is taken

$$\bar{W}'_n = \frac{\sqrt{3}}{\Delta_n}\bar{\mathcal{T}}_n\left(\bar{W}(R_{n+}) - \bar{W}(R_{n-})\right)\bar{\mathcal{T}}_n^T. \tag{C.25}$$

At this point, one only needs an expression to relate the log-derivative matrix in two adjacent sectors $n-1$ y n in R_n:

$$\bar{Y}_n(R_n) = \bar{P}_n\bar{Y}_{n-1}(R_n)\bar{P}_n^T, \tag{C.26}$$

where

$$\bar{P}_n = \bar{\mathcal{T}}_n\bar{\mathcal{T}}_{n-1}^T. \tag{C.27}$$

C.1.3   Hybrid Log-Derivative–Airy Propagator

This is possibly the best propagator in terms of convergence and computational expense. The method is based on dividing the region of integration into two parts: on the one hand, the short-range region (zone near the well and repulsive wall), where the potential varies very sharply; and on the other hand, the long-range region where the potential varies very smoothly. In the short-range region the Manolopoulos method is applied for the propagation of the logarithmic derivative, whereas in the long-range region the Airy propagator is applied.

The best way to explain the algorithm is by means of a *recipe*. Let us assume that we have already applied the log-derivative propagator of Manolopoulos up to a certain point R_n, and that from this point on, the Airy propagator takes over. The procedure:

1. A mesh in R is chosen, so that the first point coincides with the last point in the logarithmic derivative propagation area, in this case R_n. At the last point, R_N, the potential $\bar{V}(R)$ is considered to be negligible and the asymptotic conditions (Eq. (C.8) may apply.)
2. In the first sector, $\bar{W}(R)$ is evaluated at points $R_{n\pm}$, giving $\bar{k}_n^2$ and $\bar{\mathcal{T}}_n$. Likewise, the log-derivative matrix is obtained in R_n on the local base, using $\bar{Y}_n(R_n) = \bar{P}_n\bar{Y}_{n-1}(R_n)\bar{P}_n^T$, but taking into account that $\bar{\mathcal{T}}_{n-1} = \bar{I}$, since in the previous sector the logarithmic derivative was defined in the global base.

3. In each sector n', once obtained $\bar{k}^2_{n'}$ and $\bar{\mathcal{T}}_{n'}$, is evaluated $\bar{W}'_{n'}$ [Eq. (C.25)], and with these data the imbedding propagators are derived (as detailed in Sect. C.1.2). The log-derivative matrix (Eq. (C.21) is then propagated, yielding $\bar{Y}_{n'}(R_{n'+1})$. This is then transformed to the local base of the next sector, $\bar{Y}_{n'+1}(R_{n'+1})$, applying Eq. (C.26)).

4. The process is repeated until obtaining $\bar{Y}_{N-1}R_N$. Finally, the log-derivative is transformed to the global basis, applying Eq. (C.20), but taking into account that $\bar{\mathcal{T}}_N = \bar{I}$.

In addition, it should be noted that the propagation of the long-range part (Airy) is normally done with variable step size. The user defines a factor (TOLHI $\gtrsim$ 1, as given in MOLSCAT [12]) so that the relationship between two consecutive integration steps is $\Delta_i + 1 = TOLHI \times \Delta_i$. With this trick is possible to reduce the number of integration steps without losing precision during the propagation, since in the zone in which the potential is applied it becomes increasingly smoother. Moreover, Eq. (C.25) allows the propagation with a variable step size, since the derivative of the potential is well defined, for any size of step. Undoubtedly, the hybrid log-derivative–Airy method is well suited to the study of scattering at cold and ultracold energies.

References

1. Allison AC (1970) J Comput Phys 6:378
2. Johnson BR (1973) J Comput Phys 13:445
3. Manolopoulos DE (1986) J Chem Phys 85:6425
4. Alexander MH (1984) J Chem Phys 81:4510
5. Zhang JZH (1999) Theory and applications of quantum molecular dynamics. World Scientific, Singapore
6. Gordon RG (1969) J Chem Phys 51:14
7. Gordon RG (1971) Methods Comput Phys 10:81
8. Abramowitz M, Stegun IA (1972) Handbook of mathematical functions. Dover, New York
9. Zwillinger D (1997) Handbook of differential equations. Academic Press, New York
10. Polyanin AD, Zaitsev VF (2003) Handbook of exact solutions for ordinary differential equations. CRC Press, Boca Raton
11. Alexander MH, Gordon RG (1971) J Chem Phys 55:4889
12. Hutson JM, Green S (1994) MOLSCAT, collaborative Computational Project no. 6 of the UK Science and Engineering Research council, version 14

Index

A
AC Stark shift, 58
Adiabaticity parameter, 222, 228
Associative ionization, 157, 163
Auto-ionization resonance, 165

B
Beer–Lambert law, 10
Beta function, 122
Binning
    Gaussian, 220
    uniform, 220
Bipolar spherical harmonic basis, 86
Bose–Einstein condensation, 37, 39, 247

C
Classical
    capture model, 207
    collision, 4, 10
    cross section, 12
    hard-sphere collision, 13
    opacity function, 13
Compactified dimension, 240, 242
Compressible flow, 252
Coupling constant
    nonrenormalizable, 122
    renormalizable, 122
    super-renormalizable, 122
Cross section
    classical elastic three-body, 212
    elastic for atom–ion, 186
    molecular dissociation, 221
    molecular formation, 221
    quantum mechanical three-body elastic,
        212
    RA, 188

radiative charge transfer, 187
semi-classical elastic for atom–ion, 184
vibrational quenching, 221

D
Depletion of the condensate, 43
Differential cross section
    quantum mechanical, 11
Dimer field, 127
Doppler temperature limit, 65

E
Efimov state, 132
Euler equation, 251
Excess of micromotion, 177

F
Fermi pseudopotential, 145, 157
Few-body physics, 119
Fifth force, 244
Fluid
    Eulerian description, 249
    Lagrangian description, 249

G
Glory
    effect, 32
    undulations, 34
Grand angular momentum, 212
Gross–Pitaevskii equation, 41, 44, 48

H
Hierarchy problem, 239

Printed in the United States
by Baker & Taylor Publisher Services